Algae

Anatomy, Biochemistry, and Biotechnology

Algae

Anatomy, Biochemistry, and Biotechnology

Laura Barsanti
Paolo Gualtieri

Taylor & Francis
Taylor & Francis Group
Boca Raton London New York

A CRC title, part of the Taylor & Francis imprint, a member of the
Taylor & Francis Group, the academic division of T&F Informa plc.

Published in 2006 by
CRC Press
Taylor & Francis Group
6000 Broken Sound Parkway NW, Suite 300
Boca Raton, FL 33487-2742

Library of Congress Cataloging-in-Publication Data

Gualtieri, Paolo, 1952-
 Algae : anatomy, biochemistry, and biotechnology / by Laura Barsanti and Paolo Gualtieri.
  p. ; cm.
 Includes bibliographical references and index.
 ISBN-13: 978-0-8493-1467-4 (alk. paper)
 ISBN-10: 0-8493-1467-4 (alk. paper)
 1. Algae. [DNLM: 1. Algae. 2. Biotechnology. QK 566 G899a 2005] I. Barsanti, L. II. Title.

QK566.G83 2005
579.8--dc22   2005014492

Taylor & Francis Group
is the Academic Division of Informa plc.

Visit the Taylor & Francis Web site at
http://www.taylorandfrancis.com

and the CRC Press Web site at
http://www.crcpress.com

Preface

This book is an outgrowth of many years of research aimed at studying algae, especially micro-algae. Working on it, we soon realized how small an area we really knew well and how superficial our treatment of many topics was going to be. Our approach has been to try to highlight those things that we have found interesting or illuminating and to concentrate more on those areas, sacrificing completeness in so doing.

This book was written and designed for undergraduate and postgraduate students with a general scientific background, following courses on algology and aquatic biology, as well as for researchers, teachers, and professionals in the fields of phycology and applied phycology. In our intention, it is destined to serve as a means to encourage outstanding work in the field of phycology, especially the aspect of teaching, with the major commitment to arouse the curiosity of both students and teachers. It is all too easy when reviewing an intricate field to give a student new to the area the feeling that everything is now known about the subject. We would like this book to have exactly the reverse effect on the reader, stimulating by deliberately leaving many doors ajar, so as to let new ideas spring to mind by the end of each chapter.

This book covers freshwater, marine, and terrestrial forms, and includes extensive original drawings and photographic illustrations to provide detailed descriptions of algal apparatus. We have presented an overview of the classification of the algae followed by reviews of life cycles, reproductions, and phylogeny to provide conceptual framework for the chapters which follow. Levels of organization are treated from the subcellular, cellular, and morphological standpoints, together with physiology, biochemistry, culture methods and finally, the role of algae in human society. Many instances of recent new findings are provided to demonstrate that the world of algae is incompletely known and prepared investigators should be aware of this.

Each of the chapters can be read on its own as a self-containing essay, used in a course, or assigned as a supplemental reading for a course. The endeavor has been to provide a hybrid between a review and a comprehensive descriptive work, to make it possible for the student to visualize and compare algal structures and at the same time to give enough references so that the research worker can enter the literature to find out more precise details from the original sources.

The bibliography is by no means exhaustive; the papers we have quoted are the ones we have found useful and which are reasonably accessible, both very recent references and older classic references that we have judged more representative, but many excellent papers can be missing. In our opinion, too many references make the text unreadable and our intention was to put in only enough to lead the reader into the right part of the primary literature in a fairly directed manner, and we have not tempted to be comprehensive. Our intention was to highlight the more important facts, hoping that this book will complement the few specialized reviews of fine structure already published and will perhaps make some of these known to a wider audience. Our efforts were aimed at orientating the readers in the *mare magnum* of scientific literature and providing interesting and useful Web addresses.

We are grateful to the phycologists who have contributed original pictures; they are cited in the corresponding figure captions. We are also grateful to the staff at CRC Press, Boca Raton, Florida, particularly our editor, John Sulzycki, for his patience and human comprehension in addition to his unquestionable technical ability, and to the production coordinators, Erika Dery and Kari A. Budyk.

Our sincere gratitude and a special thanks to Valter Evangelista for his skillful assistance and ability in preparing the final form of all the drawings and illustrations, and for his careful attention in preparing all the technical drawings. We appreciated his efforts to keep pace with us both and to cope with our ever-changing demands without getting too upset.

We will always be grateful to Vincenzo Passarelli, who frequently smoothed a path strewn with other laboratory obligations so that we could pursue the endeavors that led up to the book, and above all because he has always tolerated the ups and downs of our moods with a smile on his face, and a witty, prompt reply. He lighted up many gloomy days with his cheerful whistling. We are sure it was not always an easy task.

For the multitudinous illustrations present in the book we are indebted to Maria Antonietta Barsanti and to Luca Barsanti, the sister and brother of Laura. When the idea of the book first arose, about four years ago, Maria Antonietta took up the challenge to realize all the drawings we had in mind for the book. But this was just a minor challenge compared with the struggle she had been engaged with against cancer since 1996. Despite all the difficulties of coping with such a disabling situation, she succeeded in preparing most of the drawings, with careful determination, interpreting even the smallest details to make them clear without wasting scientific accuracy, and still giving each drawing her unique artistic touch. She worked until the very last days, when even eating or talking were exhausting tasks, but unfortunately last February she died without seeing the outcome of her and our efforts. She will always have a very special place in our hearts and our lives. Her brother Luca Barsanti completed the drawing work in a wonderful way, making it very hard to distinguish between her artistic skill and his. His lighthearted and amusing company relieved the last and most nervous days of our work, and also for this we will always be grateful to him.

In July 2004, Mimmo Gualtieri, the only brother of Paolo, died of an unexpected heart attack. He left a huge empty room in his brother's heart.

In October 2004, our beloved friend and colleague, Dr. Patricia Lee Walne, distinguished professor of botany of the University of Tennessee in Knoxville, died after a long and serious illness.

This book is dedicated to the three of them.

About the Authors

Dr. Laura Barsanti graduated in natural science from University of Pisa, Italy. At present she is a scientist at the Biophysics Institute of the National Council of Research (CNR) in Pisa.

Dr. Paolo Gualtieri graduated in biology and computer science from University of Pisa, Italy. At present he is a senior scientist at the Biophysics Institute of the National Council of Research (CNR) in Pisa and adjunct professor at the University of Maryland, University College, College Park, Maryland.

Table of Contents

In memory of Maria Antonietta Barsanti (1957–2005), Mimmo Gualtieri (1954–2004), and Patricia Lee Walne (1932–2004), who deserved something better

1 General Overview

DEFINITION

The term *algae* has no formal taxonomic standing. It is routinely used to indicate a polyphyletic (i.e., including organisms that do not share a common origin, but follow multiple and independent evolutionary lines), noncohesive, and artificial assemblage of O_2-evolving, photosynthetic organisms (with several exceptions of colorless members undoubtedly related to pigmented forms). According to this definition, plants could be considered an algal division. Algae and plants produce the same storage compounds, use similar defense strategies against predators and parasites, and a strong morphological similarity exists between some algae and plants. Then, how to distinguish algae from plants? The answer is quite easy because the similarities between algae and plants are much fewer than their differences. Plants show a very high degree of differentiation, with roots, leaves, stems, and xylem/phloem vascular network. Their reproductive organs are surrounded by a jacket of sterile cells. They have a multicellular diploid embryo stage that remains developmentally and nutritionally dependent on the parental gametophyte for a significant period (and hence the name embryophytes is given to plants) and tissue-generating parenchymatous meristems at the shoot and root apices, producing tissues that differentiate in a wide variety of shapes. Moreover, all plants have a digenetic life cycle with an alternation between a haploid gametophyte and a diploid sporophyte. Algae do not have any of these features; they do not have roots, stems, leaves, nor well-defined vascular tissues. Even though many seaweeds are plant-like in appearance and some of them show specialization and differentiation of their vegetative cells, they do not form embryos, their reproductive structures consist of cells that are potentially fertile and lack sterile cells covering or protecting them. Parenchymatous development is present only in some groups and have both monogenetic and digenetic life cycles. Moreover, algae occur in dissimilar forms such as microscopic single cell, macroscopic multicellular loose or filmy conglomerations, matted or branched colonies, or more complex leafy or blade forms, which contrast strongly with uniformity in vascular plants. Evolution may have worked in two ways, one for shaping similarities for and the other shaping differences. The same environmental pressure led to the parallel, independent evolution of similar traits in both plants and algae, while the transition from relatively stable aquatic environment to a gaseous medium exposed plants to new physical conditions that resulted in key physiological and structural changes necessary to invade upland habitats and fully exploit them. The bottom line is that plants are a separate group with no overlapping with the algal assemblage.

The profound diversity of size ranging from picoplankton only $0.2-2.0$ μm in diameter to giant kelps with fronds up to 60 m in length, ecology and colonized habitats, cellular structure, levels of organization and morphology, pigments for photosynthesis, reserve and structural polysaccharides, and type of life history reflect the varied evolutionary origins of this heterogeneous assemblage of organisms, including both prokaryote and eukaryote species. The term algae refers to both macroalgae and a highly diversified group of microorganisms known as microalgae. The number of algal species has been estimated to be one to ten million, and most of them are microalgae.

CLASSIFICATION

No easily definable classification system acceptable to all exists for algae because taxonomy is under constant and rapid revision at all levels following every day new genetic and ultrastructural evidence. Keeping in mind that the polyphyletic nature of the algal group is somewhat inconsistent with traditional taxonomic groupings, though they are still useful to define the general character and level of organization, and the fact that taxonomic opinion may change as information accumulates, a tentative scheme of classification is adopted mainly based on the work of Van Den Hoek et al. (1995) and compared with the classifications of Bold and Wynne (1978), Margulis et al. (1990), Graham and Wilcox (2000), and South and Whittick (1987). Prokaryotic members of this assemblage are grouped into two divisions: Cyanophyta and Prochlorophyta, whereas eukaryotic members are grouped into nine divisions: Glaucophyta, Rhodophyta, Heterokontophyta, Haptophyta, Cryptophyta, Dinophyta, Euglenophyta, Chlorarachniophyta, and Chlorophyta (Table 1.1).

TABLE 1.1
Classification Scheme of the Different Algal Groups

Kingdom	Division	Class
Prokaryota eubacteria	Cyanophyta	Cyanophyceae
	Prochlorophyta	Prochlorophyceae
	Glaucophyta	Glaucophyceae
	Rhodophyta	Bangiophyceae
		Florideophyceae
	Heterokontophyta	Chrysophyceae
		Xanthophyceae
		Eustigmatophyceae
		Bacillariophyceae
		Raphidophyceae
		Dictyochophyceae
		Phaeophyceae
	Haptophyta	Haptophyceae
	Cryptophyta	Cryptophyceae
Eukaryota	Dinophyta	Dinophyceae
	Euglenophyta	Euglenophyceae
	Chlorarachniophyta	Chlorarachniophyceae
	Chlorophyta	Prasinophyceae
		Chlorophyceae
		Ulvophyceae
		Cladophorophyceae
		Bryopsidophyceae
		Zygnematophyceae
		Trentepohliophyceae
		Klebsormidiophyceæ
		Charophyceae
		Dasycladophyceae

OCCURRENCE AND DISTRIBUTION

Algae can be aquatic or subaerial, when they are exposed to the atmosphere rather than being submerged in water. Aquatic algae are found almost anywhere from freshwater spring to salt lakes, with tolerance for a broad range of pH, temperature, turbidity, and O_2 and CO_2 concentration.

They can be planktonic, like most unicellular species, living suspended throughout the lighted regions of all water bodies including under ice in polar areas. They can be also benthic, attached to the bottom or living within sediments, limited to shallow areas because of the rapid attenuation of light with depth. Benthic algae can grow attached on stones (epilithic), on mud or sand (epipelic), on other algae or plants (epiphytic), or on animals (epizoic). In the case of marine algae, various terms can be used to describe their growth habits, such as supralittoral, when they grow above the high-tide level, within the reach of waves and spray; intertidal, when they grow on shores exposed to tidal cycles: or sublittoral, when they grow in the benthic environment from the extreme low-water level to around 200 m deep, in the case of very clear water.

Oceans covering about 71% of earth's surface contain more than 5000 species of planktonic microscopic algae, the phytoplankton, which forms the base of the marine food chain and produces roughly 50% of the oxygen we inhale. However, phytoplankton is not only a cause of life but also a cause of death sometimes. When the population becomes too large in response to pollution with nutrients such as nitrogen and phosphate, these blooms can reduce the water transparency, causing the death of other photosynthetic organisms. They are often responsible for massive fish and bird kills, producing poisons and toxins. The temperate pelagic marine environment is also the realm of giant algae, the kelp. These algae have thalli up to 60 m long, and the community can be so crowded that it forms a real submerged forest; they are not limited to temperate waters, as they also form luxuriant thickets beneath polar ice sheets and can survive at very low depth. The depth record for algae is held by dark purple red algae collected at a depth of 268 m, where the faint light is blue-green and its intensity is only 0.0005% of surface light. At this depth the red part of the sunlight spectrum is filtered out from the water and sufficient energy is not available for photosynthesis. These algae can survive in the dark blue sea as they possess accessory pigments that absorb light in spectral regions different from those of the green chlorophylls a and b and channel this absorbed light energy to chlorophyll a, which is the only molecule that converts sunlight energy into chemical energy. For this reason the green of their chlorophylls is masked and they look dark purple. In contrast, algae that live in high irradiance habitat typically have pigments that protect them against the photodamages caused by singlet oxygen. It is the composition and amount of accessory and protective pigments that give algae their wide variety of colors andx for several algal groups, their common names such as brown algae, red algae, and golden and green algae. Internal freshwater environment displays a wide diversity of microalgae forms, although not exhibiting the phenomenal size range of their marine relatives. Freshwater phytoplankton and the benthic algae form the base of the aquatic food chain.

A considerable number of subaerial algae have adapted to life on land. They can occur in surprising places such as tree trunks, animal fur, snow banks, hot springs, or even embedded within desert rocks. The activities of land algae are thought to convert rock into soil to minimize soil erosion and to increase water retention and nutrient availability for plants growing nearby.

Algae also form mutually beneficial partnership with other organisms. They live with fungi to form lichens or inside the cells of reef-building corals, in both cases providing oxygen and complex nutrients to their partner and in return receiving protection and simple nutrients. This arrangement enables both partners to survive in conditions that they could not endure alone.

Table 1.2 summarizes the different types of habitat colonized by the algal divisions.

STRUCTURE OF THALLUS

Examples of the distinctive morphological characteristics within different divisions are summarized in Table 1.3.

UNICELLS AND UNICELL COLONIAL ALGAE

Many algae are solitary cells, unicells with or without flagella, hence motile or non-motile. *Nannochloropsis* (Heterokontophyta) (Figure 1.1) is an example of a non-motile unicell, while

TABLE 1.2
Distribution of Algal Divisions

Division	Common Name	Habitat			
		Marine	Freshwater	Terrestrial	Symbiotic
Cyanophyta	Blue-green algae	Yes	Yes	Yes	Yes
Prochlorophyta	n.a.	Yes	n.d.	n.d.	Yes
Glaucophyta	n.a.	n.d.	Yes	Yes	Yes
Rhodophyta	Red algae	Yes	Yes	Yes	Yes
Heterokontophyta	Golden algae Yellow-green algae Diatoms Brown algae	Yes	Yes	Yes	Yes
Haptophyta	Coccolithophorids	Yes	Yes	Yes	Yes
Cryptophyta	Cryptomonads	Yes	Yes	n.d.	Yes
Chlorarachniophyta	n.a.	Yes	n.d.	n.d.	Yes
Dinophyta	Dinoflagellates	Yes	Yes	n.d.	Yes
Euglenophyta	Euglenoids	Yes	Yes	Yes	Yes
Chlorophyta	Green algae	Yes	Yes	Yes	Yes

Note: n.a., not available; n.d., not detected.

Ochromonas (Heterokontophyta) (Figure 1.2) is an example of motile unicell. Other algae exist as aggregates of several single cells held together loosely or in a highly organized fashion, the colony. In these types of aggregates, the cell number is indefinite, growth occurs by cell division of its components, there is no division of labor, and each cell can survive on its own. *Hydrurus*

TABLE 1.3
Thallus Morphology in the Different Algal Divisions

Division	Unicellular and non-motile	Unicellular and motile	Colonial and non-motile	Colonial and motile	Filamentous	Siphonous	Parenche-matous
Cyanophyta	*Synechococcus*	n.d.	*Anacystis*	n.d.	*Calothrix*	n.d.	*Pleurocapsa*
Prochlorophyta	*Prochloron*	n.d.	n.d.	n.d.	*Prochlorothrix*	n.d.	n.d.
Glaucophyta	*Glaucocystis*	*Gloeochaete*	n.d.	n.d.	n.d.	n.d.	n.d.
Rhodophyta	*Porphyridium*	n.d.	*Cyanoderma*	n.d.	*Goniotricum*	n.d.	*Palmaria*
Heterokontophyta	*Navicula*	*Ochromonas*	*Chlorobotrys*	*Synura*	*Ectocarpus*	*Vaucheria*	*Fucus*
Haptophyta	n.d.	*Chrysochro-mulina*	n.d.	*Corym-bellus*	n.d.	n.d.	n.d.
Cryptophyta	n.d.	*Cryptomonas*	n.d.	n.d.	*Bjornbergiella*	n.d.	n.d.
Dynophyta	*Dinococcus*	*Gonyaulax*	*Gloeodinium*	n.d.	*Dinoclonium*	n.d.	n.d.
Euglenophyta	*Ascoglena*	*Euglena*	*Colacium*	n.d.	n.d.	n.d.	n.d.
Chlorarachniophyta	n.d.	*Chlorarachnion*	n.d.	n.d.	n.d.	n.d.	n.d.
Chlorophyta	*Chlorella*	*Dunaliella*	*Pseudo-sphaerocystis*	*Volvox*	*Ulothrix*	*Bryopsis*	*Ulva*

Note: n.d., not detected.

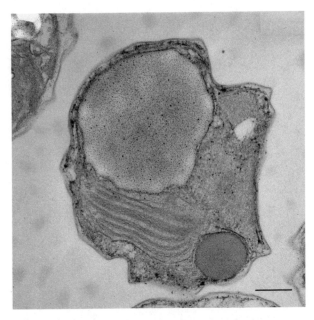

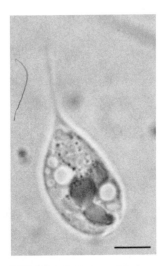

FIGURE 1.1 Transmission electron micrograph of *Nannochloropsis* sp., non-motile unicell. (Bar: 0.5 μm.)

FIGURE 1.2 *Ochromonas* sp., motile unicell. (Bar: 4 μm.)

(Heterokontophyta) (Figure 1.3) forms long and bushy non-motile colonies with cells evenly distributed throughout a gelatinous matrix, while *Synura* (Heterokontophyta) (Figure 1.4) forms free-swimming colonies composed of cells held together by their elongated posterior ends. When the number and arrangement of cells are determined at the time of origin and remain and constant during the life span of the individual colony, colony is termed coenobium. *Volvox* (Chlorophyta) (Figure 1.5) with its spherical colonies composed of up to 50,000 cells is an example of motile coenobium, and *Pediastrum* (Chlorophyta) (Figure 1.6) with its flat colonies of cells characterized by spiny protuberances is an example of non-motile coenobium.

FILAMENTOUS ALGAE

Filaments result from cell division in the plane perpendicular to the axis of the filament and have cell chains consisting of daughter cells connected to each other by their end wall. Filaments can be simple as in *Oscillatoria* (Cyanophyta) (Figure 1.7), *Spirogyra* (Chlorophyta) (Figure 1.8), or *Ulothrix* (Chlorophyta) (Figure 1.9), have false branching as in *Tolypothrix* (Cyanophyta) (Figure 1.10) or true branching as in *Cladophora* (Chlorophyta) (Figure 1.11). Filaments of *Stigonema ocellatum* (Cyanophyta) (Figure 1.12) consists of a single layer of cells and are called uniseriate, and those of *Stigonema mamillosum* (Cyanophyta) (Figure 1.13) made up of multiple layers are called multiseriate.

SIPHONOUS ALGAE

These algae are characterized by a siphonous or coenocytic construction, consisting of tubular filaments lacking transverse cell walls. These algae undergo repeated nuclear division without forming cell walls; hence they are unicellular, but multinucleate (or coenocytic). The sparsely

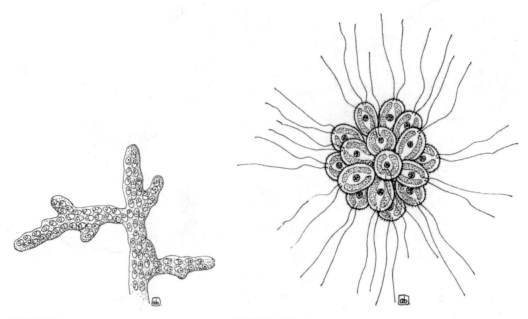

FIGURE 1.3 Non-motile colony of
Hydrurus foetidus.

FIGURE 1.4 Free-swimming colony of
Synura uvella.

branched tube of *Vaucheria* (Heterokontophyta) (Figure 1.14) is an example of coenocyte or apocyte, a single cell containing many nuclei.

PARENCHYMATOUS AND PSEUDOPARENCHYMATOUS ALGAE

These algae are mostly macroscopic with undifferentiated cells and originate from a meristem with cell division in three dimensions. In the case of parenchymatous algae, cells of the primary filament

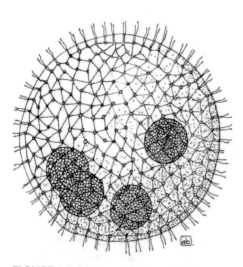

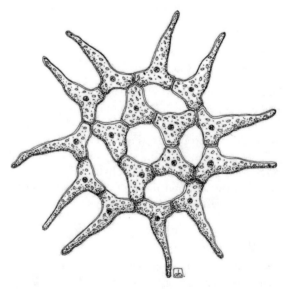

FIGURE 1.5 Motile coenobium of *Volvox
aureus.*

FIGURE 1.6 Non-motile coenobium of *Pediastrum
simplex.*

divide in all directions and any essential filamentous structure is lost. This tissue organization is found in *Ulva* (Chlorophyta) (see life cycle in Figure 1.22) and many of the brown algae. Pseudoparenchymatous algae are made up of a loose or close aggregation of numerous, intertwined, branched filaments that collectively form the thallus, held together by mucilages, especially in red algae. Thallus construction is entirely based on a filamentous construction with little or no internal cell differentiation. *Palmaria* (Rhodophyta) (Figure 1.15) is a red alga with a complex pseudoparenchymatous structure.

NUTRITION

Following our definition of the term *algae*, most algal groups are considered photoautotrophs, that is, depending entirely upon their photosynthetic apparatus for their metabolic necessities, using sunlight as the source of energy, and CO_2 as the carbon source to produce carbohydrates and ATP. Most algal divisions contain colorless heterotropic species that can obtain organic carbon from the external environment either by taking up dissolved substances (osmotrophy) or by engulfing bacteria and other cells as particulate prey (phagotrophy). Algae that cannot synthesize essential components such as the vitamins of the B_{12} complex or fatty acids also exist, and have to import them; these algae are defined auxotrophic.

However, it is widely accepted that algae use a complex spectrum of nutritional strategies, combining photoautotrophy and heterotrophy, which is referred to as mixotrophy. The relative contribution of autotrophy and heterotrophy to growth within a mixotrophic species varies along a gradient from algae whose dominant mode of nutrition is phototrophy, through those for which phototrophy or heterotrophy provides essential nutritional supplements, to those for which heterotrophy is the dominant strategy. Some mixotrophs are mainly photosynthetic and only occasionally use an organic energy source. Other mixotrophs meet most of their nutritional demand by phagotrophy, but may use some of the products of photosynthesis from sequestered prey chloroplasts. Photosynthetic fixation of carbon and use of particulate food as a source of major nutrients (nitrogen, phosphorus, and iron) and growth factors (e.g., vitamins, essential amino acids, and essential fatty acids) can enhance growth, especially in extreme environments where resources are limited. Heterotrophy is important for the acquisition of carbon when light is limiting and, conversely, autotrophy maintains a cell during periods when particulate food is scarce.

On the basis of their nutritional strategies, algae are into classified four groups:

- *Obligate heterotrophic algae.* They are primarily heterotrophic, but are capable of sustaining themselves by phototrophy when prey concentrations limit heterotrophic growth (e.g., *Gymnodium gracilentum*, Dinophyta).
- *Obligate phototrophic algae*. Their primary mode of nutrition is phototrophy, but they can supplement growth by phagotrophy and/or osmotrophy when light is limiting (e.g., *Dinobryon divergens*, Heterokontophyta).
- *Facultative mixotrophic algae*. They can grow equally well as phototrophs and as heterotrophs (e.g., *Fragilidium subglobosum*, Dinophyta).
- *Obligate mixotrophic algae*. Their primary mode of nutrition is phototrophy, but phagotrophy and/or osmotrophy provides substances essential for growth (photoauxotrophic algae can be included in this group) (e.g., *Euglena gracilis*, Euglenophyta).

REPRODUCTION

Methods of reproduction in algae may be vegetative by the division of a single cell or fragmentation of a colony, asexual by the production of motile spore, or sexual by the union of gametes.

FIGURE 1.7 Simple filament of *Oscillatoria* sp.

FIGURE 1.8 Simple filament of *Spirogyra* sp.

FIGURE 1.9 Simple filament of *Ulothrix variabilis.*

Vegetative and asexual modes allow stability of an adapted genotype within a species from a generation to the next. Both modes provide a fast and economical means of increasing the number of individuals while restricting genetic variability. Sexual mode involves plasmogamy (union of cells), karyogamy (union of nuclei), chromosome/gene association, and meiosis, resulting in genetic recombination. Sexual reproduction allows variation but is more costly because of the waste of gametes that fail to mate.

VEGETATIVE AND ASEXUAL REPRODUCTION

Binary Fission or Cellular Bisection

It is the simplest form of reproduction; the parent organism divides into two equal parts, each having the same hereditary information as the parent. In unicellular algae, cell division may be longitudinal as in *Euglena* (Euglenophyta) (Figure 1.16) or transverse. The growth of the population follows a typical curve consisting of a lag phase, an exponential or log phase, and a stationary or plateau phase, where increase in density is leveled off (see Chapter 6). In multicellular algae or in algal colonies this process eventually leads to the growth of the individual.

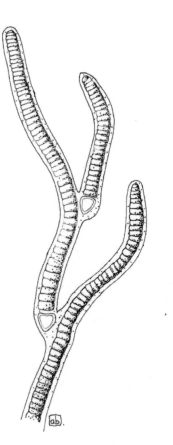

FIGURE 1.10 False branched filament of *Tolypothrix byssoidea.*

FIGURE 1.11 True branched filament of *Cladophora glomerata.*

Zoospore, Aplanospore, and Autospore

Zoospores are flagellate motile spores that may be produced within a parental vegetative cell as in *Chlamydomonas* (Chlorophyta) (Figure 1.17). Aplanospores are aflagellate spores that begin their development within the parent cell wall before being released; these cells can develop into zoospores. Autospores are aflagellate daughter cells that will be released from the ruptured wall of the original parent cell. They are almost perfect replicas of the vegetative cells that produce them and lack the capacity to develop in zoospores. Examples of autospore forming genera are *Nannochloropsis* (Heterokontophyta) and *Chlorella* (Chlorophyta). Spores may be produced within ordinary vegetative cells or within specialized cells or structures called sporangia.

Autocolony Formation

In this reproductive mode, when the coenobium/colony enters the reproductive phase, each cell within the colony can produce a new colony similar to the one to which it belongs. Cell division no longer produces unicellular individuals but multicellular groups, a sort of embryonic colony that differs from the parent in cell size but not in cell number. This mode characterizes green algae such as *Volvox* (Chlorophyta) and *Pediastrum* (Chlorophyta). In *Volvox* (Figure 1.5) division is restricted to a series of cells which produce a hollow sphere within the parent colony, and with each mitosis each cell becomes smaller. The new colony everts, its cells form flagella at their apical poles, which is released by the rupture of the parent sphere. In *Pediastrum*

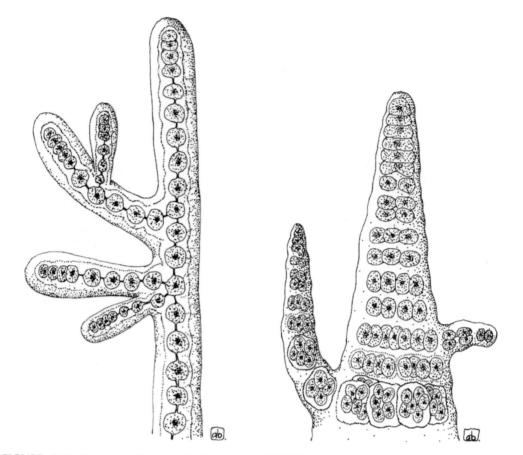

FIGURE 1.12 Uniseriate filament of *Stigonema ocellatum.*

FIGURE 1.13 Multiseriate filament of *Stigonema mamillosum.*

(Figure 1.6) the protoplast of some cells of the colony undergoes divisions to form biflagellate zoospores. These are not liberated but aggregate to form a new colony within the parent cell wall.

Fragmentation

This is a more or less random process whereby non-coenobic colonies or filaments break into two to several fragments having the capacity of developing into new individuals.

Resting Stages

Under unfavorable conditions, particularly of desiccation, many algal groups produce thick-walled resting cells, such as hypnospores, hypnozygotes, statospores, and akinetes.

Hypnospores and hypnozygotes, which have thickened walls, are produced ex novo by protoplasts that previously separated from the walls of the parental cells. Hypnospores are present in *Ulotrix* spp. (Chlorophyceae) and *Chlorococcum* spp. (Chlorophyceae), whereas hypnozygotes are present in *Spyrogyra* spp. (Chlorophyceae) and Dinophyta. Hypnospores and hypnozygotes enable these green algae to survive temporary drying out of small water bodies and also allow aerial transport from one water body to another for instance via birds. It is likely that dinophyceae cysts have a similar function.

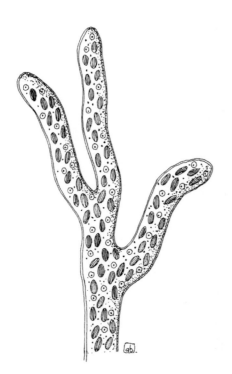

FIGURE 1.14 Siphonous thallus of *Vaucheria sessilis.*

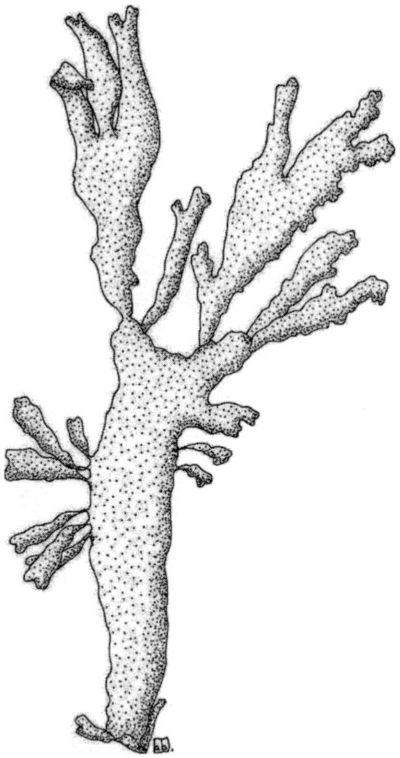

FIGURE 1.15 Pseudoparenchymatous thallus of *Palmaria palmata.*

Statospores are endogenous cysts formed within the vegetative cell by members of Chrysophyceae such as *Ochromonas* spp. The cyst walls consist predominantly of silica and so are often preserved as fossils. These statospores are spherical or ellipsoidal, often ornamented with spines or other projections. The wall is pierced by a pore, sealed by an unsilicified bung, and within the cyst lie a nucleus, chloroplasts, and abundant reserve material. After a period of dormancy the cyst germinates and liberates its contents in the form of one to several flagellated cells.

Akinetes are of widespread occurrence in the blue-green and green algae. They are essentially enlarged vegetative cells that develop a thickened wall in response to limiting environmental nutrients or limiting light. Figure 1.18 shows the akinetes of *Anabaena cylindrica* (Cyanophyta). They are extremely resistant to drying and freezing and function as a long-term anaerobic storage of the genetic material of the species. Akinetes can remain in sediments for many years, enduring very harsh conditions, and remain viable to assure the continuance of the species. When suitable conditions for vegetative growth are restored, the akinete germinates into new vegetative cells.

SEXUAL REPRODUCTION

Gametes may be morphologically identical with vegetative cells or markedly differ from them, depending on the algal group. The main difference is obviously the DNA content that is haploid instead of diploid. Different combinations of gamete types are possible. In the case of isogamy, gametes are both motile and indistinguishable. When the two gametes differ in size, we have heterogamy. This combination occurs in two types: anisogamy, where both gametes are motile, but one is small (sperm) and the other is large (egg); oogamy, where only one gamete is motile (sperm) and fuses with the other that is non-motile and very large (egg).

FIGURE 1.16 Cell division in *Euglena* sp. (Bar: 10 μm)

Algae exhibit three different life cycles with variation within different groups. The main difference is the point where meiosis occurs and the type of cells it produces, and whether there is more than one free-living stage in the life cycle.

Haplontic or Zygotic Life Cycle

This cycle is characterized by a single predominant haploid vegetative phase, with the meiosis taking place upon germination of the zygote. *Chlamydomonas* (Chlorophyta) (Figure 1.19) exhibits this type of life cycle.

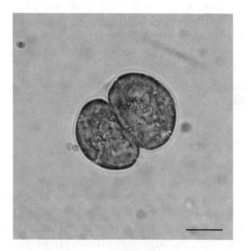

FIGURE 1.17 Zoospores of *Chlamydomonas* sp. within the parental cell wall. (Bar: 10 μm)

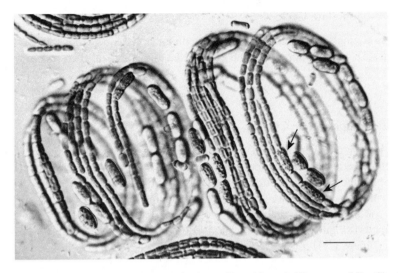

FIGURE 1.18 Akinetes (arrows) of *Anabaena cylindrica.* (Bar: 10 μm.) (Courtesy of Dr. Claudio Sili.)

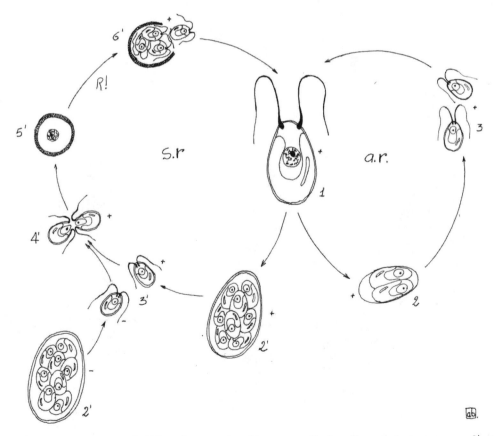

FIGURE 1.19 Life cycle of *Chlamydomonas* sp.: 1, mature cell; 2, cell producing zoospores; 2′, cell producing gametes (strain+ and strain−); 3, zoospores; 3′, gametes; 4′, fertilization; 5′, zygote; 6′, release of daughter cells. R!, meiosis; a.r., asexual reproduction; s.r., sexual reproduction.

Diplontic or Gametic Life Cycle

This cycle has a single predominant vegetative diploid phase, and the meiosis gives rise to haploid gametes. Diatoms (Figure 1.20) and *Fucus* (Heterokontophyta) (Figure 1.21) have a diplontic cycle.

Diplohaplontic or Sporic Life Cycles

These cycles present an alternation of generation between two different phases consisting in a haploid gametophyte and a diploid sporophyte. The gametophyte produces gametes by mitosis; the sporophyte produces spores through meiosis. Alternation of generation in the algae can be isomorphic, in which the two phases are morphologically identical as in *Ulva* (Chlorophyta) (Figure 1.22) or heteromorphic, with the predominance of the sporophyte as in *Laminaria* (Heterokontophyta) (Figure 1.23) or with the predominance of the gametophyte as in *Porphyra* (Rhodophyta) (Figure 1.24).

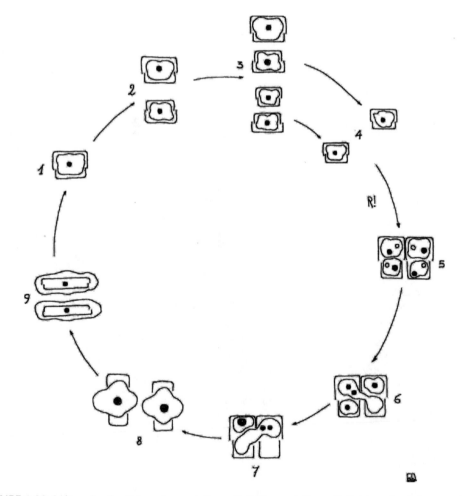

FIGURE 1.20 Life cycle of a diatom: 1, vegetative cell; 2, 3, vegetative cell division; 4, minimum cell size; 5, gametogenesis; 6, 7, fertilization; 8, auxospores; 9, initial cells. R!, meiosis.

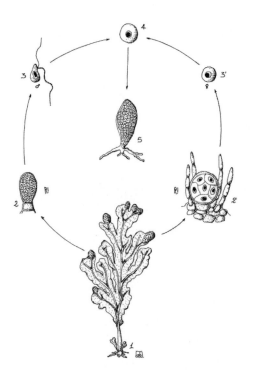

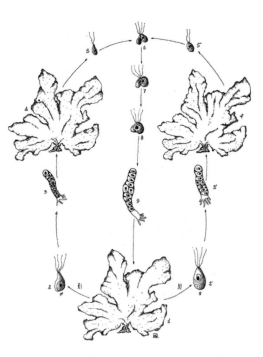

FIGURE 1.21 Life cycle of *Fucus* sp.: 1, sporophyte; 2, anteridium; 2′, oogonium; 3, sperm; 3′, egg; 4, zygote; 5, young sporophyte. R!, meiosis.

FIGURE 1.22 Life cycle of *Ulva* sp.: 1, sporophyte; 2, male zoospore; 2′, female zoospore; 3, young male gametophyte; 3′, young female gametophyte; 4, male gametophyte; 4′, female gametophyte; 5, male gamete; 5′, female gamete; 6–8, syngamy; 9, young sporophyte. R!, meiosis.

SUMMARIES OF THE TEN ALGAL DIVISIONS

The phylogenetic reconstruction adopted in this book is intended to be more or less speculative because most of the evidence has been lost and many organisms have left no trace in the fossil records. Normally, systematic groups and categories arranged in a hierarchical system on the basis of similarities between organisms replace it. Each of these natural groups consists of a set of organisms that are more closely related to each other than to organisms of a different group. This interrelationship is inferred from the fundamental similarities in their traits (homologies) and is thought to reflect fundamental similarities in their genomes, as a result of common descent. Historically, the major groups of algae are classified into Divisions (the equivalent taxon in the zoological code was the Phylum) on the basis of pigmentation, chemical nature of photosynthetic storage product, photosynthetic membranes' (thylakoids) organization and other features of the chloroplasts, chemistry and structure of cell wall, number, arrangement, and ultrastructure of flagella (if any), occurrence of any other special features, and sexual cycles. Recently, all the studies that compare the sequence of macromolecules genes and the 5S, 18S, and 28S ribosomal RNA sequences tend to assess the internal genetic coherence of the major divisions such as Cyanophyta and Procholophyta and Glaucophyta, Rhodophyta, Heterokontophyta, Haptophyta, Cryptophyta, Dinophyta, Euglenophyta, Chlorarachniophyta, and Chlorophyta. This confirms that these divisions are non-artificial, even though they were originally defined on the basis of morphology alone. Table 1.4 attempts to summarize the main characteristics of the different algal divisions.

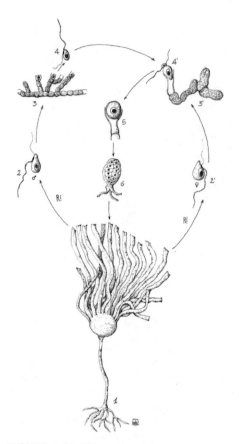

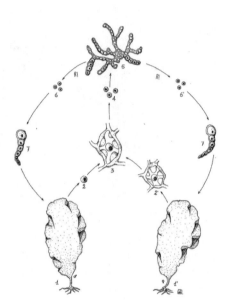

FIGURE 1.23 Life cycle of *Laminaria* sp.: 1, sporophyte; 2, male zoospore; 2′, female zoospore; 3, male gametophyte; 3′, female gametophyte; 4, sperm; 4′, egg and fertilization; 5, zygote; 6, young sporophyte. R!, meiosis.

FIGURE 1.24 Life cycle of *Porphyra* sp.: 1, male gametophyte; 1′, female gametophyte; 2, sperm; 2′, egg; 3, fertilization and zygote; 4, spores; 5, sporophyte; 6, male spore; 6′, female spores; 7, young male gametophyte; young female gametophyte. R!, meiosis.

CYANOPHYTA AND PROCHLOROPHYTA

All blue-green algae (Figure 1.25) and prochlorophytes (Figure 1.26) are non-motile Gram-negative eubacteria. In structural diversity, blue-green algae range from unicells through branched and unbranched filaments to unspecialized colonial aggregations and are possibly the most widely distributed of any group of algae. They are planktonic, occasionally forming blooms in eutrophic lakes, and are an important component of the picoplankton in both marine and freshwater systems; benthic, as dense mats on soil or in mud flats and hot springs, as the "black zone" high on the seashore, and as relatively inconspicuous components in most soils; and symbiotic in diatoms, ferns, lichens, cycads, sponges, and other systems. Numerically these organisms dominate the ocean ecosystems. There are approximately 10^{24} cyanobacterial cells in the oceans. To put that in perspective, the number of cyanobacterial cells in the oceans is two orders of magnitude more than all the stars in the sky. Pigmentation of cyanobacteria includes chlorophyll *a*, blue and red phycobilins (phycoerythrin, phycocyanin, allophycocyanin, and phycoerythrocyanin), and carotenoids. These accessory pigments lie in the phycobilisomes, located in rows on the outer surface of the thylakoids. Their thylakoids, which lie free in the cytoplasm, are not arranged in stacks, but singled

TABLE 1.4
The Main Pigments, Storage Products, and Cell Coverings of the Algal Divisions

Division	Pigments				Storage Products
	Chlorophylls	Phycobilins	Carotenoids	Xanthophylls	
Cyanophyta	*a*	*c*-Phycoerythrin *c*-Phycocyanin Allophycocyanin Phycoerythrocyanin	β-Carotene	Myxoxanthin Zeaxanthin	Cyanophycin (argine and asparagine polymer) Cyanophycean starch (α-1,4-glucan)
Prochlorophyta	*a, b*	Absent	β-Carotene	Zeaxanthin	Cyanophycean starch (α-1, 4-glucan)
Glaucophyta	*a*	*c*-Phycocyanin Allophycocyanin	β-Carotene	Zeaxanthin	Starch (α-1,4-glucan)
Rhodophyta	*a*	*r,b*-Phycoerythrin *r*-Phycocyanin Allophycocyanin	α- and β-Carotene	Lutein	Floridean starch (α-1,4-glucan)
Cryptophyta	*a, c*	Phycoerythrin-545 *r*-Phycocyanin	α-, β-, and ε-Carotene	Alloxanthin	Starch (α-1,4-glucan)
Heterokontophyta	*a, c*	Absent	α-, β-, and ε-Carotene	Fucoxanthin, Violaxanthin	Chrysolaminaran (β-1,3-glucan)
Haptophyta	*a, c*	Absent	α- and β-Carotene	Fucoxanthin	Chrysolaminaran (β-1,3-glucan)
Dinophyta	*a, b, c*	Absent	β-Carotene	Peridinin, Fucoxanthin, Diadinoxanthin Dinoxanthin Gyroxanthin	Starch (α-1,4-glucan)
Euglenophyta	*a, b*	Absent	β- and γ-Carotene	Diadinoxanthin	Paramylon (β-1,3-glucan)
Chlorarachniophyta	*a, b*	Absent	Absent	Lutein, Neoxanthin, Violaxanthin	Paramylon (β-1,3-glucan)
Chlorophyta	*a, b*	Absent	α-, β-, and γ-Carotene	Lutein Prasinoxanthin	Starch (α-1,4-glucan)

and equidistant, in contrast to prochlorophytes and most other algae, but similar to Rhodopyta and Glaucophyta.

The reserve polysaccharide is cyanophycean starch, stored in tiny granules lying between the thylakoids. In addition, these cells often contain cyanophycin granules, that is, polymer of arginine and asparagine. Some marine species also contain gas vesicles used for buoyancy regulation. In some filamentous cyanobacteria, heterocysts and akinetes are formed. Heterocysts are vegetative cells that have been drastically altered (loss of photosystem II, development of a thick, glycolipid cell wall) to provide the necessary anoxygenic environment for the process of nitrogen fixation (Figure 1.27). Some cyanobacteria produce potent hepato- and neurotoxins.

Prochlorophytes can be unicellular or filamentous, and depending on the filamentous species, they can be either branched or unbranched. They exist as free-living components of pelagic

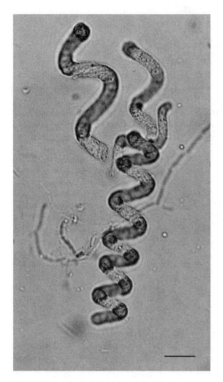

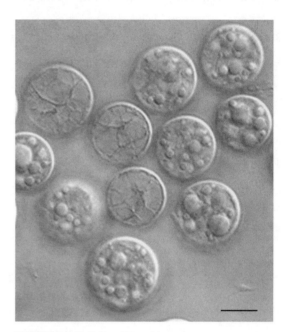

FIGURE 1.25 Trichome of *Arthrospira* sp. (Bar: 20 μm.)

FIGURE 1.26 Cells of *Prochloron* sp. (Bar: 10 μm.)

nanoplankton and obligate symbionts within marine didemnid ascidians and holothurians, and are mainly limited to living in tropical and subtropical marine environments, with optimal growth temperature at about 24°C. Prochlorophytes possess chlorophylls *a* and *b* similar to euglenoids and land plants, but lack phycobilins, and this is the most significance difference between these and cyanobacteria; other pigments are β-carotene and several xanthophylls (zeaxanthin is the principal one). Their thylakoids, which lie free in the cytoplasm, are arranged in stacks. Prochlorophytes

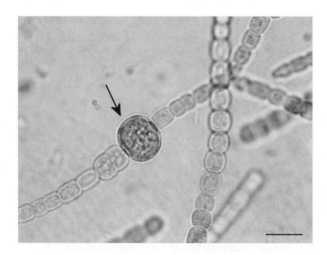

FIGURE 1.27 Heterocyst (arrow) of *Anabaena azollae*. (Bar: 10 μm.)

have a starch-like reserve polysaccharide. These prokaryotes contribute a large percentage of the total organic carbon in the global tropical oceans, making up to 25–60% of the total chlorophyll *a* biomass in the tropical and subtropical oceans. They are also able to fix nitrogen, though not in heterocysts. Both blue-green algae and prochlorophytes contain polyhedral bodies (carboxysomes) containing RuBisCo (ribulose bisphospate carboxylase/oxygenase, the enzyme that converts inorganic carbon to reduced organic carbon in all oxygen evolving photosynthetic organisms), and have similar cell walls characterized by a peptoglycan layer. Blue-green algae and Prochlorophytes can be classified as obligate photoautotrophic organisms. Reproduction in both divisions is strictly asexual, by simple cell division of fragmentation of the colony or filaments.

GLAUCOPHYTA

Glaucophytes (Figure 1.28) are basically unicellular flagellates with a dorsiventral construction; they bear two unequal flagella, which are inserted in a shallow depression just below the apex of the cell. Glaucophytes are rare freshwater inhabitants, sometimes collected also from soil samples. They posses only chlorophyll *a* and accessory pigments such as phycoerythrocyanin, phycocyanin, and allophycocyanin are organized in phycobilisomes. Carotenoids such as β-carotene and xanthophylls such as zeaxanthin are also present in their chloroplast. This unusual chloroplast lies in a special vacuole and presents a thin peptidoglycan wall located between the two plastid outer membranes. Thylakoids are not stacked. The chloroplast DNA is concentrated in the center of the chloroplast, where typically carboxysomes are present, which contain the RuBisCo enzyme. Starch is the reserve polysaccharide, which is accumulated in granular form inside the cytoplasm, but outside the chloroplast. Glaucophytes live photoautotrophically with the aid of blue-green plastids often referred to as cyanelles. Cyanelles are presumed to be phylogenetically derived from endosymbiotic cyanobacterium. Sexual reproduction is unknown in this division.

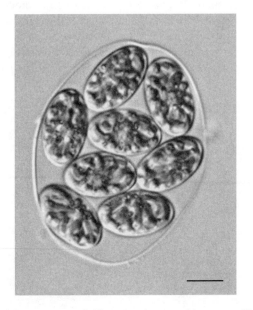

FIGURE 1.28 A group of eight autospore of *Glaucocystis nostochinearum* still retained within parent cell wall. (Bar: 10 μm.)

RHODOPHYTA

The red algae mostly consist of seaweeds but also include the genera of free-living unicellular microalgae. The class Bangiophyceae (Figure 1.24) retains morphological characters that are found in the ancestral pool of red algae and range from unicells to multicellular filaments or sheet-like thalli. The Floridophyceae (Figure 1.29) includes morphologically complex red algae and are widely considered to be a derived, monophyletic group. Rhodophyta inhabit prevalently marine ecosystems but they are also present in freshwater and terrestrial environment. The lack of any flagellate stages and the presence of accessory phycobiliproteins organized in phycobilisomes (shared with Cyanobacteria, Cryptophyta, and Glaucophyta) are unique features of this division; chlorophyll *a* is the only chlorophyll. Chloroplasts are enclosed by a double unit membrane; thylakoids do not stack at all, but lie equidistant and singly within the chloroplast. One thylakoid is present around the periphery of the chloroplast, running parallel to the chloroplast internal membrane. The chloroplastic DNA is organized in blebs scattered throughout the whole chloroplast. The most important storage product is floridean starch, an α-1,4-glucan polysaccharide. Grains of this starch are located only in the cytoplasm, unlike the starch grains produced in the Chlorophyta, which lie inside the chloroplasts. Most rhodophytes live photoautotrophically. In the great majority of red algae, cytokinesis is incomplete. Daughter cells are separated by the pit connection, a proteinaceous plug that fills the junction between cells; this connection successively becomes a plug. Species in which sexual reproduction is known generally have an isomorphic or heteromorphic diplohaplontic life cycle; haplontic life cycle is considered an exception.

HETEROKONTOPHYTA

One of the defining features of the members of this division is that when two flagella are present, they are different. Flagellate cells are termed heterokont, that is, they possess a long mastigonemate flagellum, which is directed forward during swimming, and a short smooth one that points backwards along the cell. Chrysophyceae contain single-celled individuals (Figure 1.2) as well as quite colonial forms. Xanthophyceae can be unicellular (coccoids or not) filamentous, but the most distinctive species are siphonous (Figure 1.14). All known species of Eustigmatophyceae are green coccoid unicells either single (Figure 1.1), in pairs or in colonies. Bacillariophyceae are a group of unicellular brown pigmented cells that are encased by a unique type of silica wall, composed of two overlapping frustules that fit together like a box and lid (Figure 1.30 and Figure 1.31). Raphidophyceae are unicellular wall-less heterokonts (Figure 1.32). Dictyochophyceae, known as silicoflagellates, are unicells that bear a single flagellum with mastigonemes

FIGURE 1.29 Frond of *Rhodophyllis acanthocarpa.* (Bar: 5 cm.)

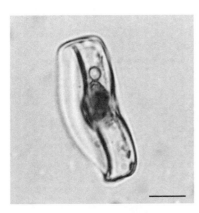

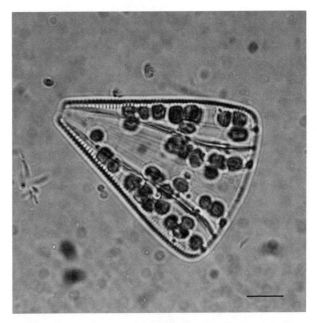

FIGURE 1.30 Marine diatom.   **FIGURE 1.31** Freshwater diatom. (Bar: 20 μm.)
(Bar: 10 μm.)

(Figure 1.33). Phaeophyceae are multicellular, from branched filaments to massive and complex kelp (Figure 1.34). Other groups of algae have been described as belonging to this division, such as Pelagophyceans and Sarcinochrysidaleans *sensu* Graham and Wilkox (2000) and Parmales *sensu* Van den Hoek et al. (1995). Heterokontophyta are mostly marine; but they can be found also in freshwater and terrestrial habitats. They show a preponderance of carotenoids over chlorophylls that result in all groups having golden rather than grass green hue typical of other major algal divisions. The members of this division possess chlorophylls a, c_1, c_2, and c_3 with the exception of the Eustigmatophyceae that have only chlorophyll a. The principal accessory pigments are β-carotene, fucoxanthin, and vaucheriaxanthin. The thylakoids are grouped into stacks of three, called lamellae. One lamella usually runs along the whole periphery of the chloroplast, which is termed girdle lamella, absent only in the Eustigmatophyceae. The chloroplasts are enclosed in their own double membrane and also by a fold of the endoplasmic reticulum. The chloroplastic DNA is usually arranged in a ring-shaped nucleoid. Dictyochophyceae species possess several nucleoids scattered inside the chloroplast. The main reserve polysaccharide is chrysolaminarin, a β-1,3-glucan, located inside the cytoplasm in special vacuoles. The eyespot consists of a layer of globules, enclosed within the chloroplast, and together with the photoreceptor, located in the smooth flagellum, forms the photoreceptive apparatus. The members of this division can grow photoautotrophically but can also combine different nutritional strategies such as heterotrophy. The Heterokontophyta species that reproduce sexually have a haplontic (Chrysophyceae), diplontic (Bacillariophyceae) or diplohaplontic (Phaeophyceae) life cycle.

HAPTOPHYTA

The great majority of Haptophyta are unicellular, motile, palmelloid, or coccoid (Figure 1.35), but a few form colonies or short filaments. These algae are generally found in marine habitats, although there are a number of records from freshwater and terrestrial environments. Flagellate cells bear two naked flagella, inserted either laterally or apically, which may have different length. A structure apparently found only in algae of this division is the haptonema, typically a long thin organelle

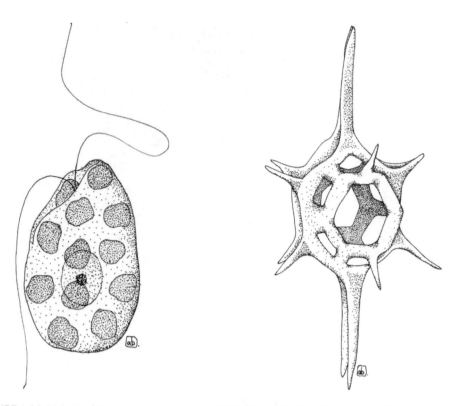

FIGURE 1.32 Unicell of *Heterosigma akashiwo*.   **FIGURE 1.33** The silicoflagellate *Distephanus speculum*.

reminiscent of a flagellum but with a different ultrastructure. The chloroplast contains only chloro-phylls *a,* c_1, and c_2. The golden yellow brown appearance of the chloroplast is due to accessory pigments such as fucoxanthin, β-carotene, and other xanthins. Each chloroplast is enclosed

FIGURE 1.34 Frond of *Bellotia eriophorum*. (Bar: 5 cm.)

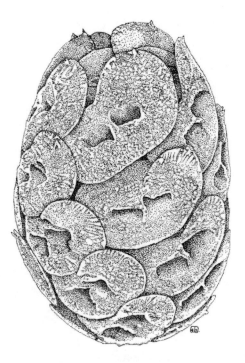

FIGURE 1.35 Unicell of *Helicosphaera carteri*.

within a fold of endoplasmic reticulum, which is continuous with the nuclear envelope. Thylakoids are stacked in threes, and there are no girdle lamellae. The nucleic DNA is scattered throughout the chloroplast as numerous nucleoids. When present as in *Pavlova*, the eyespot consists in a row of spherical globules inside the chloroplast; no associated flagellar swelling is present. The most important storage product is the polysaccharide chrysolaminarine. The cell surface is typically covered with tiny cellulosic scales or calcified scales bearing spoke-like fibrils radially arranged. Most haptophytes are photosynthetic, but heterotrophic nutrition is also possible. Phagotropy is present in the forms that lack a cell covering. A heteromorphic diplohaplontic life cycle has been reported, in which a diploid planktonic flagellate stage alternates with a haploid benthic filamentous stage.

CRYPTOPHYTA

The unicellular flagellates belonging to the division Cryptophyta are asymmetric cells dorsiventrally constructed (Figure 1.36). They bear two unequal, hairy flagella, subapically inserted, emerging from above a deep gullet located on the ventral side of the cell. The wall of this gullet is lined by numerous ejectosomes similar to trichocysts. Cryptophytes are typically free-swimming in freshwater and marine habitats; palmelloid phases can also be formed, and some members are known to be zooxanthellae in host invertebrates or within certain marine ciliates. Cryptophyta possess only chlorophylls *a* and c_2. Phycobilins are present in the thylakoid lumen rather than in phycobilisomes. The chloroplasts, one or two per cell, are surrounded by a fold of the endoplasmic reticulum. In the space between these membranes a peculiar organelle, the nucleomorph, is located. This organelle can be interpreted as the vestigial nucleus of the red algal endosymbiont that gave rise to the chloroplasts of the Cryptophyta. Thylakoids are arranged in pairs, with no girdle lamellae. The pyrenoid projects out from the inner side of the chloroplast. The chloroplast DNA is condensed in small nucleoids scattered inside the chloroplast. The reserve polysaccharide accumulates

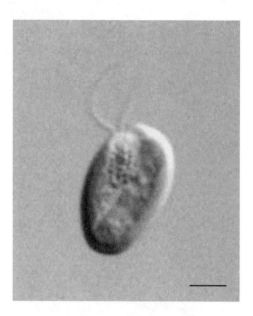

FIGURE 1.36 Unicell of *Cryptomonas* sp. (Bar: 6 μm.)

in the periplastidial space as starch granules. Sometimes an eyespot formed by spherical globules is present inside the plastid, but it is not associated with the flagella. The cell is enclosed in a stiff, proteinaceous periplast, made of polygonal plates. Most forms are photosynthetic, but heterotrophic nutrition also occurs. The primary method of reproduction is simply by longitudinal cell division, but sexual reproduction has recently been documented.

DINOPHYTA

The members of this division are typical unicellular flagellates (Figure 1.37) but can be also non-flagellate, ameboid, coccoid, palmelloid, or filamentous. Dinoflagellates have two flagella with independent beating pattern, one training and the other girdling that confers characteristic rotatory swimming whirling motion. Flagella are apically inserted (desmokont type) or emerge from a region close to the midpoint of the ventral side of the cell (dinokont type). Most dinoflagellates are characterized by cell-covering components that lie beneath the cell membrane. Around the cell there is a superficial layer of flat, polygonal vesicles, which can be empty or filled with cellulose

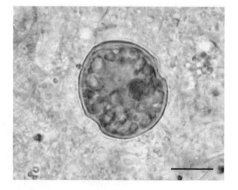

FIGURE 1.37 A marine dinoflagellate. (Bar: 30 μm.)

plates. In dinokont type dinoflagellates, these thecal plates generally form a bi-partite armor, consisting of an upper, anterior half and a lower, posterior half, separated by a groove known as cingulum where the transversal flagellum is located (Figure 1.38). A smaller groove, the sulcus, extends posteriorly from the cingulum, and hosts the longitudinal flagellum. The two flagella emerge from a pore located at the intersection of the two grooves. Very often they are important components of the microplankton of freshwater and marine habitats. Though most are too large (2–2000 μm) to be consumed by filter feeders, they are readily eaten by larger protozoa, rotifer, and planktivorous fishes. Some Dinoflagellates are invertebrate parasites, others are endosymbionts (zooxanthellae) of tropical corals. Dinoflagellates possess chlorophylls a, b, c_1, and c_2, fucoxanthin, other carotenoids, and xanthophylls such as peridinin, gyroxanthin diester, dinoxanthin, diadinoxantin, and fucoxanthin. The chloroplasts, if present, are surrounded by three membranes. Within the chloroplasts the thylakoids are for the most part united in a stack of three. The chloroplast DNA is localized in small nodules scattered in the whole chloroplast, with typical pyrenoids. A really complex photoreceptive system is present in the dinophytes such as *Warnowia polyphemus*, *Warnowia pulchra*, or *Erythropsidinium agile* consisting of a "compound eye" composed of a lens and a retinoid. Most dinoflagellates are distinguished by a dinokaryon, a special eukaryotic nucleus involving fibrillar chromosomes that remain condensed during the mitotic cycles. The principal reserve polysaccharide is starch, located as grains in the cytoplasm, but oil droplets are present in some genera. At the surface of the cell there are trichocysts which discharge explosively when stimulated. Besides photoautotrophy, dinoflagellates exhibit an amazing diversity of nutritional types because about half of the known species lack plastids and are therefore obligate heterotrophic. Some are notorious for nuisance blooms and toxin production, and many exhibit bioluminescence. Dinophyceae have generally a haplontic life history.

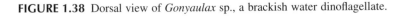

FIGURE 1.38 Dorsal view of *Gonyaulax* sp., a brackish water dinoflagellate.

EUGLENOPHYTA

Euglenophyta include mostly unicellular flagellates (Figure 1.39) although colonial species are common. They are widely distributed, occurring in freshwater, brackish and marine waters, most soils, and mud. They are especially abundant in highly heterotrophic environments. The flagella arise from the bottom of a cavity called reservoir, located in the anterior end of the cell. Cells can also ooze their way through mud or sand by a process known as metaboly, a series of flowing movements made possible by the presence of the pellicle, a proteinaceous wall which lies inside the cytoplasm. The pellicle can have a spiral construction and can be ornamented. The members of this division share their pigmentation with prochlorophytes, green algae, and land plants, because they have chlorophylls a and b, β- and γ-carotenes, and xanthins. However, plastids could be colorless or absent in some species. As in the Dinophyta the chloroplast envelope consists of three membranes. Within the chloroplasts, the thylakoids are usually in groups of three, without a girdle lamella and pyrenoids may be present. The chloroplast DNA occurs as a fine skein of tiny granules. The photoreceptive system consisting of an orange eyespot located free in the cytoplasm and the true photoreceptor located at the base of the flagellum can be considered unique among unicellular algae. The reserve polysaccaccharide is paramylon, β-1,3-glucan, stored in the granules scattered inside the cytoplasm and not in the chloroplasts like the starch of the Chlorophyta. Though these possess algae chlorophylls, they are not photoautotrophic but rather obligate mixotrophic, because they require one or more vitamins of the B group. Some color-less genera are phagotrophic, with specialized cellular organelle for capture and ingestion of prey; some others are osmotrophic. Some of the pigmented genera are facultatively heterotrophic. Only asexual reproduction is known in this division. Euglenophyta posses unique cellular and biochemi-cal features that place these microorganisms closer to trypanosomes than to any other algal group.

CHLORARACHNIOPHYTA

They are naked, uninucleate cells that form a net-like plasmodium via filopodia (Figure 1.40). The basic life cycle of these algae comprises ameboid, coccoid, and flagellate cell stages. The ovoid

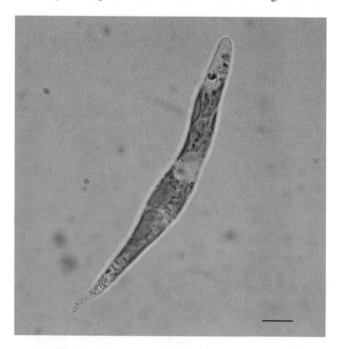

FIGURE 1.39 Unicell of *Euglena mutabilis*. (Bar: 10 μm.)

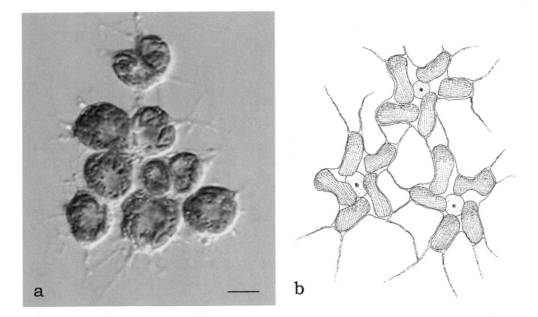

FIGURE 1.40 Plasmodial reticulum of *Chlorarachnion*, bright field microscope image (a) and schematic drawing (b). (Bar: 4 μm.)

zoospores bear a single flagellum that during swimming wraps around the cell. Chlorarachniophytes are marine. They posses chlorophylls *a* and *b*. Each chloroplast has a prominent projecting pyrenoid and is surrounded by four-membrane envelope. Thylakoids are grouped in stacks of one to three. A nucleomorph is present between the second and third membranes of the chloroplast envelope. The origin of this organelle is different from the origin of the cryptophytes nucleomorph, because the chlorarachniophytes originated from a green algal endosymbiont. Paramylon (β-1,3-glucan) is the storage carbohydrate. They are phototrophic and phagotrophic engulfing bacteria, flagellates, and eukaryotic algae. Asexual reproduction is carried out by either normal mitotic cell division or zoospore formation. Sexual reproduction characterized by heterogamy has been reported for only two species.

CHLOROPHYTA

A great range of somatic differentiation occurs within the Chlorophyta, ranging from flagellates to complex multicellular thalli differentiated into macroscopic organs. The different level of thallus organization (unicellular, colonial, filamentous, siphonous, and parenchimatous) have traditionally served as the basis of classification of this division. Prasinophyceae are unicellular motile algae covered on their cell body and flagella by non-mineralized organic scales (Figure 1.41). The class Chlorophyceae comprises flagellated cells even naked or covered by a cell wall termed theca (Figure 1.42). All Ulvophyceae known to date are sessile organisms having walled vegetative cells. Except for a small group of species, the thalli are usually multicellular or coenocytic during at least some part of the life history. Many species have microscopic, filamentous thalli, but most are macroscopic seaweeds, capable of considerable morphological differentiation (Figure 1.22). Cladophorophyceae take the form of branched or unbranched filaments of multinucleate cells with periodic cross walls (Figure 1.11). The organization of the thallus in the class of Briopsidophyceae is always syphonous; syphonous thalli can combine to form fairly complex tissues (Figure 1.43). The Zygnematophyceae species are either coccoids or filamentous (Figure 1.8). In all the Trentepohliophyceae the thalli consist of branched or unbranched filaments with uninucleate cells

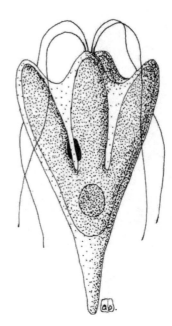

FIGURE 1.41 Unicell of *Pyramimonas longicauda.*

(Figure 1.44). Klebsormidiophyceae have coccoid and branched or unbranched filamentous forms (Figure 1.45). Charophyceae have macroscopic thalli, which exhibit the characteristic of both the filamentous and syphonous levels of organization (Figure 1.46). Dasycladophyceae have syphonous thalli in many species encrusted with calcium carbonate (Figure 1.47). Chlorophytes show a wide diversity in the number and arrangements of flagella associated with individual cells (one or up to eight in the apical or subapical region). Flagellated cells are isokont, which means the flagella are similar in structure, but could differ in length. These algae are ubiquitous in freshwater, marine, and terrestrial habitats. Chlorophyta possess chlorophylls *a* and *b*, β- and γ-carotene, and several xanthophylls as accessory pigments. Chloroplasts are surrounded by a two-membrane envelope without any endoplasmic reticulum membrane. Within the chloroplasts, thylakoids are

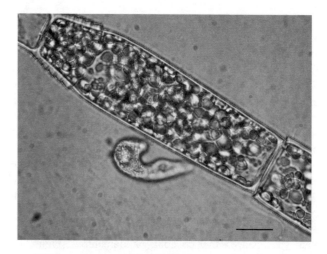

FIGURE 1.42 Filament of *Oedogonium* sp., with a *Peranema* sp. cell. (Bar: 20 μm.)

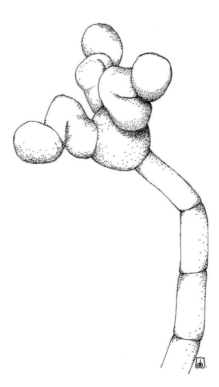

FIGURE 1.43 Thallus of *Codium* sp. (Bar: 2 cm.)　　**FIGURE 1.44** Thallus of *Trentepohlia arborum.*

stacked to form grana. Pyrenoids, if present, are embedded within the chloroplast and often penetrated by thylakoids. The circular molecules of chloroplast DNA are concentrated in numerous
small blobs (1–2 μm in diameter). The most important reserve polysaccharide is starch, which
occurs as a grain inside the chloroplasts; glucan is present in the cell wall of Cladophorophyceae
and Bryopsidophyceae and β-1,4 mannan in Dasycladophyceae. Eyespot, if present, is located
inside the chloroplast, and consists of a layer of carotenoid-containing lipid droplets between the
chloroplast envelope and the outermost thylakoids. Chlorophyta are photoautotropic but can be
also heterotropic. No sexuality is known in Prasinophyceae but the genus *Nephroselmis* has a haplontic life cycle. In Chlorophyceae, reproduction is usually brought about through the formation of
flagellate reproductive cells. The life cycle is haplontic. In Ulvophyceae the life cycle is haplontic,
isomorphic, and diplohaplontic. In Cladophorophyceae and Trentepohliophyceae, the life cycle of
reproductive species are diplohaplontic and isomomorphic. In Bryopsidophyceae, Klebsormidiophyceae, Charophyceae, Zygnematophyceae, and Dasycladophyceae life cycle is haplontic. As
the advanced land plants and the "modest" Trentepohliophyceae class possess the same mechanism
of cell division, that is, using the phragmoplast disc where the cells will divide, plant evolution
researchers believe that the land plants derived directly from this fresh-water algae class.

ENDOSYMBIOSIS AND ORIGIN OF EUKARYOTIC ALGAE

Within the algae, different evolutionary lineages are discernable. Three major eukaryotic photosynthetic groups have descended from a common prokaryotic ancestor, through an endosymbiotic
event. Therefore, these algae possess primary plastid, that is, derived directly from the prokaryotic
ancestor. Other algal groups have acquired their plastids via secondary (or tertiary) endosymbiosis,
where a eukaryote already equipped with plastids is preyed upon by a second eukaryotic cell.

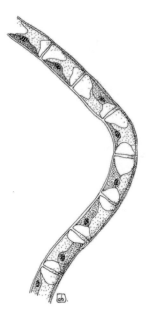

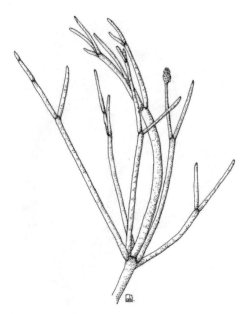

FIGURE 1.45 Filament of *Klebsormidium* sp.   **FIGURE 1.46** Thallus of *Nitella* sp.

Endosymbiotic process produced nested cellular compartments one inside the other, which can give information about the evolutionary history of the algae containing them.

Cyanobacteria evolved more than 2.8 billion years ago and have played fundamental roles in driving much of the ocean carbon, oxygen, and nitrogen fluxes from that time to present. The evolution of cyanobacteria was a major turning point in biogeochemistry of Earth. Prior to the

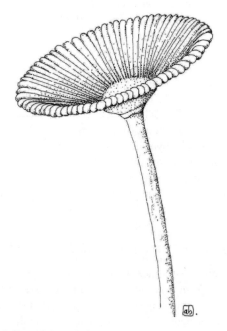

FIGURE 1.47 Portion of the thallus of *Acetabularia* sp.

appearance of these organisms, all photosynthetic organisms were anaerobic bacteria that used light to couple the reduction of carbon dioxide to the oxidation of low free energy molecules, such as H_2S or preformed organics. Cyanobacteria developed a metabolic process, the photosynthesis, which exploits the energy of visible light to oxidize water and simultaneously reduces CO_2 to organic carbon represented by $(CH_2O)n$ using light energy as a substrate and chlorophyll a as a requisite catalytic agent. Formally oxygenic photosynthesis can be summarized as:

$$CO_2 + H_2O + \text{light} \xrightarrow{\text{Chlorophyll } a} (CH_2O)n + O_2$$

All other oxygen producing algae are eukaryotic, that is, they contain internal organelles, including a nucleus, one or more chloroplasts, one or more mitochondria, and, most importantly, in many cases they contain a membrane-bound storage compartment or vacuole. The three major algal lineages of primary plastids are the Glaucophyta lineage, the Chlorophyta lineage, and the Rhodopyta lineage (Figure 1.48).

Glaucophyta lineage occupies a key position in the evolution of plastids. Unlike other plastids, the plastids of glaucophytes retain the remnant of a Gram-negative bacterial cell wall of the type found in cyanobacteria, with a thin peptidoglycan cell wall and cyanobacterium-like pigmentation that clearly indicate its cyanobacterial ancestry. In fact, the *Cyanophora paradoxa* plastid genome shows the same reduction as other plastids when compared with free-living cyanobacteria (it is 136 kb and contains 191 genes). The peptidoglycan cell wall of the plastid is thus a feature retained from their free-living cyanobacterial ancestor. In this context, the Glaucophyta are remarkable only for their retention of an ancestral character present in neither green nor red plastids. No certain case of a secondary plastid derived from Glaucophyta is known.

Green algae (Chlorophyta) constitute the second lineage of primary plastids. The simple two-membrane system surrounding the plastid, the congruence of phylogenies based on nuclear and organellar genes, and the antiquity of the green algae in the fossil record all indicate that the green algal plastid is of primary origin. In these chloroplasts, chlorophyll b was synthesized as a secondary pigment and phycobiliproteins were lost. Another hypothesis is that the photosynthetic ancestor of green lineage was a prochlorophyte that possessed chlorophylls a and b and lacked phycobiliproteins.

The green lineage played a major role in oceanic food webs and the carbon cycle from about 2.2 billion years ago until the end-Permian extinction, approximately 250 million years ago. It was this similarity to the pigments of plants that led to the inference that the ancestors of land plants (i.e., embryophytes) would be among the green algae, and is clear that phylogenetically plants are a group of green algae adapted to life on land. Euglenophyta and Chlorarachniophyta are derived from this primary plastid lineage by secondary endosymbiosis; the green algal plastid present in Euglenophyta is bounded by three membranes, while the green algal plastid present in the Chlorarachniophyta is bound by four membranes.

Since that time, however, a second group of eukaryotes has risen to ecological prominence; that group is commonly called the "red lineage." The plastids of the red algae (Rhodophyta) constitute the third primary plastid lineage. Like the green algae, the red algae are an ancient group in the fossil record, and some of the oldest fossils interpreted as being of eukaryotic origin are often referred to the red algae, although clearly these organisms were very different from any extant alga. Like those of green algae, the plastids of red algae are surrounded by two membranes. However, they are pigmented with chlorophyll a and phycobiliproteins, which are organized into phycobilisomes. Phycobilisomes are relatively large light-harvesting pigment/protein complexes that are water-soluble and attached to the surface of the thylakoid membrane. Thylakoids with phycobilisomes do not form stacks like those in other plastids, and consequently the plastids of red algae (and glaucophytes) bear an obvious ultrastructural resemblance to cyanobacteria.

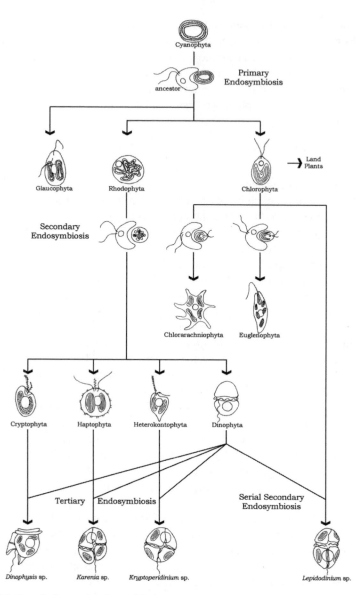

FIGURE 1.48 Algal evolution and endosymbiotic events.

A number of algal groups have secondary plastids derived from those of red algae, including several with distinctive pigmentation. The cryptomonads (Cryptophyta) were the first group in which secondary plastids were recognized on the basis of their complex four membrane structure. Like red algae, they have chlorophyll *a* and phycobiliproteins, but these are distributed in the intrathylakoidal space rather than in the phycobilisomes found in red algae, Glaucophyta, and Cyanophyta. In addition, cryptomonads possess a second type of chlorophyll, chlorophyll *c*, which is found in the remaining red lineage plastids. These groups, which include the Heterokontophyta (including kelps, diatoms, chrysophytes, and related groups), Haptophyta (the coccolithophorids), and probably those dinoflagellates (Dinophyta) pigmented with peridinin, have chlorophylls *a* and *c*, along with a variety of carotenoids, for pigmentation. Stacked thylakoids are found in those lineages (including the cryptomonads) that lack phycobilisomes. The derivation of chlorophyll *c* containing plastids from the red algal lineage is still somewhat conjectural, but recent analyses of both gene sequences and gene content are consistent with this conclusion.

A few groups of dinoflagellates have plastids now recognized to be derived from serial secondary endosymbiosis (the uptake of a new primary plastid-containing endosymbiont) such as *Lepidodinium* spp. or tertiary endosymbiosis (the uptake of the secondary plastid-containing endosymbiont), such as *Dinophysis, Karenia*, and *Kryptoperidinium.*

All these groups are comparatively modern organisms; indeed, the rise of dinoflagellates and coccolithophorids approximately parallels the rise of dinosaurs, while the rise of diatoms approximates the rise of mammals in the Cenozoic. The burial and subsequent diagenesis of organic carbon produced primarily by members of the red lineage in shallow seas in the Jurassic period provide the source rocks for most of the petroleum reservoirs that have been exploited for the past century by humans.

SUGGESTED READING

Andersen, R. A., Biology and systematics of heterokont and haptophyte algae, *American Journal of Botany*, 91, 1508–1522, 2004.

Bhattacharya, D. and Medlin, L. K., Dating algal origin using molecular clock methods, *Protist*, 155(1), 9–10, 2004.

Bold, H. C. and Wynne, M. J., *Introduction to the Algae — Structure and Reproduction*, Englewood Cliffs, NJ, Prentice-Hall, Inc., 1978.

Bourrelly, P., *Les Algue d'eau douce. Initiation à la Systèmatique. Les algues verdes*, Paris, Editions N. Boubèe, 1966.

Bourrelly, P., *Les Algue d'eau douce. Initiation à la Systèmatique. Les algues jaunes et brunes*, Paris, Editions N. Boubèe, 1968.

Bourrelly, P., *Les Algue d'eau douce. Initiation à la Systèmatique. Eugleniens, Peridinies, Algues rouges et Algues bleues*, Paris, Editions N. Boubèe, 1970.

Bravo-Sierra, E. and Hernandez-Becerril, D. U., Parmales from the Gulf of Tehuantepec, Mexico, including the description of a new species, *Tetraparma insecta*, and a proposal to the taxonomy of the group, *Journal of Phycology*, 39, 577–583, 2003.

Carr, N. G. and Whitton, B. A., *The Biology of Cyanobacteria*, Oxford, Blackwell Scientific, 1982.

Cavalier-Smith, T., Chloroplast evolution: secondary symbiogenesis and multiple losses, *Current Biology*, 12, R62–R64, 2002.

Dyall, S. D., Brown, M. T. and Johnson, P. J., Ancient invasions: from endosymbionts to organelles, *Science*, 304, 253–257, 2004.

Graham, L. E. and Wilcox, L. W., *Algae*, Upper Saddle River, NJ, Prentice-Hall, 2000.

Hackett, J. D., Anderson, D. M., Erdner, D. L., and Bhattacharya, D., Dinoflagellates: a remarkable evolutionary experiment, *American Journal of Botany*, 91, 1523–1534, 2004.

Jones, R. I., Mixotrophy in planktonic protists: an overview, *Freshwater Biology*, 45, 219–226, 2000.

Keeling, P. J., Diversity and evolutionary history of plastids and their hosts, *American Journal of Botany*, 91, 1481–1493, 2004.

Lewin, R. A., Prochlorophyta, a matter of class distinctions, *Photosynthesis Research*, 73, 59–61, 2002.

Lewis, L. A. and McCourt, R. M., Green algae and the origin of land plants, *American Journal of Botany*, 91, 1535–1556, 2004.

Margulis, L., Corliss, J. O., Melkonian, M., and Chapman, D. J. (Eds.). *Handbook of Protoctista*, Boston, Jones and Bartlett Publishers, 1990.

Marin, B., Origin and fate of chloroplasts in the euglenoida, *Protist*, 155(1), 13–14, 2004.

McFadden, G. I. and van Dooren, G. G., Evolution: red algal genome affirms a common origin of all plastids, *Current Biology*, 14(13), R514–R516, 2004.

Nozaki, H., Matsuzaki, M., Misumi, O., Kuroiwa, H., Hasegawa, M., Higashiyama, T., Shini, T., Kohara, Y., Ogasawara, N., and Kuroiwa, T., Cyanobacterial genes transmitted to the nucleus before divergence of red algae in the chromista, *Journal of Molecular Evolution*, 59(1), 103–113, 2004.

Patterson, D. J., The diversity of Eukaryotes, *The American Naturalist*, 65, S96–S124, 1999.

Saunders, G. W. and Hommersand, M. H., Assessing red algal supraordinal diversity and taxonomy in the context of contemporary systematic data, *American Journal of Botany*, 91, 1494–1507, 2004.

Schoonhoven, E. *Ecophysiology of mixotrophs.* http://www.bio.vu.nl/thb/education/Scho2000.pdf, 2000.

South, G. R. and Whittick, A., *Introduction to Phycology*, Oxford, Blackwell, 1987.

Tilden, J. E., *The Algae and Their Life Relations. Fundamental of Phycology*, Minneapolis, The University of Minnesota Press, 1935.

Van den Hoek, C., Mann, D. G., and Jahns, H. M., *Algae — An Introduction to Phycology*, Cambridge, U.K., Cambridge University Press, 1995.

2 Anatomy

CYTOMORPHOLOGY AND ULTRASTRUCTURE

The description of the algal cell will proceed from the outside structures to the inside components. Details will be given only for those structures that are not comparable with analogue structures found in most animals and plants. The reader is referred to a general cell biology textbook for the structure not described in the following.

OUTSIDE THE CELL

Cell surface forms the border between the external word and the inside of the cell. It serves a number of basic functions, including species identification, uptake and excretion/secretion of various compounds, protection against desiccation, pathogens, and predators, cell signaling and cell–cell interaction. It serves as an osmotic barrier, preventing free flow of material, and as a selective barrier for the specific transport of molecules. Algae, besides naked membranes more typical of animal cells and cell walls similar to those of higher plant cells, possess a wide variety of cell surfaces. The terminology used to describe cell surface structures of algae is sometimes confusing; to avoid this confusion, or at least to reduce it, we will adopt a terminology mainly based on that of Presig et al. (1994).

Cell surface structures can be grouped into four different basic types:

- Simple cell membrane (Type 1)
- Cell membrane with additional extracellular material (Type 2)
- Cell membrane with additional intracellular material in vesicles (Type 3)
- Cell membrane with additional intracellular and extracellular material (Type 4)

Type 1: Simple Cell Membrane

This cell surface consists of a simple or modified plasma membrane. The unit membrane is a lipid bilayer, 7–8 nm thick, rich of integral and peripheral proteins. Several domains exist in the membrane, each distinguished by its own molecular structure. Some domains have characteristic carbohydrate coat enveloping the unit membrane. The carbohydrate side chains of the membrane glycolipids and glycoproteins form the carbohydrate coat. Difference in thickness of plasma membrane may reflect differences in the distribution of phospholipids, glycolipids, and glycoproteins (Figure 2.1).

A simple plasma membrane is present in the zoospores and gametes of Chlorophyceae, Xanthophyceae (Heterokontophyta), and Phaeophyceae (Heterokontophyta), in the zoospores of the Eustigmatophyceae (Heterokontophyta), and in the spermatozoids of Bacillariophyceae (Heterokontophyta). This type of cell surface usually characterizes very short-lived stages and, in this transitory naked phase, the naked condition is usually rapidly lost once zoospores or gametes have ceased swimming and have become attached to the substrate, as wall formation rapidly ensues. A simple cell membrane covers the uninucleate cells that form the net-like plasmodium of the Chlorarachniophyta during all their life history. Most Chrysophyceae occur as naked cells, whose plasma membrane is in direct contact with water, but in *Ochromonas*, the membrane is covered with both a carbohydrate coat and surface blebs and vesicles, which may serve to trap bacteria and other particles that are subsequently engulfed as food. The properties of the membrane or its domains may change from one stage in the life cycle to the next.

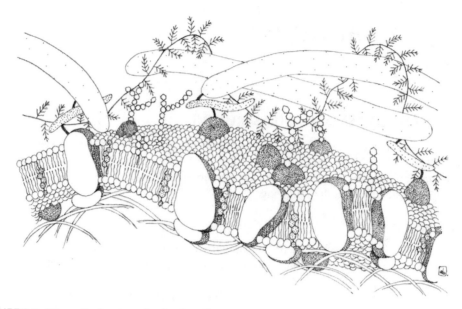

FIGURE 2.1 Schematic drawing of a simple cell membrane.

Type 2: Cell Surface with Additional Extracellular Material

Extracellular matrices occur in various forms and include mucilage and sheaths, scales, frustule, cell walls, loricas, and skeleta. The terminology used to describe this membrane-associated material is quite confusing, and unrelated structures such as the frustule of diatoms, the fused scaled covering of some prasynophyceae, and the amphiesma of dinoflagellates have been given the same name, that is, theca. Our attempt has been to organize the matter in a less confusing way (at least in our opinion).

Mucilages and Sheaths

These are general terms for some sort of outer gelatinous covering present in both prokaryotic and eukaryotic algae. Mucilages are always present and we can observe a degree of development of a sheath that is associated with the type of the substrate the cells contact (Figure 2.2). All cyanobacteria secrete a gelatinous material, which, in most species, tends to accumulate around the cells or trichome in the form of an envelope or sheath. Coccoid species are thus held together to form colonies; in some filamentous species, the sheath may function in a similar manner, as in the formation of *Nostoc* balls, or in development of the firm, gelatinous emispherical domes of the marine *Phormidium crosbyanum*. Most commonly, the sheath material in filamentous species forms a thick coating or tube through which motile trichomes move readily. Sheath production is a continuous process in cyanobacteria, and variation in this investment may reflect different physiological stages or levels of adaptation to the environment. Under some environmental conditions the sheath may become pigmented, although it is ordinarily colorless and transparent. Ferric hydroxide or other iron or metallic salts may accumulate in the sheath, as well as pigments originating within the cell. Only a few cyanobacterial exopolysaccharides have been defined structurally; the sheath of *Nostoc commune* contains cellulose-like glucan fibrils cross-linked with minor monosaccharides, and that of *Mycrocystis flos-aquae* consists mainly of galacturonic acid, with a composition similar to that of pectin. Cyanobacterial sheaths appear as a major component of soil crusts found throughout the world, from hot desert to polar regions, protecting soil from erosion, favoring water retention and nutrient bio-mobilization, and affecting chemical weathering of the environment they colonize.

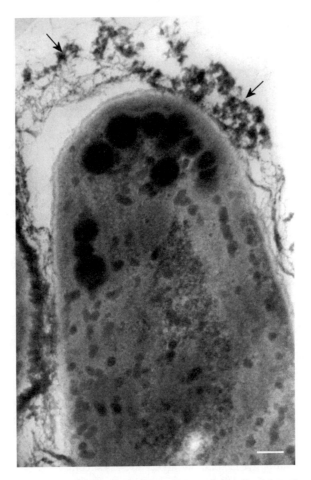

FIGURE 2.2 Transmission electron microscopy image of the apical cell of *Leptolyngbya* spp. trichome in longitudinal section. The arrows point to the mucilaginous sheath of this cyanobacterium. Inside the cell osmiophylic eyespot globules are present. (Bar: 0.15 μm). (Courtesy of Dr. Patrizia Albertano.)

In eukaryotic algae, mucilages and sheaths are present in diverse divisions. The most common occurence of this extracellular material is in the algae palmelloid phases, in which non-motile cells are embedded in a thick, more or less stratified sheath of mucilage. This phase is so-called because it occurs in the genus *Palmella* (Chlorophyceae), but it occurs also in other members of the same class, such as *Asterococcus* sp., *Hormotila* sp., *Spirogyra* sp., and *Gleocystis* sp. A palmelloid phase is present also in *Chroomonas* sp. (Cryptophyceae) and in *Gleodinium montanum* vegetative cells (Dynophyceae) and in *Euglena gracilis* (Euglenophyceae) (Figure 2.3). Less common are the cases in which filaments are covered by continuous tubular layers of mucilages and sheath. It occurs in the filaments of *Geminella* sp. (Chlorophyceae). A more specific covering exists in the filaments of *Phaeothamnion* sp. (Chrysophyceae), because under certain growth conditions, cells of the filaments dissociate and produce a thick mucilage that surround them in a sort of colony resembling the palmelloid phase.

Scales

Scales can be defined as organic or inorganic surface structures of distinct size and shape. Scales can be distributed individually or arranged in a pattern sometimes forming an envelope around

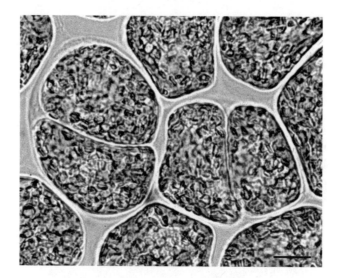

FIGURE 2.3 Palmelloid phase of *Euglena gracilis*. (Bar: 10 μm.)

the cell. They occur only in eukaryotic algae, in the divisions of Heterokontophyta, Haptophyta, and Chlorophyta. They can be as large as the scales of Haptophyta (1 μm), but also as small as the scales of Prasynophyceae (Chlorophyta) (50 nm). There at least three distinct types of scales: non-mineralized scales, made up entirely of organic matter, primarily polysaccharides, which are present in the Prasynophyceae (Chlorophyta); scales consisting of calcium carbonate crystallized onto an organic matrix, as the coccoliths produced by many Haptophyta; and scales constructed of silica deposited on a glycoprotein matrix, formed by some members of the Heterokontophyta.

Most taxa of the Prasinophyceae (Chlorophyta) possess several scale types per cell, arranged in 1–5 layers on the surface of the cell body and flagella, those of each layer having a unique morphology for that taxon. These scales consist mainly of acidic polysaccharides involving unusual 2-keto sugar acids, with glycoproteins as minor components. Members of the order Pyramimonadales exhibit one of the most complex scaly covering among the Prasinophyceae. It consists of three layers of scales. The innermost scales are small, square, or pentagonal; the intermediate scales are either naviculoid, spiderweb-shaped, or box shaped (Figure 2.4); the outer layer consists of large basket or crown-shaped scales. It is generally accepted that scales of the Prasinophyceae are synthesized within the Golgi apparatus; developing scales are transported through the Golgi apparatus by cisternal progression to the cell surface and released by exocytosis. In some Prasynophyceae genera such as *Tetraselmis* and *Scherffelia*, the cell body is covered entirely by fused scales. The scale composition consists mainly of acidic polysaccharides. These scales are produced only during cell division. They are formed in the Golgi apparatus and their development follow the route already described for the scales. After secretion, scales coalesce extracellularly inside the parental covering to form a new cell wall.

In the Haptophyta, cells are typically covered with external scales of varying degree of complexity, which may be unmineralized or calcified. The unmineralized scales consist largely of complex carbohydrates, including pectin-like sulfated and carboxylated polysaccharides, and cellulose-like polymers. The structure of these scales varies from simple plates to elaborate, spectacular spines and protuberances, as in *Chrysochromulina* sp. (Figure 2.5) or to the unusual spherical or clavate knobs present in some species of *Pavlova*.

Calcified scales termed coccoliths are produced by the coccolithophorids, a large group of species within the Haptophyta. In terms of ultrastructure and biomineralization processes, two

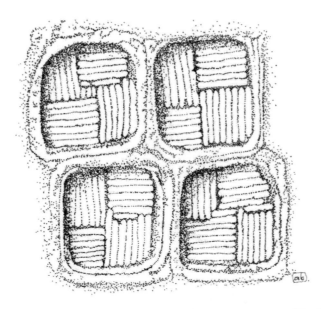

FIGURE 2.4 Box shaped scales of the intermediate layer of *Pyramimonas* sp. cell body covering.

very different types of coccoliths are formed by these algae: heterococcoliths, (Figure 2.6) and holococcoliths (Figure 2.7). Some life cycles include both heterococcolith and holococcolith-producing forms. In addition, there are a few haptophytes that produce calcareous structures that do not appear to have either heterococcolith or holococcolith ultrastructure. These may be products of further biomineralization processes, and the general term nannolith is applied to them.

Heterococcoliths are the most common coccolith type, which mainly consist of radial arrays of complex crystal units. The sequence of heterococcolith development has been described in detail in *Pleurochrysis carterae*, *Emiliana huxleyi*, and the non-motile heterococcolith phase of *Coccolithus pelagicus*. Despite the significant diversity in these observations, a clear overall pattern is discernible in all cases. The process commences with formation of a precursor organic scale inside Golgi-derived vesicles; calcification occurs within these vesicles with nucleation of a protococcolith ring

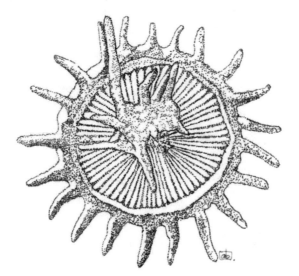

FIGURE 2.5 Elaborate body scale of *Chrysochromulina* sp.

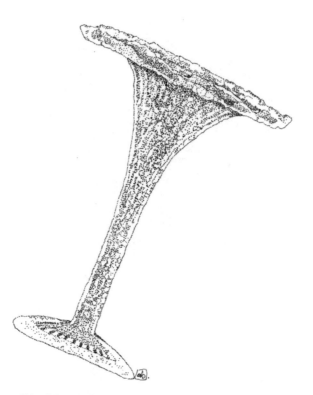

FIGURE 2.6 Heterococcolith of *Discosphaera tubifera*.

of simple crystals around the rim of the precursor base-plate scale. This is followed by growth of these crystals in various directions to form complex crystal units. After completion of the coccolith, the vesicle dilates, its membrane fuses with the cell membrane and exocytosis occur. Outside the cell, the coccolith joins other coccoliths to form the coccosphere, that is the layer of coccoliths surrounding the cell (cf. Chapter 1, Figure 1.35).

Holococcoliths consist of large numbers of minute morphologically simple crystals. Studies have been performed on two holococcolith-forming species, the motile holococcolith phase of *Coccolithus pelagicus* and *Calyptrosphaera sphaeroidea*. Similar to the heteroccoliths, the holococcoliths are

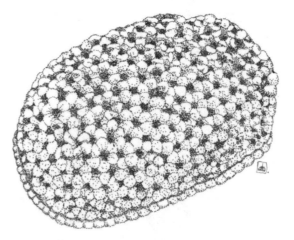

FIGURE 2.7 Holococcolith of *Syracosphaera oblonga*.

underlain by base-plate organic scales formed inside Golgi vesicles. However, holococcolith calcification is an extracellular process. Experimental evidences revealed that calcification occurs in a single
highly regulated space outside the cell membrane, but directly above the stack of Golgi vesicles. This
extracellular compartment is covered by a delicate organic envelope or "skin." The cell secretes
calcite that fills the space between the skin and the base-plate scales. The coccosphere grows progressively outward from this position. As a consequence of the different biomineralization strategies,
heterococcoliths are more robust than the smaller and more delicate holococcoliths.

Coccolithophorids, together with corals and foraminifera, are responsible for the bulk of
oceanic calcification. Their role in the formation of marine sediment and the impact their
blooms may exert on climate change will be discussed in Chapter 4.

Members of the Chrysophyceae (Heterokontophyta) such as *Synura* sp. and *Mallomonas* sp. are
covered by armor of silica scales, with a very complicated structure. *Synura* scales consists of a
perforated basal plate provided with ribs, spines, and other ornamentation (Figure 2.8). In
Mallomonas, scales may bear long, complicated bristles (Figure 2.9). Several scale types are produced in the same cell and deposited on the surface in a definite sequence, following an imbricate,
often screw-like pattern. Silica scales are produced internally in deposition vesicles formed by the
chrysoplast endoplasmic reticulum, which function as moulds for the scales. Golgi body vesicles
transporting material fuse with the scale-producing vesicles. Once formed the scale is extruded
from the cell and brought into correct position on the cell surface.

Frustule

This structure is present only in the Bacillariophyceae (Heterokontophyta). The frustule is an ornate
cell membrane made of amorphous hydrated silica, which displays intricate patterns and designs
unique to each species. This silicified envelope consists of two overlapping valves, an epitheca
and a slightly smaller hypotheca. Each theca comprises a highly patterned valve and one or
more girdle bands (cingula) that extend around the circumference of the cell, forming the region
of theca overlay. Extracellular organic coats envelop the plasma membrane under the siliceous

FIGURE 2.8 Ornamented body scale of *Synura petersenii.*

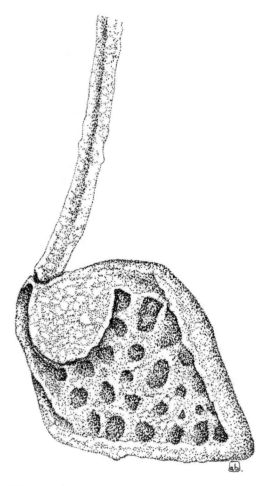

FIGURE 2.9 Body scale of *Mallomonas crassisquama.*

frustule. They exist in the form of both thick mucilaginous capsules and thin tightly bound organic sheaths. The formation of the frustule has place in the silica deposition vesicles, derived from the Golgi apparatus, wherein the silica is deposited. The vesicles eventually secrete their finished product onto the cell surface in a precise position.

Diatoms can be divided artificially in centric and pennate because of the symmetry of their frustule. In centric diatoms, the symmetry is radial, that is, the structure of the valve is arranged in reference to a central point (Figure 2.10). However, within the centric series, there are also oval, triradiate, quadrate, and pentagonal variation of this symmetry, with a valve arranged in reference to two, three, or more points. Pennate diatoms are bilaterally symmetrical about two axes, apical and trans-apical, or only in one axis, (Figure 2.11); some genera possess rotational symmetry, (cf. Chapter 1, Figure 1.30). Valves of some pennate diatoms are characterized by an elongated fissure, the raphe, which can be placed centrally, or run along one of the edges. At each end of the raphe and at its center there are thickenings called polar and central nodules. Addiction details in the morphology of the frustule are the stria, lines composed of areolae, and pores through the valve that can go straight through the structure, or can be constricted at one side. Striae can be separated by thickened areas called costae. Areolae are passageways for the gases, nutrients exchanges, and mucilage secretion for movement and attachment to substrates or other cells of colony. Other pores, also known as portules, are present on the surface of the valve.

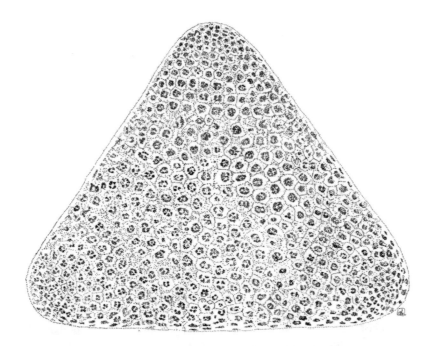

FIGURE 2.10 *Triceratium* sp., a centric diatom.

There are two types of portules: fultoportulae (Figure 2.12) found only in the order Thalassiosirales and rimoportulae (Figure 2.13), which are universal. The structure of the fultoportulae is an external opening on the surface of the valve extended or not into a protruding structure (Figure 2.12). The other end penetrates the silica matrix and is supported with two to five satellite pores. The portules

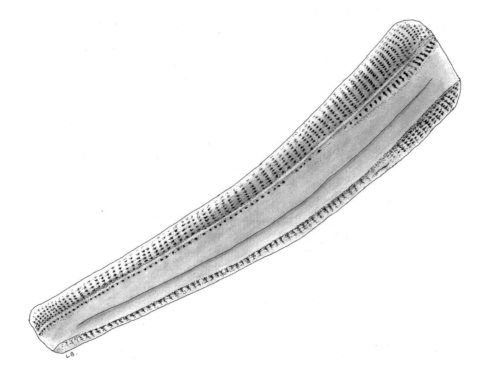

FIGURE 2.11 *Rhoicosphenia* sp., a pennate diatom.

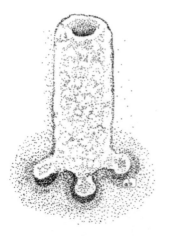

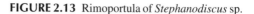

FIGURE 2.12 Fultoportula of *Thalassiosira* sp.   **FIGURE 2.13** Rimoportula of *Stephanodiscus* sp.

function in the excretion of several materials, among them are β-chitin fibrils. These fibrils are manufactured in the conical invaginations in the matrix, under the portule. This may be the anchoring site for the protoplast. The rimoportula is similar to the fultoportulae, except that it has a simpler inner structure. The rimoportula does not have satellite pores in the inner matrix. However, the rimoportula does have some elaborate outer structures that bend, have slits, or are capped. Sometimes the valve can outgrow beyond its margin in structures called setae that help link adjacent cells into linear colonies as in *Chaetoceros* spp., or possess protuberances as in *Biddulphia* spp. that allow the cells to gather in zig-zag chains (Figure 2.14). In other genera such as *Skeletonema* the valve presents a marginal ridge along its periphery consisting of long, straight spines, which make contact between adjacent cells, and unite them into filaments. Some genera also possess a labiate process, a tube through the valve with internally thickened sides that may be flat or elevated.

Diatoms are by far the most significant producer of biogenic silica, dominating the marine silicon cycle. It is estimated that over 30 million km^2 of ocean floor are covered with sedimentary deposits of diatom frustules. The geological and economical importance of these silica coverings as well as the mechanism of silica deposition will be discussed in Chapter 4.

Cell Wall

A cell wall, defined as a rigid, homogeneous and often multilayered structure, is present in both prokaryotic and eukaryotic algae.

In the Cyanophyta the cell wall lies between the plasma membrane and the mucilaginous sheath; the fine structure of the cell wall is of Gram-negative type. The innermost layer, the electron-opaque layer or peptidoglycan layer, overlays the plasma membrane, and in most cyanobacteria its width varies between 1 and 10 nm, but can reach 200 nm in some *Oscillatoria* species. Regularly arranged discontinuities are present in the peptidoglycan layer of many cyanobacteria; pores are located in single rows on either side of every cross wall, and are also uniformly distributed over the cell surface. The outer membrane of the cell wall appears as a double track structure tightly connected with the peptidoglycan layer; this membrane exhibits a number of evaginations representing sites of extrusion of material from the cytoplasm through the wall into the slime. The cell wall of Prochlorophyta is comparable to that of the cyanobacteria in structure and contains muramic acid.

Eukaryotic algal cell wall is always formed outside the plasmalemma, and is in many respects comparable to that of higher plants. It is present in the Rhodophyta, Eustigmatophyceae (Figure 2.15a and 2.15b), Phaeophyceae (Heterokontophyta), Xanthophyceae (Heterokontophyta),

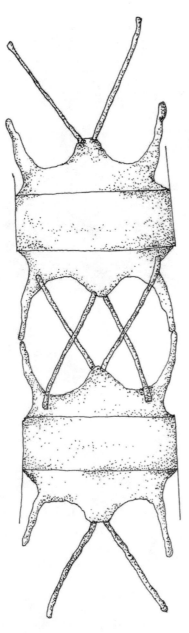

FIGURE 2.14 Cells of *Biddulphia* sp.

Chlorophyceae, and Charophyceae (Chlorophyta). Generally, cell walls are made up of two components, a microfibrillar framework embedded in an amorphous mucilaginous material composed of polysaccharides, lipids, and proteins. Encrusting substances such as silica, calcium carbonate, or sporopollenin may be also present. In the formation of algal cell walls the materials required are mainly collected into Golgi vesicles that then pass it through the plasma membrane, where enzyme complexes are responsible for the synthesis of microfibrils, in a pre-determinate direction.

In the Floridophyceae (Rhodophyta) the cell wall consists of more than 70% of water-soluble sulfated galactans such as agars and carrageenans, commercially very important in food and pharmaceutical industry, for their ability to form gels. In the Phaeophyceae (Heterokontophyta) cell wall mucilagine is primarily composed of alginic acid; the salts of this acid have valuable

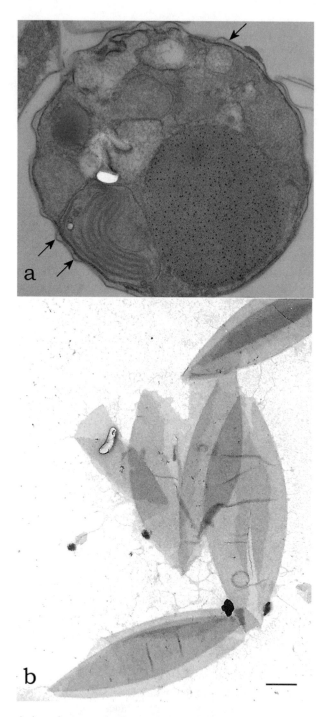

FIGURE 2.15 Transmission electron microscopy image of *Nannochloropsis* sp. in transversal section. (a) Arrows point to the cell wall. Negative staining of the shed cell walls (b). (Bar: 0.5 μm.)

emulsifying and stabilizing properties. In the Xanthophyceae (Heterokontophyta) the composition of the wall is mainly cellulosic, while in the Chlorophyceae (Chlorophyta) xylose, mannose, and chitin may be present in addition to cellulose. Some members of the Chlorophyceae (Chlorophyta) and Charophyceae (Chlorophyta) have calcified walls.

Lorica

These enveloping structures are present in some members of the class Chrysophyceae (Hetero-kontophyta) such as *Dinobryon* sp. or *Chrysococcus* sp. and in some genera of the Chlorophy-ceae, such as *Phacotus, Pteromonas,* and *Dysmorphococcus.* These loricas are vase-shaped structures with a more or less wide apical opening, where the flagella emerge. These structures can be colorless, or dark and opaque due to manganese and iron compound impregnation. We can expect different shapes corresponding to different species. In *Dinobryon* sp., the lorica is an interwoven system of fine cellulose or chitin fibrils (Figure 2.16). In *Chrysococcus* sp., it can consist of imbricate scales. In *Phacotus*, the lorica is calcified, ornamented, and is composed

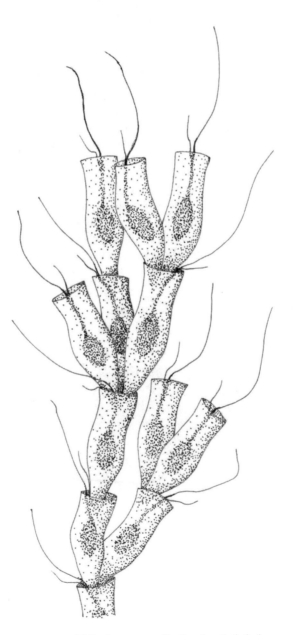

FIGURE 2.16 Tree-like arrangement of *Dinobryon* sp. cells showing their loricas.

of two cup-shaped parts that separate at reproduction. In *Pteromonas*, the lorica extend into a projecting wing around the cell and is composed of two shell-like portions joined at the wings (Figure 2.17).

Skeleton

A siliceous skeleton is present in a small group of marine organisms called silicoflagellates, belonging to the division of Heterokonthophyta. This skeleton is placed outside the plasma membrane; it is a three-dimensional structure resembling a flat basket, which consists of a system of branched tubular elements bearing spinose endings, (cf. Chapter 1, Figure 1.33). The protoplast is contained inside the basket and has a spongy or frothy appearance, with a central dense region containing the nucleus and the perinuclear dictyosomes and numerous cytoplasmic pseudopodia extending outward, containing the plastids. Sometimes a delicate cell covering of mucilage can be detected.

Type 3: Cell Surface with Additional Intracellular Material in Vesicles

In this type of cell surface, the plasma is underlined by a system of flattened vesicles. An example is the complex outer region of dinoflagellates termed amphiesma. Beneath the cell membrane that bounds dinoflagellate motile cells, a single layer of vesicles (amphiesmal vesicles) is almost invariably present. The vesicles may contain cellulosic plates (thecal plates) in taxa that are thus termed thecate or armored; or the vesicles may lack thecal plates, such taxa being termed athecate, unarmored, or naked. In athecate taxa, the amphiesmal vesicles play a structural role. In thecate taxa, thecal plates, one of which occurs in each vesicle, adjoin one another tightly along linear plate sutures, usually with the margin of one plate overlapping the margin of the adjacent plate. Cellulosic plates vary from very thin to thick, and can be heavily ornamented by reticula or striae; trichocyst pores, which may lie in pits termed areolae, penetrate most of them.

A separate layer internal to the amphiesmal vesicles may develop. It is termed pellicle, though in the case of dinoflagellates the term "pellicle" refers to a surface component completely different

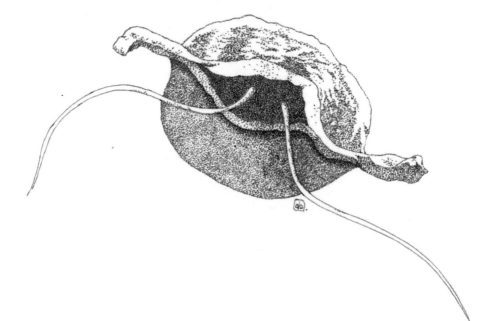

FIGURE 2.17 Lorica of *Pteromonas protracta*.

from the euglenoid pellicle, hence with a completely different accepted meaning, and in our opinion its use should be avoided. The layer consists primarily of cellulose, sometimes with a dinosporine component, a complex organic polymer similar to sporopollenin that make these algae fossilizable. In some athecate genera, such as *Noctiluca* sp., this layer reinforces the amphiesma, and the cells are termed pelliculate. This layer is sometimes present beneath the amphiesma, as in *Alexandrium* sp., or *Scrippsiella* sp., and forms the wall of temporary cysts.

According to Dodge and Crawford (1970), the amphiesma construction falls into eight reasonably distinct categories: (1) simple membrane underlain by a single layer of vesicles 600–800 nm in length, rather flattened, circular, or irregular in shape, with a gap of at least 40 nm between adjacent vesicles that may contain dense granular material; beneath the vesicles are parallel rows of microtubules which lie in groups of three; this simple arrangement is present in *Oxyrrhis marina*; (2) simple membrane underlain by closely packed polygonal (generally hexagonal) vesicles 0.8–1.2 μm in length, frequently containing fuzzy material; these vesicles and the cell membrane are occasionally perforated by trichocyst pores; beneath the vesicles lie microtubules in rows of variable number; this type of amphiesma has been found in *Amphidinium carteri*; (3) as in category (2), but with plug-like structures associated with the inner side of the vesicles; these plugs are cylindrical structures 120 nm long, and are arranged in single lines between single or paired microtubules; an example of this arrangement is present in *Gymnodinium venefi-cum*; (4) as in category (2), but with thin (about 20 nm) plate-like structure in the flattened vesicles; this amphiesma characterizes *Aureodinium pigmentosum*; (5) in this group the vesicles contain plates of medium thickness (60 nm), which slightly overlap; in *Woloszynskia coronata* the plates are perforated by trichocyst pores; (6) the plates are thicker (up to 150 nm), reduced in number with a marked diversity of form; each plate has two or more sides bearing ridges and the remaining sides have tapered flanges; where the plates join, one plate bears a ridge and the opposite bears a flange; *Glenodinium foliaceum* belongs to this category; (7) the plates can be up to 25 μm large and up to 1.8 μm thick; they bear a corrugated flange on two or more sides, and a thick rim with small projections on the opposing edges; these plates may overlap to a considerable extent, and their surface may be covered by a pattern of reticulations; a distinctive member of this category is *Ceratium* sp.; and (8) amphiesma consisting of two large plates, with one or more small plates in the vicinity of the flagellar pores at the anterior end of the cell; plates can be very thin and perforated by two or three simple trichocyst pores as in *Prorocentrum nanum*, or thick and with a very large number (up to 60) of trichocyst pores as in *Prorocentrum micans* (Figure 2.18).

The arrangement of thecal plates is termed tabulation, and it is of critical importance in the taxonomy of dinoflagellates. Tabulation can also be conceived of as the arrangement of amphiesmal vesicles with or without thecal plates. The American planktologist and parasitologist Charles Kofoid developed a tabulation system allowing reference to the shape, size, and location of a particular plate; plates were recognized as being in series relative to particular landmarks such as the apex, cingulum (girdle), sulcus. His formulas (i.e., the listing of the total number of plates in each series) were especially useful for most gonyaulacoid and peridinioid dinoflagellates. Apart from some minor changes introduced afterwards, the Kofoid System is still the standard in the description of new taxa. Plates are numbered consecutively from that closest to the midventral position, continuing around to the cell left. A system of superscripts and other marks are used to designate the plate series. Two complete transverse series of plates are present in the epitheca: apical (′) and precingular (″), counted from the ventral side in a clockwise sequence. Also the hypotheca is divided into two transverse series: postcingular (‴) and antapical (⁗). Some genera possess also an incomplete series of plates on the dorsal surface of the epitheca, termed anterior intercalary plates (a), and on the hypotheca, termed posterior intercalary plates (p). Cingular (C) and sulcal (S) plates are also identified (Figure 2.19). Thus, for example, the dinoflagellate *Proteperidinium steinii* has a formula 4′, 3a, 7″, 3C, 6S, 5‴, 2⁗, which indicates four apical plates, three anterior intercalary plates, seven precingular plates, three cingular plates, six sulcal plates, five postcingular plates, and two antapical plates.

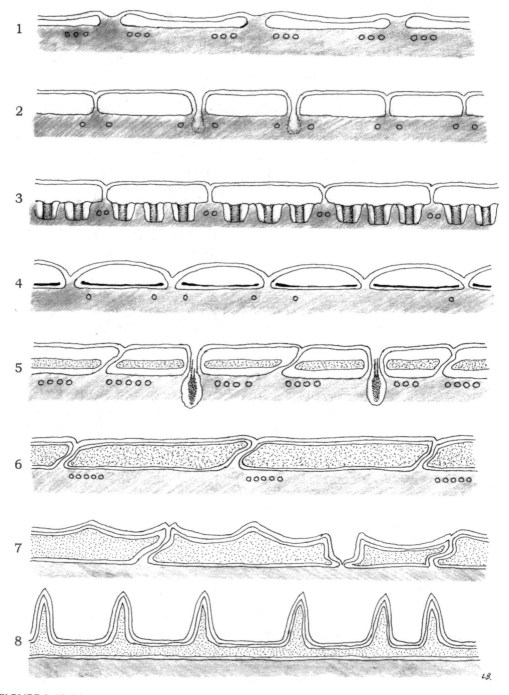

FIGURE 2.18 Diagram of the eight distinct categories of the dinoflagellate amphiesma.

Type 4: Cell Surface with Additional Extracellular and Intracellular Material

Both the surface structure of the Cryptophyta and that of the Euglenophyta can be grouped under this type. The main diagnostic feature of the members of the Cryptophyta is their distinctive kind of cell surface, colloquially termed Periplast. Examples are *Chroomonas* (Figure 2.20) and *Cryptomonas*; in these algae the covering consists of outer and inner components, present on both sides of the

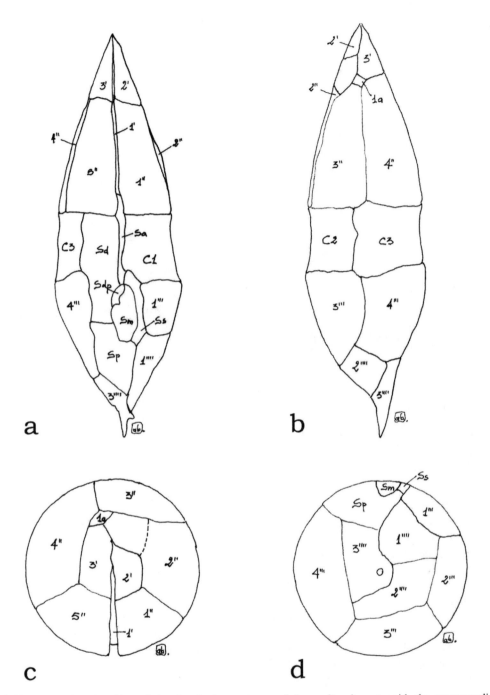

FIGURE 2.19 Line drawings of the thecal plate patterns of *Lessardia elongata* with the corresponding numeration. Ventral view (a), dorsal view (b), apical view (c), and antiapical view (d).

membrane, variable in their composition. The inner component comprises protein and may consist of fibril material, a single sheet or multiple plates having various shapes, hexagonal, rectangular, oval, or round. The outer component may have plates, heptagonal scales, mucilage, or a combination of any of these. The pattern of these plates can be observed on the cell surface when viewed with SEM and freeze-fracture TEM, but it is not obvious in light microscopy view.

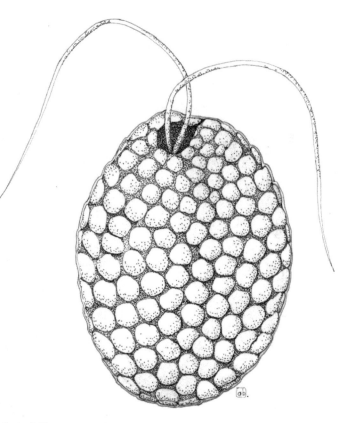

FIGURE 2.20 Periplast of *Chroomonas* sp.

Euglenophyta possess an unusual membrane complex called the pellicle, consisting of the plasma membrane overlying an electron-opaque semicontinuous proteic layer made up of overlapping strips. These strips or striae that can be described as long ribbons that usually arise in the flagellar pocket and extend from the cell apex to the posterior. Each strip is curved at both its edges, and in transverse section it shows a notch, an arched or slightly concave ridge, a convex groove, and a heel region where adjacent strips interlock and articulate. The strips can be arranged helically or longitudinally; the first arrangement, very elastic, is present in the "plastic euglenids" (e.g., *Euglena*, *Peranema*, and *Distigma*), either heterotrophic or phototrophic, where the strips are more than 16. Their relational sliding over one another along the articulation edges permits the cells to undergo "euglenoid movement" or "metaboly." This movement is a sort of peristaltic movement consisting of a cytoplasmic dilation forming at the front of the cell and passing to the rear. The return movement of the cytoplasm is brought about without dilation. The more rigid longitudinal arrangement is present in the "aplastic euglenids" (e.g., *Petalomonas*, *Pleotia*, and *Entosiphon*), all heterotrophic, where the strips are usually less than 12. These euglenids are nor capable of metaboly.

The ultrastructure of the pellicular complex shows three different structural levels (Figure 2.21):

- The plasma membrane with its mucilage coating (first level)
- An electron-opaque layer organized in ridges and grooves (second level)
- The microtubular system (third level)

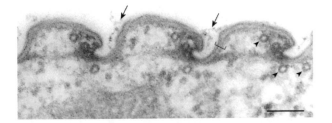

FIGURE 2.21 Transmission electron microscopy image of the surface of *Euglena gracilis* in transverse section, showing the three different structural levels of the pellicle. Arrows point to the first level (mucus coating); a square bracket localizes the second level (ridges and grooves); arrowheads point the third level (microtubules). (Bar: 0.10 μm.)

First Level

A dense irregular layer of mucilaginous glycoprotein covers the external surface of the cell. It has a fuzzy texture that, however, has a somehow ordered structure of orientated threads. Mucilage bodies present beneath the cell surface secret the mucilaginous glycoproteins. The consolidation of the secretory products and their arrangement at one pole or around the periphery of the cell leads to the formation of peduncles (stalks of fixation) and other enveloping structures homologous to the loricas of Chrysophyceae and Chlorophyceae. Peduncles are present in *Colacium*, an euglenophyte that forms small arborescent colonies (Figure 2.22). Its cells, with reduced flagella, are attached by their anterior pole by a peduncle consisting of an axis of neutral polysaccharides and a cortex of acid polysaccharides. Loricas are present in *Trachelomonas* sp. (Figure 2.23), *Strombomonas verrucosa* (Figure 2.24), and *Ascoglena*; they are very rigid, made up of mucilaginous filaments impregnated with ferric hydroxide or manganese compounds which confer an

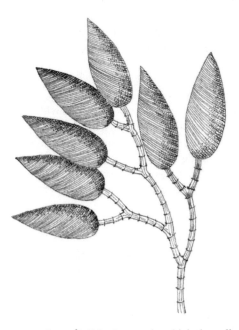

FIGURE 2.22 A small arborescent colony of *Colacium* sp. in which the cells are joined to one another by mucilaginous stalks.

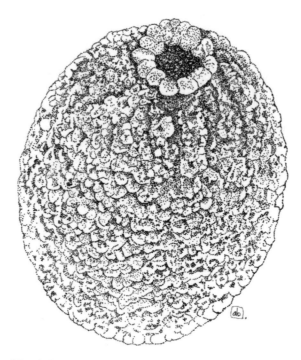

FIGURE 2.23 Lorica of *Trachelomonas* sp.

orange, brown to black coloration to the structure. These loricas fit loosely over the body proper of the cell. They possess a sharply defined collar that tapers to a more or less wide apical opening, where the flagella emerge, or possess a wide opening in one pole and attached to a substrate at the other pole, as in *Ascoglena*.

Beneath the mucus coating, there is the plasma membrane (Figure 2.25). This cell membrane is continuous and covers the ridges and grooves on the whole cell and can be considered the external surface of the cell. The protoplasmic face (PF) of the plasma membrane shows that the strips are covered with numerous peripheral membrane proteins of about 10 nm.

Second Level

This peripheral cytoplasmic layer has a thickness that varies with the species. It consists of roughly twisted proteic fibers with a diameter from 10 to 15 nm arranged with an order texture or parallel striation (Figure 2.26a). The overall structure resembles the wired soul present in the tires, which gives the tire its resistance to tearing forces. Transversal fibers are detectable in some euglenoids, which connect the two longitudinal edges of the ridge of each strip (Figure 2.26b).

Third Level

There is a consistent number and arrangement of microtubules associated with each pellicular strip, which are continuous with those that line the flagellar canal and extend into the region of the reservoir. Within the ridge in the region of the notch there are three to five, usually four, microtubules about 25 nm diameter running parallel along each strip. Two of these are always close together and are located immediately adjacent to the notch adhering to the membrane (Figure 2.21).

The lack of protein organization in the groove regions gives higher plasticity to these zones, and together with presence of parallel microtubules in the ridge regions gives the characteristic pellicular pattern to the surface of euglenoids.

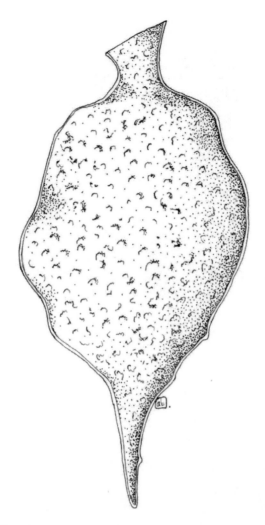

FIGURE 2.24 Lorica of *Strombomonas verrucosa.*

The solid structure of the pellicle confers a very high degree of flexibility and resistance to the cells. Our experience with *E. gracilis* allow us to say that this alga possesses one of the strongest covering present in these microorganisms. A pressure of more than 2000 psi is necessary to break the pellicular structure of this alga.

FLAGELLA AND ASSOCIATED STRUCTURES

Flagella can be defined as motile cylindrical appendages found in widely divergent cell types throughout the plant and animal kingdom, which either move the cell through its environment or move the environment relative to the cell.

Motile algal cell are typically biflagellate, although quadriflagellate types are commonly found in green algae; it is generally believed that the latter have been derived from the former, and a convincing example of this derivation is *Polytomella agilis* from Chlamydomonas. A triflagellate type of zoospore such as that of *Acrochaete wittrockii* (Chlorophyceae, Chlorophyta) may have originated from a quadriflagellate ancestor by reduction, whereas the few uniflagellate forms are most likely descendant of biflagellated cells. Intermediate cases exist, which carry a short

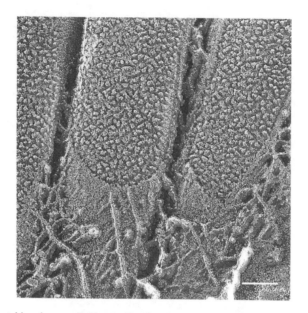

FIGURE 2.25 Deep-etching image of *E. gracilis* showing the mucus coating of the cell surface and the protoplasmic fracture of the cell membrane. (Bar: 0.10 μm.) (Courtesy of Dr. Pietro Lupetti.)

second flagellum, as in *Mantoniella squamata* (Prasinophyceae, Chlorophyta) or *Euglena gracilis*, where one flagellum is reduced to a stub (Figure 2.27); in some species, one flagellum of the pair is reduced to a nonfunctional basal body attached to the functional one, as in the uniflagellate swarmer of *Dictyota dichotoma* (Phaeophyceae, Heterokontophyta). A special case of multiflagellate alga is

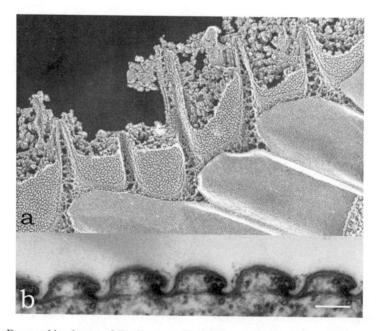

FIGURE 2.26 Deep-etching image of *Euglena gracilis* showing the second structural level of the pellicular complex, showing the regular texture of the internal face of the pellicle stripes (a). Transmission electron microscopy image of the pellicle of *E. gracilis* in transverse section showing the transversal fibers connecting the edges of successive ridges (b). (Bar: 0.10 μm.)

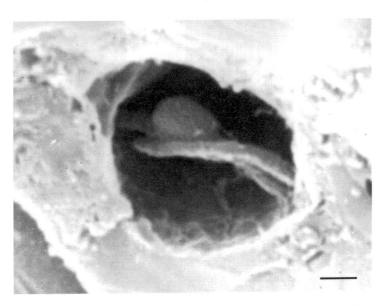

FIGURE 2.27 Scanning electron microscopy image of the reservoir of *E. gracilis* in longitudinal section showing the locomotory emerging flagellum bearing the photoreceptor and the nonemerging flagellum reduced to a stub. (Bar: 0.50 μm.) (Courtesy of Dr. Franco Verni.)

the naked zoospore of *Oedogonium*, where the numerous flagella form a ring or crown around the apical portion of the cell (stephanokont zoospore).

The characteristics of the flagella in a pair, that is, relative length and surface features, have led to a specific nomenclature. When the two flagella differ in length and surface features, one being hairy and the other smooth, they are termed "heterokont." This term applies to all the members of the division Heterokontophyta. When the two flagella are equal in length and appearance, the term "isokont" is used (Figure 2.28), which applies to the algae of the division Haptophyta and to green

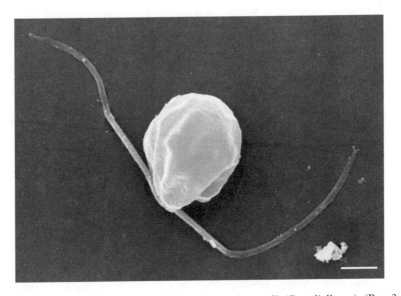

FIGURE 2.28 Scanning electron microscopy image of an isokont cell. (*Dunaliella* sp.). (Bar: 3 μm.)

algae such as Chlorophyceae and Charopyceae. Within this group, there are few genera whose flagella differ in length, which are termed "anisokont."

Description of flagella anatomy will proceed from outside to the inside, from the surface features and components to the axoneme and additional inclusions to the structures anchoring the flagella to the cell.

Flagellar Shape and Surface Features

Deviations from the cylindrical shape are rare among the algae. Usually the flagellar membrane fits smoothly around the axoneme and a total diameter of 0.25–0.35 μm, excluding scales, hairs, etc., holds for most species. If extra material is present between the axoneme and the flagellar membrane, the flagellum diameter increases either locally as in the case of flagellar swellings, or through almost the entire length as in the case of paraxial rods. Minor deviations from the cylindrical shape are caused by small extensions of the membrane to form one or more longitudinal keels running the length of the flagellum. Greater extension of the membrane forms a ribbon or wing supported along the edge by a paraxial rod. More variations are present in the flagellar tip, because flagella can possess a hairpoint, that is, their distal part is thinner with respect to the rest of the flagellum or blunt-tipped, with an abundance of intermediates between these two types.

Flagellar surface is smooth in many algae, where only a simple plasma membrane envelopes the axoneme. Sometimes, however, a distinct, apparently homogeneous dense layer covers the flagellar membrane throughout (Figure 2.29). One of the two flagella of Heterokontophyta is smooth, and smooth flagella are present in members of the Haptophyta, such as *Chrysochromulina parva*, and in many Chlorophyta, such as *Chlamydomonas reinhardtii*.

Flagellar Scales

Flagella may bear a high variety of coverings and ornamentation, which often represent a taxonomic feature. The occurrence of flagellar scale follows that of cell body scales, because they are present only in eukaryotic algae, in the divisions of Heterokontophyta, Haptophyta, and Chlorophyta. As for the cell body scales, they have a silica-based composition in the Heterokonthophyta, a mixed

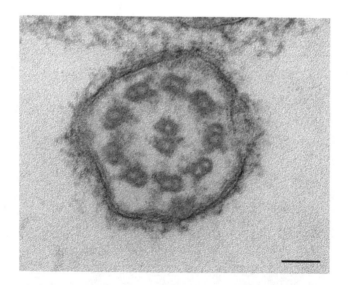

FIGURE 2.29 Transmission electron microscopy image of a *Dunaliella* sp. flagellum in transverse section, showing the homogeneous fuzzy coating of its membrane. (Bar: 0.10 μm.)

structure of calcium carbonate and organic matter in the Haptophyta, and a completely organic nature in the Chlorophyta.

Members of the Chrysophyceae with flagellar scales (Heterokontophyta) fall into two groups: one possessing exactly the same type of scale on both flagellar and body surface, the other showing flagellar scale different in structure and arrangement from body scales. Example of the first group is *Sphaleromantis* sp., whose flagella and cell body are closely packed with scales of very peculiar appearance, resembling the branched structure of a tree. Examples of the second group are *Mallomonas* sp. and *Synura* sp.; in both genera, flagellar scales are not arranged in a regular pattern, are very small (under 300 nm) and possess different morphological types, the most characteristic being the annular type. As the body scales, flagellar scales are produced in deposition vesicles, extruded from the cell and brought into correct position in relation to the other scales and the cell surface.

As described earlier, flagella of the Haptophyta are usually equal in length and appearance (isokont), however, members of the genus *Pavlova* possess two markedly unequal flagella, the anterior much longer than the posterior, and carrying small, dense scales in the form of spherical or clavate knobs. These scales are often arranged in regular rows longitudinally, or can be randomly disposed on the flagellum. Scales are formed inside the Golgi apparatus, and then released to the cell surface by fusion of the plasmalemma and the cisternal membrane.

Flagellar scales are known from almost all the genera of the class Prasinophyceae (Chlorophyta). These algae possess non-mineralized organic scales on their cell body and flagella, the same type of scale being rarely present on both surfaces. On the flagella, the scales are precisely arranged in parallel longitudinal rows, sometimes in one layer, two layers, or even three layers on top of each other. Each layer usually contains only one type of scales. The four flagella of *Tetraselmis* sp. are covered by different types of scales: pentagonal scales attached to the flagellar membrane (Figure 2.30), rod-shaped scales covering the pentagonal scales, and hair scales organized in two rows on opposite sides of the flagellum. A fourth type termed "knotted scales" is present only in some strains, but their precise arrangement is not known. In *Nephroselmis spinosa* the flagellar surface is coated by two different types of scales arranged in two distinct layers. Scales of the inner layer, deposited directly on the membrane, are small and square, 40 nm across (Figure 2.31); scales of the outer layer are rod-shaped, 30–40 nm long, and are deposited atop the inner scales. As in *Tetraselmis*, hair scales of at least two different types are also present covering the flagella. In *Pyramimonas* sp., the scales are extremely complex in structure and ornamentation, and belong to three different types. Minute pentagonal scales, 40 nm wide, form the layer covering the membrane, which in turn is covered by limuloid scales, 313 nm long and 190 nm wide, arranged in nine rows (Figure 2.32); each flagellum also bears two rows of almost opposite tubular hair scales, 1.3 μm long. Spider web scales with an ellipsoid outline are

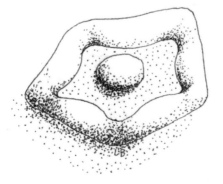

FIGURE 2.30 Pentagonal scale of the flagellar membrane of *Tetraselmis* sp.

FIGURE 2.31 Square scale of the flagellar membrane of *Nephroselmis spinosa*.

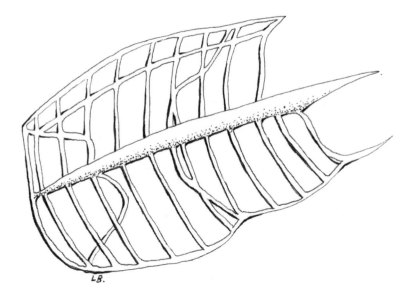

FIGURE 2.32 Limuloid scale of the flagellar membrane of *Pyramimonas* sp.

present in *Mamiella gilva*, which are ornamented by a radial spoke elongated into a conspicuous spine (Figure 2.33).

The scales are synthesized within the Golgi vesicles. The vesicles then migrate to the base of the flagella and from here are extruded and arranged on the flagella.

Flagellar Hairs

Flagellar hairs can be grouped into two types: tubular and non-tubular (simple) hairs. Tubular hairs consist of two or more distinct regions, at least one of which is thick and tubular, while the distal elements may be simpler. This type of hairs is further divided into cryptophycean hairs, tripartite hairs, and prasinophycean hairs.

The cryptophycean hairs are unique for arrangement to the Cryptophyceae (Cryptophyta), being attached in two opposite rows on the longer flagellum, and on a single row on the shorter one. On the long flagellum the hairs consist of a tubular proximal part, 1.5–2.5 μm long, and a non-tubular distal filament, 1 μm long, while the hairs on the shorter flagellum are shorter, 1–1.5 μm long, with a distal filament 1 μm long.

Tripartite hairs are the hair type of the Heterokontophyta, (Figure 2.34a and 2.34b). These hairs consist of three morphological regions, that is, a short basal region, a tubular hollow shaft, and a distal region. The basal part is 0.2–0.3 μm long and tapers towards the site of attachment to the flagellar membrane, at which point dense structures are present that connect the hairs to the peripheral axoneme microtubules. The hollow shaft shows a range of length from 0.7–0.8 to 2 μm, and a diameter of about 16 nm. The distal parts of each hair, called terminal filaments or fibers, are extremely fragile, hence difficult to detect because readily shed during electron microscopy preparation. In some cases, they are organized in a 2+1 structure, that is, two short filaments 0.3 μm long, and one long filament 1 μm long, however, differences exist in their number, length, and diameter.

Cells of the Prasinophyceae carry hairs on all their flagella, whether one, two, four, or eight, which are very diverse in morphology. They can vary in length from 0.5 to 3 μm, and a single flagellum may carry more than one hair type. An example is *Mantoniella* sp., bearing hairs on the flagellar tip which are longer than those on the side. In *Pyramimonas orientalis* both lateral and apical hairs are bipartite and of the same length, with the lateral hairs divided into a short, thick base of

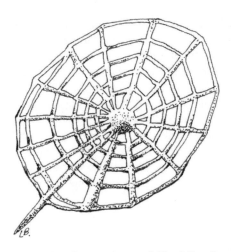

FIGURE 2.33 Spider web scale of the flagellar membrane of *Mamiella gilva*.

160 nm, and a long, thin distal part of 650 nm, and the apical hairs possessing a long thick base and a very short, thin tip. In *Nephroselmis* and *Tetraselmis* there is a single hair type divided into two regions of roughly the same length (0.5 μm).

Unlike tubular hairs, simple, nontubular hairs are not differentiated into regions; they are thin and very delicate, probably consisting of a single row of subunits. These hairs occur in a variety of groups, but are unique in the two divisions of Euglenophyta and Dinophyta, whose hairy coverings share certain features not known to occur in any other algal group.

In Euglenophyta, long, simple hairs are arranged in a single row on the emergent part of the flagella. In genera with two emergent flagella, the hair covering is similar on the two flagella. In *Euglena gracilis* these long hairs consist of a single filament 3–4 μm long, with a diameter of 10 nm, while in *Eutreptiella gymnastica*, they are 4–5 μm long and assembled in unilateral bundles. In addition to these long hairs, euglenoid flagella carry a dense felt of shorter hairs, which in *Euglena* are approximately half as long and half as thick as the long hairs. These short hairs, precisely positioned with respect to each other and to axonemal components,

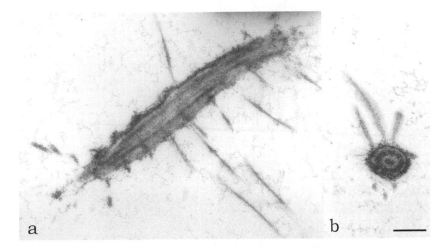

FIGURE 2.34 Transmission electron microscopy image of the long trailing flagellum of *Ochromonas danica* in both longitudinal (a) and transverse sections (b), showing the tripartite hairs. (Bar: 0.25 μm.)

consist of a sheath about 240–300 nm in length, which represent the basic unit. The units, each formed by loops, side arms and filaments, lie parallel to each other in the longitudinal direction of the flagellum (Figure 2.35); two groups of short hairs are arranged helically on each narrow side of the flagellum, separated from each other by two membrane areas without hair attachments. In Dinophyta, both the longitudinal and the transverse flagellum carry hairs, but unlike Eugleno-phyta, the hairy coverings on the two flagella are different. The transverse flagellum carries uni-lateral hairs except in the proximal part; they are 2–4 μm long and arranged in bundles, each bundle consisting of differently sized hairs. In *Oxyrrhis marina*, hairs are of three different lengths, the longest in the middle. Hairs on the longitudinal flagellum are shorter than those on the transverse flagellum (0.4–0.75 μm), but similar in diameter (10 nm). Simple, non-tubular hairs are present also in some Glaucophyta and Chlorophyta.

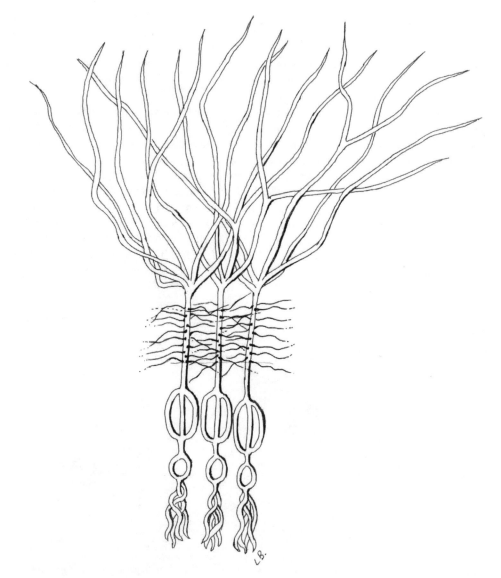

FIGURE 2.35 Short hairs of *Euglena* flagellum.

Flagellar Spines

Flagellar spines are a peculiarity of unknown function confined to male gametes of a few oogamous brown algae. The spermatozoids of *Dictyota* sp. are unique in possessing a longitudinal row of 12 very short spines on their single hairy flagella (these spermatozoids are basically biflagellate, but the second flagellum is reduced to its basal body only). Spines are absent on the distal 2.5–3 μm of the flagellum, and on the proximal 10 μm. In some Fucales (*Himanthalia*, *Xiphophora*, and *Hormosira*), spermatozoids possess only a single spine, up to 1.0 μm long. In all these algae, each spine is made up of electron-dense material, located between the flagellar membrane and the peripheral axonemal doublets.

INTERNAL FEATURES

Axoneme

The movements of the flagella are generated by a single functional unit, the axoneme, which consists of a long cylinder, from 10 to 100 μm long, with a 0.2 μm diameter. Its structure, as seen in cross-sections by electron microscopy, is almost ubiquitous: it is made of nine equally spaced outer microtubule doublets (A and B) approximately 40 nm in diameter surrounding two central microtubules, the central pair (Figure 2.29). This arrangement is maintained by a delicate series of linkages to give the classical 9+2 pattern. The nine outer doublets are numbered starting from number 1 located in the plane orthogonal to the plane including the central pair and counting clockwise when looking from the tip of the flagellum. The former plane allows the definition of the curvature directions during beating as left or right relative to it. Doublets are transiently linked by outer/inner Dynein arms (ODA and IDA) that represent the flagellar motor, and permanently interconnected by nexin links; the radial spokes connect the central pair to the peripheral microtubules of the outer doublets. Divergence from the basic 9+2 pattern are rare, but include the spermatozoid of some centric diatoms (9+0) and the chlorophyta *Golenkinia minutissima* (9+1), as well as the haptonema of the Prymnesiophyceae (Figure 2.36). This structure develops between the two flagella of these algae, and it is sometimes longer than the flagella themselves. It resembles a flagellum, but contains a central shaft of six to eight microtubules arranged in a cylinder, with no doublets. In transverse section, the microtubules are disposed in an arc of a circle or in a ring and are surrounded by a limb of the smooth endoplasmic reticulum. The distal part of the haptonema is fairly straightforward. It is surrounded by plasma membrane, which is continuous over the tip of the haptonema and may be smooth, drawn into a tip or form a spathulate projection.

The bulk of axonemal proteins (70%) is made of tubulins, the building blocks (heterodimers) that polymerize linearly to form microtubules. Those tubulins, which constitute the wall of microtubules belong to the *a* and *b* families, whose sequences have been conserved during evolution (other families, *g*, *d*, and *e*, are responsible for microtubule nucleation at the level of the basal bodies/centrosomes). A large molecular diversity among tubulins is generated by a series of post-translational modifications such as acetylation, detyrosylation, polyglutamylation, or polyglycylation. Tektin filaments are present at the junction between the A and B microtubules of each doublet. The internal and external arms that graft to the peripheral doublets represent 10–15% of the global protein mass of axonemes and are essentially formed by the "dynein-ATPases" motor (the Greek word "dyne" means force). Microtubular dyneins are large multimolecular complexes with a pseudo-bouquet shape, and a molecular mass ranging from 1.4 (bouquets with two heads) to 1.9 MDa (bouquets with three heads) for the whole molecule, and approximately 500 kDa for the largest subunits containing the ATP hydrolysis site. The size of both outer and inner dynein arms is approximately 50 nm. Among the 250 different polypeptides present in the axoneme, as estimated by bidimensional electrophoresis, only a few have been associated with a function.

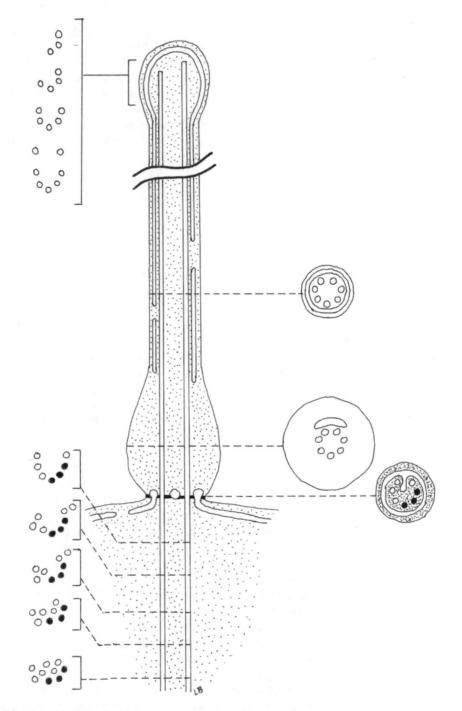

FIGURE 2.36 Schematic drawing of the generalized representation of the structure of the haptonema of *Chrysochromulina*.

Paraxial (Paraxonemal) Rod

In addition to the 9+2 axoneme, algal flagella may contain a number of other structures within the flagellar membrane. Among those extending through the entire length of the flagella, there is the paraflagellar rod (PFR), which is known only in the order of Pedinellales (Chrysophyceae, Hetero-kontophyta), and in members of the Euglenophyta and Dinophyta. PFRs are complex and highly organized lattice-like structures that run parallel to the axoneme. Among the different groups, PFRs are very similar structurally and biochemically.

In the Pedinellales, the membrane of the single emergent flagellum is expanded into a sheath or fin supported along the edge by the rod, which is cross-banded. Owing to the presence of the fins, these algae are excellent swimmers. Their axoneme beats in a distinct planar wave. Paraflagellar structures of characteristic appearance are present in some dinoflagellates. A hollow cylinder with the wall composed of helically arranged filaments is present in the single emergent (longitudi-nal) flagellum of *Noctiluca* gametes, and in the longitudinal flagellum of both *Gyrodinium lebour-iae* and *Oxyrrhis marina*. This rod is about as big as the axoneme in diameter and runs along the axoneme to which it is attached for almost its entire length. The thin filaments (nanofilaments with a 2–4 nm diameter) of its highly geometrically organized network are periodically attached to some of the outer doublet microtubules of the axoneme, which in *O. marina* is doublet no. 4. A different type of paraxial rod is present in the complex-shaped transverse flagellum of dino-flagellates. This rod takes a nearly straight path along the inner wall of the sulcus and is regularly banded in the transverse direction. In contrast, the axoneme itself is distinctly helical.

A rod is present in most members of Euglenophyta. In genera with two emergent flagella, such as *Eutreptia* and *Eutreptiella*, both flagella carry a paraxial rod. In *Eutreptiella*, where the two flagella differ in length, the rods differ from each other in thickness and fine structure, the longer flagellum carrying a more complex rod than the shorter flagellum. In genera with a flagellar appar-atus reduced to one long emergent flagellum and one short flagellum not extending beyond the reservoir region, only the emergent flagellum retains the rod, which extends its entire length. An example is *Euglena gracilis*; in this alga, the rod arises just above the flagellar transition zone and is located latero-ventrally with respect to the axoneme and the cell body. In cross-thin sections of isolated and demembranated flagella, the rod appears hollow with an outer diameter of 90 nm. Images obtained from negative staining preparations show that the rod is made up of several coiled filaments, with a diameter of 22 nm, forming a seven-start left-handed helix with a pitch of 45° and a periodicity of 54 nm. Extending from the surface of the rod a series of goblet-like projections can be observed, which form the point of attachment between the rod and one of the axonemal doublet microtubules (Figure 2.37a and 2.37b). The PFR does not assume any consistent orientation with respect to the central-pair microtubules of the emergent flagellum.

Two major protein components of the PFR have been identified in euglenoids, and dinoflagel-lates with a number of possible minor protein constituents. These major proteins (referred from now as PFR_1 and PFR_2) migrate in the SDS–PAGE as a doublet of similar abundance. Depending on the organism, the mobility for PFR_1 ranges from 70 to 80 kDa, and for PFR_2 from 62 to 70 kDa. Coiled-coils are a common structural motif in the filament formed using the PFR1 and PFR2 proteins. Database searches reveal a 41-residue conserved region of the PFR_1/PFR_2 family that bears a sig-nificant relationship to a conserved motif within the central coiled-coil rod of tropomyosin.

Other Intraflagellar Accessory Structures

Other intraflagellar accessory structures are present in dinoflagellates besides the PFR, the so-called R fiber (Rf), and the so-called striated fiber (Sf). These structures do not show any kind of lattice or precisely organized structure. However, they deserve mentioning because they run for long dis-tances along the axonemal structure and therefore are also candidate for modulating the axonemal beating, possibly by Ca^{2+}-dependent contraction or through dipole–dipole forces.

The Rf is made of thin filaments 2–4 nm in diameter. It may be as large as the axoneme or the PFR in diameter (300–500 nm), runs along the major part of the axoneme and is attached to it via the PFR, the PFR linking the Rf to the axoneme. The Rf may contract and shows transversal stria-tions of variable periodicities and thickness only during its contraction, but not in its relaxed or fully

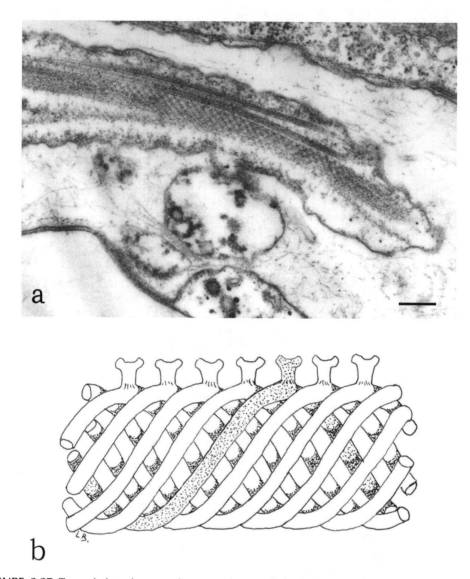

FIGURE 2.37 Transmission electron microscopy image of the locomotory flagellum of *E. gracilis* in longitudinal section showing the PFR (a). Schematic drawing of the PFR showing the coiled filaments and goblet-like projections (b). (Bar: 0.40 μm.)

contracted state (Figure 2.38). The precise structure of the Rf varies according to the fixation conditions for TEM, mostly depending on the Ca^{2+} concentration, suggesting that its contractility is Ca^{2+}-dependent. In some dinoflagellates such as *Ceratium furca*, the Rf is a good candidate for the induction of the complete retraction of the longitudinal flagellum in the flagellar pocket, a movement that cannot be explained by the axoneme structure itself. In this retracted state, the Rf is contracted and the axoneme is highly folded (more tightly than during the usual flagellar beating). Therefore, the Rf could modulate the properties of the PFR and of axoneme motility through constraints imposed to the PFR.

The so-called Sf is also made of thin filaments. It is much smaller than the PFR or the axoneme in diameter (about 35 nm), and runs along three fourths of the axoneme. Its transversal striations suggest its implication in contractile processes that could modulate axonemal motility through gradual changes in the axonemal wavelength or amplitude.

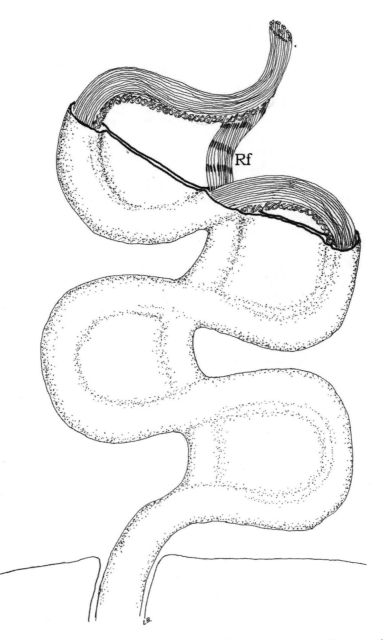

FIGURE 2.38 Schematic drawing of longitudinal flagellum of *Ceratium furca* during its contraction showing the striated Rf.

Transition Zone

While both the flagellar axoneme and the basal body that continues it are very constant in morphology among the different algae, the transition region where they meet varies considerably, and is considered one of the most useful indicators of phylogenetic relationships. This region sometimes contains particular structures such as helices, star-shaped bodies, and transverse partitions referred to as basal plates. Five main types of transition zones may be distinguished, taking into account that secondary variations can appear within each.

Type 1 (Figure 2.39) appears the simplest, with only one basal plate situated at the level of the point of inflexion of the flagellar membrane. Immediately above it, the flagellum shows a slight narrowing. Radial fibers connect the peripheral doublets with submembrane swellings. This type of transition zone is present in the Phaeophyceae (Heterokontophyta).

In Type 2 (Figure 2.40) the basal plate is replaced by a sort of plug located beneath the point of inflexion of the flagellar membrane, far from the point of origin of the central pair of doublets. In this transition zone the nine outer doublets show a strong dilatation, which leaves room for a fibrillar spiral suspended on the doublets by short spokes. The spiral can reach the central doublet, or be much shorter. Each outer doublet is associated via other spokes with a thickening of the membrane, whose folding forms a star with nine characteristics arms (Figure 2.41a and 2.41b). This transition region is typical of the Euglenophyta. Among these algae, *Entosiphon* sulcatum is unique for its long transition region, the spiral of which surrounds the proximal 1 μm of the central doublet.

Type 3 transition regions possess a double system of complex plates. Variations exist due to the distance separating the plates or to the presence of interposition material between them. In Dinophyta, the basal plate is duplicated and ring-shaped; above this there are two median discs, the uppermost supporting the central doublet of the axoneme, (Figure 2.42a). In Cryptophyta, an apical plate is located at the level of the narrowing of the flagellum above the point of inflexion of the flagellar membrane; the central doublet is right above this plate. A second plate bearing a ring-shaped thickening is located beneath it. In the Glaucophyta of the genus *Cyanophora* the apical plate has the same localization present in Cryptophyta, but it is ring-shaped and traversed by the central doublet that continues towards the basal plate situated at the level of the point of inflexion of the flagellar membrane (Figure 2.42b). Further variations of this type are present in Haptophyta, where two widely spaced plates are present, the apical below the central doublet. Each plate corresponds to a flagellar constriction, and the space between them possibly contains fibrillar material; in cross-section a stellate structure is visible, the arm of which connects with the a-tubules of the peripheral doublets (Figure 2.42c).

In Type 4 (Figure 2.43) there is only one basal plate situated at the point where the flagellum emerges, but this type of transition region is characterized by a very peculiar structure called "transitional helix." In longitudinal section this appears as a double row of punctae equidistant from the doublets, representing the four to six turns of a helix (Figure 2.44a and b). Some variations occur in

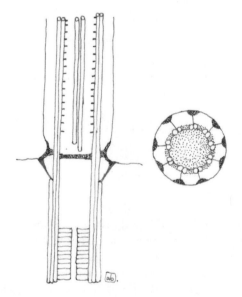

FIGURE 2.39 Type 1 transition zone (Phaeophyceae, Heterokontophyta).

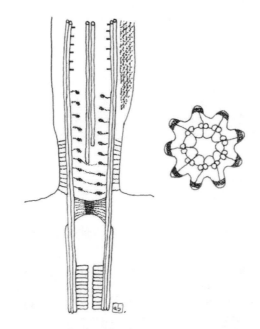

FIGURE 2.40 Type 2 transition zone (Euglenophyta).

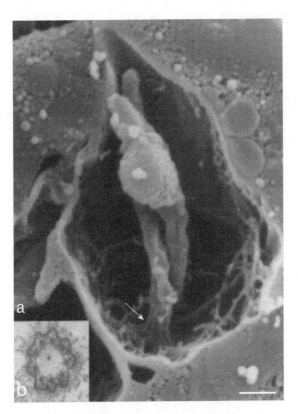

FIGURE 2.41 Scanning electron microscopy image of the reservoir region of *Euglena gracilis*, showing the two flagella arising from the bottom of the reservoir. The arrow points to the thickened folding of the flagellum membrane, typical of the transition zone (a). Transmission electron microscopy image of the transition zone in transverse section, showing the characteristic star configuration (b). (Bar: 0.50 μm.)

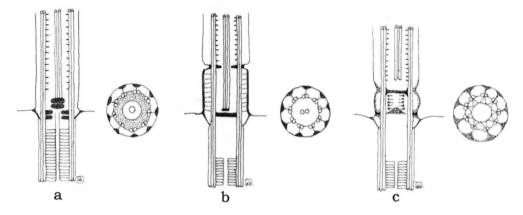

FIGURE 2.42 Type 3 transition zone of Dinophyta (a), Glaucophyta (b), and Haptophyta (c).

the number of gyres, which in a short flagellum may be as low as one. This helix is present in Chrysophyceae, Xanthophyceae, and Eustigmatophyceae (Heterokontophyta).

Type 5 (Figure 2.45) is characterized by the so-called "stellate pattern"; typically, it is divided into a longer distal and a shorter proximal part, separated by a basal plate. In longitudinal section, the structure resembles an H, with the cross-bar located a short distance above the cell surface (Figure 2.46). Transversely, the cross-bar may or may not extend to the peripheral doublets of the axoneme. Variations regard the length of the proximal part, the location of the plate, and the appearance of additional rings. This transition zone is typical of Chlorophyta.

Basal Bodies

A flagellum cannot be dissociated from its base, the basal body, or kinetosome. This structure has a cylindrical form, with an average diameter of 0.2 μm and a variable height (average 0.5 μm). The wall of the cylinder is discontinuous, and consists of nine microtubular triplets tilted to the radii at an angle of 130° and interconnected by transverse desmosomes. The complete tubule A consists of

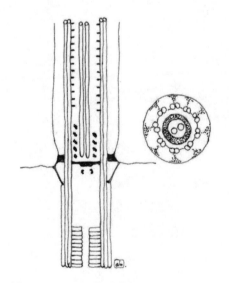

FIGURE 2.43 Type 4 transition zone of Chrysophyceae, Xantophyceae, and Eustigmatophyceae (Heterokontophyta).

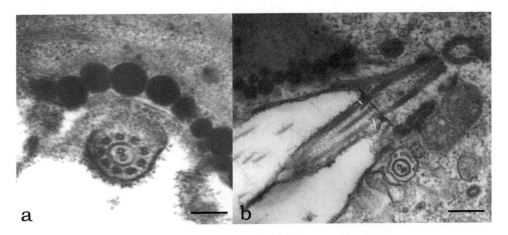

FIGURE 2.44 Transmission electron microscopy image of the transition zone of *Ochromonas danica* in transverse section (a) (Bar: 0.20 μm); longitudinal section of the short flagellum of *Ochromonas* (b); arrows point to the double rows of punctae representing the turn of the helix. (Bar: 0.40 μm.)

13 protofilaments and the incomplete tubules B and C have 10 protofilaments. The proximal part of the basal body contains a fibrogranular structure termed the cartwheel, composed of a longitudinal central tubule and nine series of spokes joined to the triplets. This structure seems to be present in nearly all species of algae, with variations reported mainly for the length of the basal body. While most green algae usually possess very short basal bodies (250–450 nm), those of the Prasino-phyceae are often twice as long (560–690 nm). Some members of the Haptophyta, such as *Chrysochromulina*, also possess very long basal bodies (850–875 nm), but this length can reach a value of 1300 nm in some Euglenophyta such as *Entosiphon*.

In some Chlorophyta two structures are present in association with the proximal ends of the basal bodies: the terminal cap and the proximal sheaths. The terminal cap is a more or less prominent electron-dense flap located on the anterior surface of the basal body, which folds over and covers in part its proximal end. The proximal sheath is located posterior to the proximal end of the basal body, and can have a half-cylindrical shape or be wedge-shaped, narrow proximally, and broad distally.

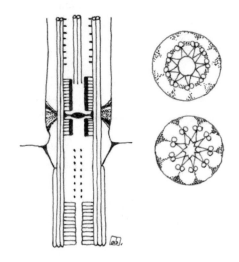

FIGURE 2.45 Type 5 transition zone (Chlorophyta).

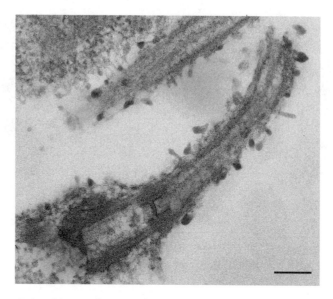

FIGURE 2.46 Transmission electron microscopy image of the transition zone of *Dunaliella* sp. in longitudinal section showing the "H" structure. (Bar: 0.40 μm.)

Variations in the number of basal bodies mainly reflect variations in the number of flagella. It is possible to distinguish:

- Cells with only one basal body, carrying one flagellum; this situation is very rare and is present in the euglenophyte of uncertain affinity, *Scytomonas pusilla*, and in the gametes of the diatoms *Lithodesmium* and *Biddulphia*
- Cells with two neighboring basal bodies, forming the so-called primordial pair indicated as 2A
- Cells with numerous basal bodies

In the case of two basal bodies, two possibilities exist: only one basal body, generally situated close to the nucleus or linked to it, possesses a flagellum, as in the spores of *Hydrurus*; or both basal bodies possess a flagellum. The latter is the most frequently occurring situation in flagellate algae. Most often the basal bodies are inclined towards each other, sometimes they are perpendicular, or they may also be parallel facing the same directions or antiparallel facing opposite directions. In the Glaucophyta the two basal bodies are inclined towards each other; in the Heterokontophyta they are parallel in the Chrysophyceae *Mallomonas* or *Synura* or almost perpendicular in *Ochromonas*, lie at an obtuse angle to each other in the Xanthophyceae and Phaeophyceae, and are inclined to each other at an angle of about 90° in the Eustigmatophyceae and Raphidophyceae; in the Haptophyta the angle between the two basal bodies can be acute or obtuse; in both Cryptophyta and Euglenophyta the basal bodies are almost parallel, while in the Dinophyta they lie at almost 180° to each other and slightly overlap. In the Chlorophyta the angle between the two basal bodies can vary (about 90° in Chlamydomonas, 180° in *Acrosiphonia* zoid) and the couple can assume three different configurations (Figure 2.47):

- 12 o'clock–6 o'clock configuration, in which the basal bodies are in line with each other
- 11 o'clock–5 o'clock configuration, in which the basal bodies are anticlockwise rotated relative to the first configuration
- 1 o'clock–7 o'clock, in which the basal bodies are clockwise rotated relative to the first configuration

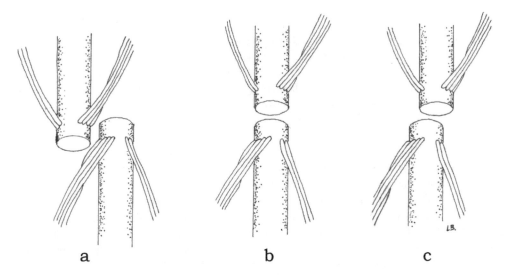

FIGURE 2.47 Schematic drawing of cruciate arrangements of basal bodies and roots of Chlorophyta. 11 o'clock–5 o'clock type (a), 12 o'clock–6 o'clock type (b), 1 o'clock–7 o'clock type (c).

The 11 o'clock–5 o'clock configuration is present in Ulvophyceae, Cladophorophyceae, Bryopsidophyceae, Trentepohliophyceae, and Dasycladophyceae, while both the 12 o'clock–6 o'clock and 1 o'clock–7 o'clock configurations are present in the Chlorophyceae.

In the case of many basal bodies, the increase in their number and the corresponding increase in the number of flagella can result from:

- A replica of each basal body of the pair, as in the stephanokont zoospore of *Oedogonium* (Chlorophyceae, Chlorophyta)
- Replica of the primordial pair, indicated by the formula $X \cdot 2A$, as in the dinoflagellate *Phaeopolykrikos* sp. (Dinophyta) where the primordial pair has undergone four replicas $(4 \cdot 2A)$
- Addition of new basal bodies in the surroundings of the primordial pair without a replication of it, indicated by the formula $2A + N$; algae with four flagella, as *Tetraselmis suecica* or *Pyramimonas lunata* (Prasinophyceae, Chlorophyta), have a formula $2A + 2N$, while algae with eight flagella as *Pyramimonas octopus* (Prasinophyceae, Chlorophyta), are indicated by the formula $2A + 6N$

Root System

All algal flagella appear to possess flagellar roots, that is, microtubular or fibrillar, often cross-banded structures, which extend from the basal bodies into the cytoplasm either underlying the plasma membrane or projecting into the cell and making contact with other organelles such as the nucleus, the mitochondria, the Golgi apparatus, or the chloroplasts. Diverse functions have been assigned to the flagellar roots: anchoring devices or stress absorbers, sensory transducers, and skeletal and organizational structures for morphogenetic processes.

The great diversity of morphology and arrangement of the flagellar root systems among the different algae make it necessary to describe them separately in each division. As described earlier, Cyanophyta, Prochlorophyta, and Rhodophyta lack any flagellar apparatus, hence they will not be considered in the following.

Glaucophyta

In the cyanelle-containing genera *Gloeochaete* and *Glaucocystis*, four roots are arranged in a cruciate manner, that is, four roots spreading out more or less evenly from the basal bodies. All four roots appear identical, each containing about 20 microtubules in *Glaucocystis*, and about 50 in *Gloeo-chaete*. In *Cyanophora* the roots are two. In all the genera each root contains a multilayered structure consisting of a band of microtubules, which overlies several layers of parallel plates (Figure 2.48).

Heterokontophyta

The root system will be described for each class of this division, selecting when possible a genus representing the morphological cell type within the class.

The root system of *Ochromonas* can be considered the basic type of the Chrysophyceae. A single large cross-banded contractile root, termed rhizoplast, is typically present, associated with the basal body of the longer pleuronematic flagellum. After leaving the basal body, this fibrous root passes closely against the edge of the chloroplast and reaches the tip of the pyriform nucleus. It then splits into several branches, ramifying over the nuclear surface, with some branches located in the narrow space between the nucleus and the associated Golgi body. Four microtubular roots, R1, R2, R3, and R4, anchor the two flagella in the cell, R1 and R2 are associated with the basal body of the long pleuronematic flagellum, and R3 and R4 are associated with the basal body of the short smooth flagellum. The three-stranded R1 describes an arc in the anterior part of the cell just beneath the cell membrane. The two-stranded R2 originates at the opposite side of the basal body, running along the cell membrane. R3 and R4, consisting of a species-specific number of microtubules, arise from the opposite sides of the basal body of the short flagellum,

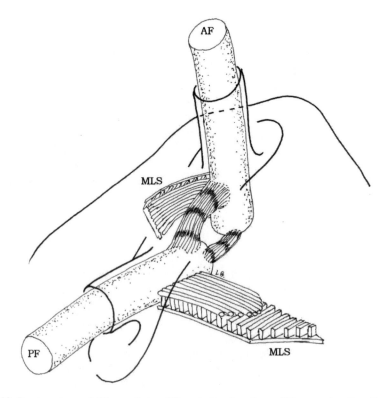

FIGURE 2.48 Root system of Glaucophyta. AF, anterior flagellum; PF, posterior flagellum; and MLS, multilayered structure.

and form a loop around and under this flagellum. R1 forms the base of attachment of numerous microtubules running towards the posterior end of the cell, with cytoskeletal function (Figure 2.49).

As yet no complete analysis of the flagellar root systems of Xanthophyceae exists. The typical root system as presently understood appears to consist of three different types of structures: a descending root originating near the basal bodies, resembling the rhizoplast in extending from the basal bodies along the nuclear surface, between the nucleus and the Golgi body, but differing in being unbranched, and in consisting of a succession of rectangular blocks rather than fibers as in the Chrysophyceae; a cross-banded fibrillar root composed of slightly curved bands, originating together with, but at right angles to the descending root, and terminating at the cell membrane; microtubular roots in various number near the basal bodies, each of which containing three or four microtubules. In *Vaucheria* the system is completely different; neither a descending root nor a cross-banded fiber are present, but its anterior protrusion is supported by a single broad microtubular root of 8–9 microtubules arranged in a row. This root originates near the base of the anterior pleuronematic flagellum, and from here passes forward along the cell membrane to the tip of the protrusion, turns around and runs back on the opposite side of the cell, again along the membrane.

Only few data are available on the flagellar roots of Eustigmatophyceae. Both microtubular and cross-banded fibrillar types are present. The microtubular structures, consisting of 2–5 microtubules, arise close to the basal bodies and pass anteriorly and posteriorly in the cell. The cross-banded roots are narrow and pass from the region of the flagellar bases along the anterior flattened face of the nucleus.

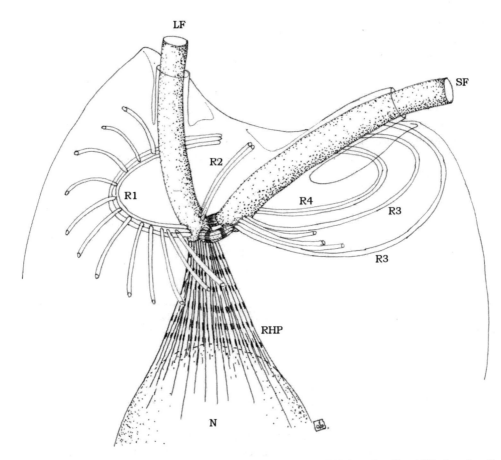

FIGURE 2.49 Root system of *Ochromonas* sp. (Heterokontophyta). LF, long flagellum; SF, short flagellum; R1, R2, R3, and R4, microtubular roots; RHP, rhizoplast; and N, nucleus.

The ultrastructure of the flagellar root system has been described only in one species of Bacillariophyceae, *Biddulphia laevis*. In the spindle-shaped spermatoids of this centric diatom a system of microtubules radiates from the only basal body present into the cell, to form a cone on the anterior part of the nucleus, and at least some of these microtubules extend throughout the length of the cell. They appear to maintain the elongated shape of the nucleus, and thus probably of the cell.

An elaborate root system connects the basal bodies to the anterior surface of the nucleus in the Raphidophyceae. In *Chattonella* two or three roots have been described, one multi-layered associated with the pleuronematic flagellum, and one with the smooth flagellum. The latter is a band of 9–10 microtubules which extends from the anterior part of the cell, passing between the basal bodies, and joining with other microtubules before extending along the nucleus in a depression of its surface. A large number of distinctly cross-banded fibrous roots are also present, extending from the flagellar bases to ensheathe the anterior cone of the nucleus.

In the Phaeophyceae the structure of the flagellar root system appears remarkably uniform. The main characteristic of the system is the root supporting the flat, elastic proboscis in spermatozoids of brown algae. In *Fucus* this root is very broad, contains usually 7 microtubules (but there can be up to 15 microtubules), which travel from the base of the anterior flagellum along the plasmalemma to the anterior end of the cell, where the proboscis is located. Here, they bend back and run along the cell membrane on the opposite side. A bypassing microtubular root, of about five microtubules, originates in the anterior part of the cell, runs along the proboscis root, bypasses the basal bodies, without contact, and continues towards the posterior part of the cell. Other two minor microtubular roots, consisting of only one microtubule, are present, one extending anteriorly, the other posteriorly. The basal bodies are interlinked by three cross-banded connectives, the deltoid, the strap-shaped, and the button-shaped bands (Figure 2.50).

Haptophyta

In the Haptophyta there are different types of flagellar roots. Members of the order Pavlovales, such as *Pavlova* and *Diacronema*, possess a fibrous root, non-striated, which extends from the base of the anterior flagellum, and passes into the cell along the inner face of the nucleus, becoming progressively wider. In some species, another fibrous root originates at the base of the haptonema. Two microtubular roots extend from the base of the posterior flagellum: a seven-stranded root which runs under the periplast, and a two-stranded root arising almost at right angles to the seven-stranded root, running inside the cell opposite to the haptonema. Fibrous connecting bands are present between the basal bodies (Figure 2.51). In the algae of the order Isochrysidales, such as *Pleurochrysis*, the structure of the flagellar root system is more complex. Three main microtubular roots are associated with the two basal bodies, two broad roots, no. 1 and no. 2, arising near the left flagellum, and a smaller root, no. 3, arising near the right flagellum. Root no. 1 extends from the basal body up towards the cell apex and then curves backwards to run inside the cell. A fibrous root is associated with root no. 1. Closely packed microtubules organized in a bundle branch off perpendicularly from both root no. 1 and root no. 2. The basal bodies are connected to each other by distal, intermediate, and proximal connecting bands. Accessory connecting bands link the haptonema to the basal bodies and the left basal body to the broad microtubular root no. 1 (Figure 2.52).

Cryptophyta

The flagellar roots of these algae include two characteristic components, the rhizostyle and the compound rootlet. The rhizostyle is a posteriorly directed microtubular structure. It originates alongside the basal body of the dorsal flagellum, extends deep into the cell, parallel to the gullet, behind the layer of trichocysts, and ends in the posterior part of the cell. On the way, it runs through a groove in the nuclear surface. In many cryptomonads, such as *Chilomonas*, each of the rhizostyle microtubule bears a wing-like lamellar projection. The compound rootlet consists of a cross-banded fibrous band and microtubular roots. The fibrous band originates from the basal bodies of the dorsal flagellum,

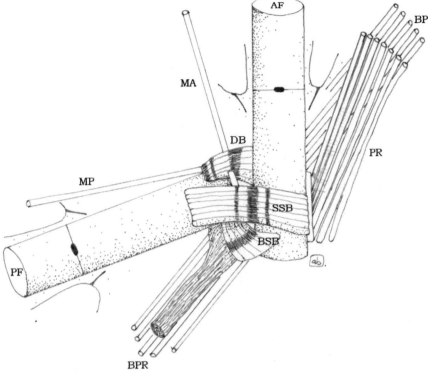

FIGURE 2.50 Root system of *Fucus* sp. (Heterokontophyta). AF, anterior flagellum; PF, posterior flagellum; PR, proboscis root; BPR, bypassing microtubular root; MA, minor anterior root; MP, minor posterior root; DB, deltoid band; SSB, strap-shaped band; and BSB, button-shaped band.

but perpendicularly to the rhizostyle. A microtubular root arises near the rhizostyle and passes between the basal bodies in close association with the fibrous band, a second microtubular root extends dorso-laterally in a curved path, and a third root, which can be very short, originates near the rhizostyle and extends anteriorly. In addition to this rootlet, a conspicuous twelve-stranded microtubular root is present in *Chilomonas*, together with a mitochondrion-associated lamella root, while a delicated cross-banded anchoring fiber connecting one of the basal bodies to the ventral groove is present in *Cryptomonas*. The striated components of the root system have been shown to contain the contractile protein centrin (Figure 2.53).

Dinophyta

The root apparatus of these algae is quite complex for number and appearance of ancillary structures associated to the microtubular roots, and for the spatial relationship between roots and other cell organelles. Minor features can be considered specie-specific, whereas major components are common to almost all the dinoflagellates. The longitudinal basal body and the transverse basal body are interconnected by a small striated connective band. A multimembered microtubular root, the longitudinal root originates on left side of the longitudinal basal body and continues posteriorly along the sulcus. A cross-striated fibrous root, the transverse striated root, emanates from the left side of the transverse basal body and runs along the transverse flagellar canal and the cingulum. Striated connectives link the transverse striated root to both the longitudinal root and the longitudinal basal body, and the proximal portion of the longitudinal basal body to the longitudinal root. The most distinct connective is a large electron-opaque fiber termed the nuclear fibrous

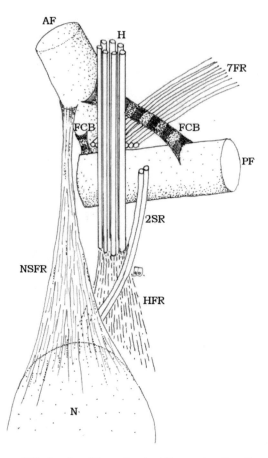

FIGURE 2.51 Root system of Pavlovales (Haptophyta). AF, anterior flagellum; PF, posterior flagellum; NSFR, nonstriated fibrous root; H, haptonema; HFR, haptonema fibrous root; 2SR, two-stranded root; 7SR, seven-stranded root; FCB, fibrous connecting band; and N, nucleus.

connective, which links the proximal parts of the longitudinal root, the transverse striated root, and the transverse and longitudinal basal bodies with the nucleus. Fibrous rings with cross-striations, called fibrous collars, encircle the flagellar canals, the longitudinal striated collar less conspicuous than the transverse striated collar. A non-striated fiber interconnects the two collars. The fibrous elements of this complex root apparatus are likely to contain centrin (Figure 2.54).

Euglenophyta

Euglenoids show a rather uniform root structure, with one microtubular root opposite each basal body and a single microtubular root in between, termed dorsal, intermediate, and ventral roots (Figure 2.55). The dorsal root is anchored to the dorsal basal body at the side furthest from the ventral basal body; the intermediate and ventral roots are both associated with the ventral basal body at its dorsal and ventral sides, respectively. In *Euglena mutabilis*, the dorsal and the intermediate roots consist of three microtubules, while the ventral root consists of five microtubules; in *E. gracilis*, the ventral root is formed by five microtubules as in *E. mutabilis*, while the microtubules of the dorsal root are more numerous. These roots extend from the basal bodies along the reservoir and into the cytoplasm, usually along the cell periphery, but in some species towards the nucleus. Flagellar roots are believed to play an important role in maintaining the cell shape. In some species the two flagellar basal bodies are connected by a conspicuous transversely striated connective.

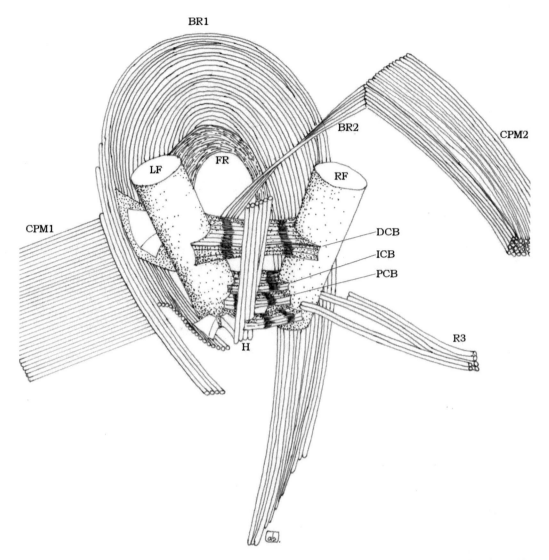

FIGURE 2.52 Root system of *Pleurochrysis* sp. (Haptophyta). LF, left flagellum; RF, right flagellum; H, haptonema; BR1 and BR2, broad microtubular roots; FR, fibrous root; R3, small microtubule root; CPM1, CPM2, closely packed microtubules; DCB, distal connecting band; ICB, intermediate connecting band; and PCB, proximal connecting band.

Besides the intraflagellar accessory structures described above, euglenoids with two emergent flagella possess extraflagellar accessory structures, the so-called rootlets. Two major classes of rootlets have been described: microtubular rootlets and filamentous rootlets. Microtubular rootlets are made of a usually complex and species-specific network (geometry, number of elements) anchoring the flagellar apparatus in the whole cell body. However, due to known functions of microtubules in other systems, these microtubules could not only play a role of anchor, but also provide a stable oriented network along which cargos could be transported from inside the cell body to the flagellar apparatus.

The filamentous rootlets are made of filaments that can contain centrin. These appendages are attached to the pair of basal bodies located at the base of the flagella and extend inside the cytoplasm, usually in the direction of the nucleus or along the plasma membrane. Classically, they

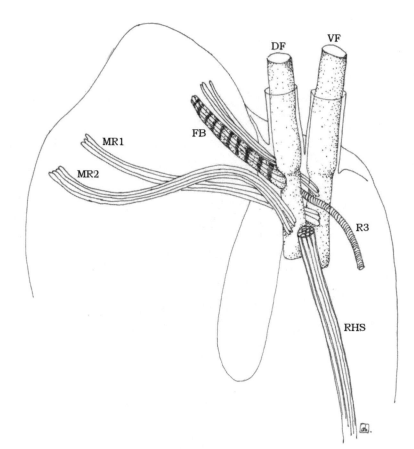

FIGURE 2.53 Root system of Cryptophyta. DF, dorsal flagellum; VF, ventral flagellum; RHS, rhizostyle; FB, fibrous band; MR1 and MR2, microtubular roots; and R3, root.

have been considered as anchoring structures providing the role of roots to the whole flagellar apparatus. However, they are contractile structures as well, which may show transversal striations (centrin type, not assembling type) whose periodicity varies with their contraction state.

Chlorarachniophyta

The single flagellum of the ovoid zoospores of *Chlorarachmon reptans* possesses a root system that consists of a microtubular component appearing as a 3+1 structure near the level of emergence of the flagellum, soon increasing to 8 + 1, and a second root with a homogeneous substructure that occupies a distinct concavity in the nuclear envelope.

In contrast to *C. reptans*, cells of *Bigelowiella natans* are basically biflagellate. The second flagellum, however, is exceptionally short and represented only by a barren basal body inserted at approximately right angle to the emergent flagellum. In this alga the four flagellar roots are represented by microtubular structures only with no cross-banded roots. A microtubular root is present on either sides of long flagellum basal body; the largest and most conspicuous root, which attaches to the outside of the long flagellum basal body, is five-stranded and forms an "L" in the area between the nucleous and the plasma membrane. Just before terminating, this root became two-stranded. The second microtubular root associated with the long flagellum basal body emerges from the corner between the two basal bodies and is one-stranded. The third flagellar root emerges on the outside of the short basal body and it is three-stranded. The fourth microtubular

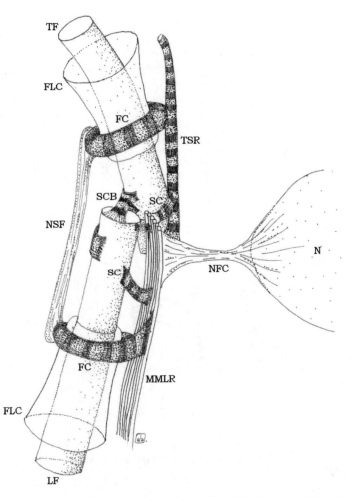

FIGURE 2.54 Root system of Dinophyta. TF, transverse flagellum; LF, longitudinal flagellum; SCB, striated connective band; MMLR, multimembered longitudinal root; TSR, transverse striated root; SC, striated connectives; NFC, nuclear fibrous connective; N, nucleus; FC, fibrous collars; FC, flagellar canals; and NSF, nonstriated fiber.

root is the most unusual; it is one-stranded and it seems to emerge within the lumen of the short basal body next to the cartwheel structure (Figure 2.56).

Chlorophyta

Flagellar apparatus of most green algae are characterized by a cruciate root system. A cruciate root system consists of four roots spreading out more or less evenly from the basal bodies, and with opposite roots usually possessing identical numbers of microtubules. As in most Chlorophyta two of the roots are two-stranded, the general arrangement of microtubular root follows the X-2-X-2 system, with X varying from three to eight microtubules. The root system will be described for each class of this division, selecting when possible a genus representing the morphological cell type within the class.

The diversity of flagellar apparatuses of Prasinophyceae is unique among the Chlorophyta. The root system is cruciate with a 4-2-4-2 system in the tetraflagellate *Pyramimonas* and *Tetraselmis*, a 4-2-0-0 system in *Mantoniella squamata*, due to the reduction of one of the two flagella, and to its

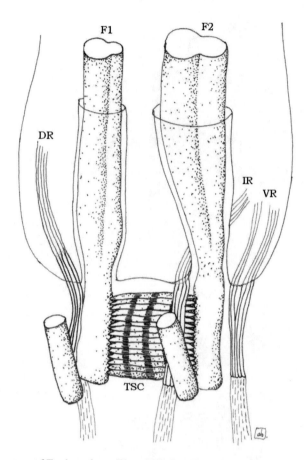

FIGURE 2.55 Root system of Euglenophyta. F1 and F2: flagella; DR, dorsal root; IR, intermediate root; VR, ventral root; and TSC, transversely striated connective.

lack of roots, a 4-3-0-(7-12) system in *Nephroselmis*, and a 4-6-4-6 system in *Mesostigma*. When four microtubular roots are present, as in *Pyramimonas*, they extend up the sides of the flagellar pit towards the anterior of the cell, where they join the microtubules of the cytoskeleton, which radiate from the flagellar region below the membrane. An elaborate system of non-striated fibers connects the basal bodies, which are associated with one or more rhizoplasts. These contractile structures extend from the proximal ends of the basal bodies down to the chloroplast, where they branch over the chloroplast surface and get in contact with it. Some tetraflagellate members of this class possess a synistosome, a fibrous band longitudinally striated connecting two of the basal bodies, and an asymmetrical structure termed lateral fibrous band, which forms an arc on one side of the four basal bodies. Proximal connective fibers may be present between the basal bodies.

The microtubular root system of flagellate Chlorophyceae has the X-2-X-2 pattern except in the stephanokont reproductive cells of Oedogoniales. The roots diverge from the basal bodies and run beneath the cell membrane towards the posterior of the cell. Fibrous roots are generally present and associated with the two-membered microtubular roots. Rhizoplasts extend from the basal bodies to the nucleus. The basal bodies are connected by a robust upper striated connective and two lower striated connectives (Figure 2.57). In the stephanokont zoospore of *Oedogonium* sp. the flagellar bases are connected by a transversely striated fibrous band running around the top of the zoospore above the flagella. Three-membered microtubular roots perpendicular to the basal bodies depart from them and extend towards the posterior of the cell. Other striated components are present in close association with the microtubular roots.

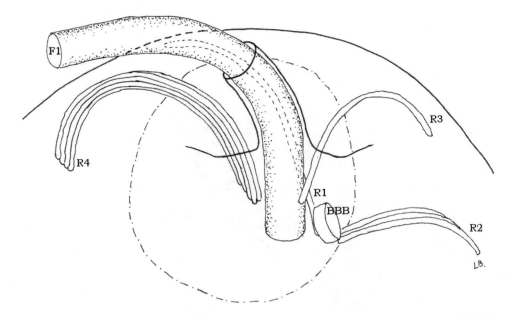

FIGURE 2.56 Root system of *Bigelowiella natans* (Chlorarachniophyta). F1, main flagellum; BBB, barren basal body; R1, R2, R3, and R4, microtubular roots.

The microtubular rootlets of the flagellate reproductive cells of Ulvophyceae follow the X-2-X-2 arrangement of most green algae, with X = 4 in *Ulva* and *Enteromorpha*, and X = 5 in *Ulothrix*. Striated bands connect the rootlets to the basal bodies, which are connected anteriorly by an upper connective; in some genera, additional striated bands between the basal bodies and striated component associated with the two-membered root can be detected.

Members of the Cladophorophyceae generally conform to the X-2-X-2 system, with X = 3 in *Chaetomorpha* and X = 4 in *Cladophora*. The microtubular rootlet system of the biflagellate cells extend posteriorly nearly parallel to one another and to the basal body from which they arise. In the X-membered rootlets, an electron-dense strut or wing connects one of the uppermost microtubules to the subtending singlet, and the entire rootlet is usually subtended by a massive, more or less striated structure. Also the two-membered rootlets may be accompanied by such structures. An upper transversely striated connective links the basal bodies.

The flagellar apparatus of the reproductive cells of the Bryopsidophyceae is anchored in the cell by four microtubular roots following the usual X-2-X-2 pattern. Each of the microtubules in the rootlet may be subtended by an electron-dense wing. The basal bodies are connected anteriorly by a non-striated upper connective, with a typically pronounced arched appearance. Transversely striated bands connecting the rootlets to the basal bodies are also present. Species exist in this class, which produce stephanokont zoospores with more than 30 flagella. In these cells, the basal bodies are connected by a non-striated fibrous upper ring, which can be considered the result of the fusion of many non-striated upper connectives. The proximal ends of the basal bodies are partially enclosed in a second lower ring of amorphous material. Four- and six-membered microtubular roots depart from between the basal bodies.

No flagellate reproductive cells are present in the Zygnematophyceae.

The microtubular root system of the Trentepohliophyceae motile cells (tetraflagellate zoospores and biflagellate gametes) do not follow the X-2-X-2 pattern, but show a 6-4-6-4 arrangement in *Trentepohlia* sp. The dorsiventrally compressed form of the flagellate cells forces the basal bodies and the root system components in a flattened arrangement. Basal bodies are aligned perpendicular to the long axis of the cell, with a parallel or antiparallel arrangement, and the microtubular

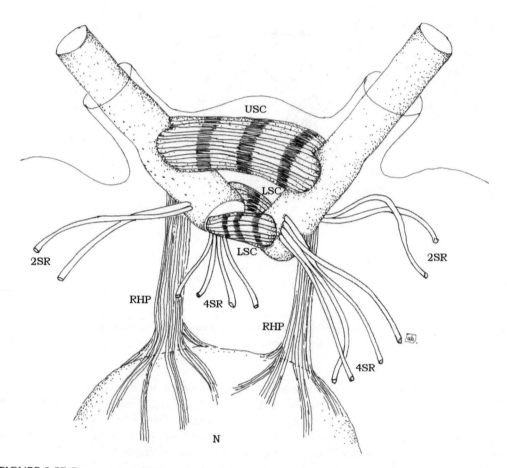

FIGURE 2.57 Root system of Chlorophyceae (Chlorophyta). 2SR, two-stranded roots; 4SR, four-stranded roots; RHP, rhizoplasts; N, nucleus; USC, upper striated connective; and LSC, lower striated connectives.

rootlets, especially in the gametes, extend posteriorly nearly parallel to one another and to the basal body from which they arise. Two of the rootlets (the four-membered rootlets in *Trentepohlia* sp.) are associated with a complex columnar structure resembling the multilayered structure of the Charophyceae.

Biflagellate cells of Klebsormidiophyceae such as *Chaetosphaeridium* sp. and *Coleochaete* sp. are characterized by a unilateral construction, in which the two equal flagella emerge on one side of the cell, below the apex. A transversely striated connective links the basal bodies. The system anchoring the flagella in the cell consists of a single lateral root of about 60 microtubules, which extend from the basal bodies along the cell side down to the posterior. At the level of the basal bodies this broad root enters a multilayered structure in which the microtubules are located between two laminate plates.

In the biflagellate male gametes of the Charophyceae the basal bodies are connected to each other by a conspicuous fibrous linkage; the root system consists of a main broad band of microtubules and a small secondary root.

The root system and the associated structures of the biflagellate motile cells of Dasycladophyceae is scarcely distinguishable from those of the Cladophorophyceae. The X-2-X-2 pattern is present, with the basal bodies and microtubular rootlets showing a flattened arrangement. A striated distal fiber connects the proximal ends of the basal bodies, from which two prominent rhizoplasts depart.

A strongly suggested reading on this topic is the review by Moenstrup (1982).

How Algae Move

Cytoplasm, cell walls, and skeletons of algae have a density greater than the medium these organisms dwell in. The density of fresh water is 1.0 g cm^{-3} and that of sea water ranges from 1.021 to 1.028 g cm^{-3}, but most cytoplasmic components have a density between 1.03 and 1.10 g cm^{-3}, the silica forming the diatom frustule and the scales of Chrysophyceae have a density of 2.6 g cm^{-3}, and both calcite and aragonite of Haptophyta coccoliths reach an even higher value of 2.7 g cm^{-3}. With this density values, algae must inevitably sink. Therefore, one of the problems facing planktonic organisms (organisms that wander in the water or are carried about by the movements of the water rather than by their own ability to swim) is how to keep afloat in a suitable attitude between whatever levels are suitable for their life. The phytoplankton must obviously remain floating quite close to the surface because only here is there a sufficient illumination for photosynthesis. There are broadly two solutions by which algae can keep afloat and regulate their orientation and depth: a dynamic solution, obtaining lift by swimming; and a static solution, by buoyancy control, or through adaptations reducing sinking rates. In many cases the two solutions function together.

Swimming

What does it mean to swim? It means that an organism immersed in a liquid environment is allowed to deform its body in the same manner. The algae are all good movers or better good swimmers. They swim more or less continuously and control their level chiefly by this means. For example dinoflagellates, which can achieve speeds of 200–500 μm sec^{-1}, are said to maintain themselves near the surface by repeated bursts of upward swimming, alternating with short intervals of rest during which they slowly sink. Motility is present in unicellular algae or colonies that are propelled by flagella; in some classes it is confined only to gametes and asexual zoospores provided with flagella, which are used as motor system for their displacement in the fluid medium.

In order to move through a fluid the swimming cell must use its motor system to push a portion of the fluid medium in the direction opposite to that in which the movement is to take place. Forward movement of a swimming alga is resisted by two things: the inertial resistance of the fluid that must be displaced, which depends on the density of the fluid and the viscous drag experienced by the moving organism, that is, the rearward force exerted on the organism by the fluid molecules adhering to its surface when it passes through the viscous fluid. The ratio of inertial and viscous forces is the Reynolds number (R), which depends on the size of the organism (related to the linear dimension, (l)), its velocity (u), and to the density (ρ) and viscosity (η) of the fluid medium according to the equation:

$$R = \text{(Inertial forces)} * \text{(Viscous forces)}^{-1} = l * u * \rho * \eta^{-1} \tag{2.1}$$

As the ratio between the viscosity and the density is the kinematic viscosity (v in cm^2 sec^{-1}), Equation (2.1) can be written as:

$$R = l * u * v^{-1} \tag{2.2}$$

In water the kinematic viscosity is 10^{-2} cm^2 sec^{-1}.

An algae 50 μm long swimming at 10 μm sec^{-1} has the minuscule Reynold's number of 5×10^{-4}, hence inertial effects are vanishing and the major constraint is the viscous drag; this means that what an algae is doing at the moment is entirely determined by the forces that are exerted on it *at that moment* and by nothing in the past. Therefore, when the flagellum stops, forward movement of the cell will cease abruptly without gradual deceleration.

Reynolds number might get up to 10^5 for a fish long 10 cm swimming at 1 m sec^{-1}. If the same fish would be swimming at the same Reynolds number of an alga, it would be as if it was swimming inside molasses.

Another funny thing about motion at low Reynolds number is reciprocal motion. As time does not matter the deformation that produces the swimming must be asymmetrical. Therefore, the pattern of flagellar beating must be three-dimensional and asymmetric, that is, the forward stroke should be different from the reverse.

For optimum propulsive efficiency, cell body size should be 15–40 times the flagellum radius (about 0.1 μm), and this ratio is present in many algae. When the cell body size is larger than predicted, as in *Euglena*, the effective radius of the flagellum is modified by simple, non-tubular hairs.

Beat patterns of most smooth flagella (i.e., without hairs) are three-dimensional, and the analysis of the motion is far from straightforward. However, it is clear that the direction in which the microorganism moves is opposite to the direction in which the waves are propagated along the length of the flagellum, so that in almost all cases, when the cell body is to be pushed along, a wave must be initiated at the base of the smooth flagellum. Although basal initiation is more common than distal, both are known. The velocity of forward movement is always a small fraction of the velocity of the wave running along the flagellum, and its propulsive efficiency depends on the ratio of its amplitude and wavelength. Unlike smooth flagella, the propulsive force generated by a flagellum bearing tubular hairs is in the same direction as the direction of wave propagation. These stiff hairs remain perpendicular to the axis of the flagellum as it bends (Figure 2.58). A wave moving away from the cell body will cause the hairs to act as oars and the overall effect will be to propel the cell flagellum first.

Control characteristics, and thus behavioral peculiarities, are connected with the functioning of the propelling structure of the cell. If the cell is asymmetric, it advances spinning along its axis; it can correct its trajectory only by sudden steering obtained by changing the insertion angle of flagella, or by the stiffening of internal structures. This behavior can be attributed to all heterokont or uniflagellate algae. In the case of a symmetric cell, it can accomplish a gradual smooth correction of its trajectory going forward without spinning (or rotating with a very long period), and displacing the barycenter of the motor couple. This behavior can be attributed to all isokont cells.

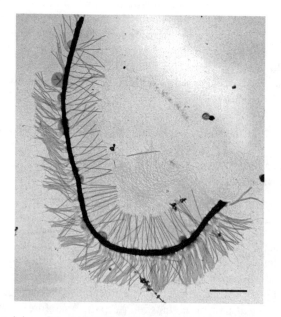

FIGURE 2.58 Negative staining of the trailing flagellum of *Ochromonas danica*. (Bar: 1 μm.)

Examples of main swimming patterns among algae will be described as follows.

In *Ochromonas* sp. (Heterokontophyta) only the flagellum bearing hairs seems to be active during swimming. It is directed forward and executes simultaneous undulatory and helical waves that travel from its base to the tip. The resulting flagellum movements cause the whole body of the cell to rotate as it moves forwards. The shorter flagellum trails backward passively, lying against the cell; it is capable of acting as a rudder to steer the cells. The two rows of stiff hairs cause a reversal of the flagellum thrust. Water is propelled along the flagellum from the tip to the base, so that the cell is towed forward in the direction of the flagellum (Figure 2.59).

In desmokont dinoflagellates such as *Prorocentrum* sp., the longitudinal flagellum, which extends apically, beats with an anterior-to-posterior whipping action, generating a wave in a tip-to-base mode. The second flagellum, perpendicular to the first, is coiled and attached to the cell body except for the tip, which beats in a whiplash motion, while the attached part undulates (Figure 2.60). In dinokont dinoflagellates, such as *Peridinium* sp. or *Gymnodinium* sp., the two flagella emerge at the intersection of the cingulum (transverse furrow) and sulcus (longitudinal sulcus). The longitudinal flagellum extends apically running in the sulcus and is the propelling and steering flagellum, while the ribbon-shaped transverse flagellum is coiled, lies perpendicular to the first and runs around the cell in the cingulum. It is thought to be responsible for driving the cell forward and it also brings about rotation. The longitudinal flagellum beats with a planar waveform, which contributes to forward movement. The longitudinal flagellum can also reverse the swimming direction: it stops beating, points in a different direction by bending, and then resumes beating. This steering ability is related to change in orientation of basal bodies and contraction of the structures associated to the axoneme. In the transverse flagellum, a spiral wave generated in a base-to-tip mode is propagated along the axoneme, bringing about backward thrust and rotation at the same time (Figure 2.61).

As described earlier, the emergent flagellum of *Euglena* sp. (Euglenophyta), bears simple hairs 3–4 μm long. These long hairs are arranged in tufts of three to four and form a single row that runs along the flagellum spirally with a low pitch. The flagellar hairs increase the thrust of the flagellum against the surrounding water. During swimming the long flagellum trails beside the cell body and performs helical waves, generated in a base-to-tip mode (Figure 2.62).

A peculiar swimming pattern is present in the ovoid zoospores of *Chlorarachnion reptans* and *Bigelowiella natans* (Chlorarachniophyta), which bear a single flagellum inserted a little below the cell apex. This flagellum bears very delicate hairs markedly different from the tubular hairs of Heterokontophyta. During swimming the flagellum wraps back around the cell in a downward spiral, lying in a groove along the cell body. The cells rotate around the longitudinal axis during swimming and the anterior or posterior end of the cell moves in either narrow or wide helical path which appears as a side-to-side roking or wobbling (Figure 2.63).

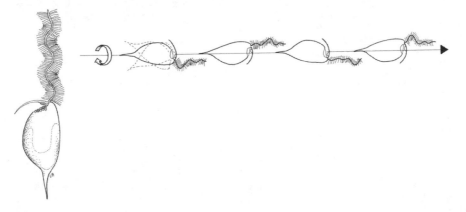

FIGURE 2.59 Swimming pattern of *Ochromonas danica*.

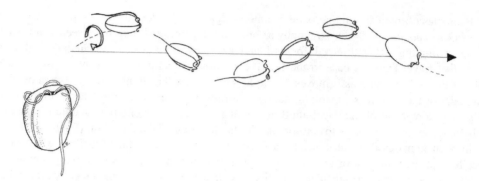

FIGURE 2.60 A desmokont dinoflagellate (*Prorocentrum* sp.) and its swimming pattern.

In isokont biflagellate algae such as *Chlamydomonas* or *Dunaliella* (Chlorophyta), during the effective stroke the flagella bend only at the base, push more water backwards than adhereing to them during the forward recovery stroke, thus bringing about net forward movement. While swimming these cells also rotate. Speed ranging from 100 to 200 μm sec^{-1} can be reached by these cells during forward swimming. Backward swimming is also possible, during which the flagella perform undulatory movement (Figure 2.64).

An interesting question is why the algae swim. All algae in an aquatic environment have a need to exchange molecules such as O_2, CO_2, and NH_3 with environment. As all solid boundaries in a liquid medium have associated with them a boundary layer in which water movement is reduced (due to the no-slip boundary), this layer will impede the nutrient uptake of the organisms by creating a small depleted layer around them. Turbulence is very ineffective in transporting nutrients towards such small organisms as the smallest length scale of turbulent eddies are of the order of several millimeters. Therefore, algae must rely on molecular diffusion to overcome the nutrient gradient across the boundary layer. Diffusion, that is the slow mixing caused by the random motion of molecules, is important in the world of low Reynolds number, because here stirring is not any good. The alga's problem is not its energy supply; its problem is its environment. At low Reynolds number you cannot shake off your environment. If you move, you take it along; it only gradually falls behind.

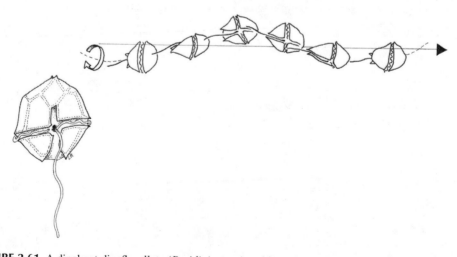

FIGURE 2.61 A dinokont dinoflagellate (*Peridinium* sp.) and its swimming pattern.

FIGURE 2.62 Swimming pattern of *Euglena gracilis*.

Algae use their motility (be it sinking or swimming) to generate movement relative to the water and hence replenish the boundary layer with nutrients. Depending on the size of the organism, the motive for swimming must differ, however, because its effects differ significantly. For small algae in the $1-10$ μm range, diffusion is about 100 times more effective in supplying nutrients than movement. This is often expressed as the Sherwood number (S):

$$S = (\text{Time for transport by diffusion}) * (\text{Time for transport by movement})^{-1}$$

$$= (L^2 * D^{-1}) * (L * u^{-1})^{-1} = (L * u)^{-1} * D$$

(2.3)

where L is the distance over which the nutrient is to be transported, u the water velocity, and D the diffusion constant.

For scale of the order of 1 μm the ratio is $\approx 10^{-2}$. Diffusion is about 100 times faster than movement. Hence, in this world of low Reynolds numbers, nothing is gained by trying to reduce the diffusion barrier by generating turbulent advection. In this context the only possible advantage to the alga of undertaking locomotion is that it might encounter nutrients in a higher concentration. For this purpose, a helical swimming path is more useful than a straight one inspite of the longer distance for the same

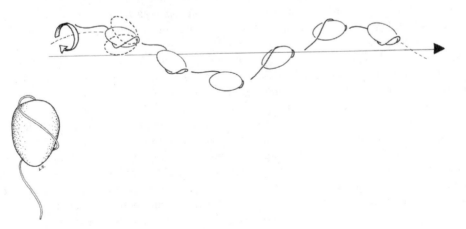

FIGURE 2.63 Swimming pattern of *Bigelowiella* sp.

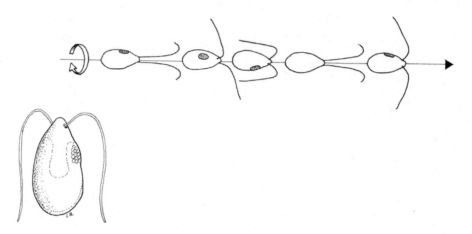

FIGURE 2.64 Swimming pattern of isokont biflagellate algae (*Dunaliella salina*).

displacement. This is because a helical swimming path enables the detection of three-dimensional component of a gradient, whereas the straight path allows detection of only one dimension.

Purcell (1977) summarized it by saying that the organism does not move like a cow that is grazing on pasture, it moves to find greener pasture.

Only the species that swim very fast such as the dinoflagellates (about 500 μm sec^{-1}) can overcome the diffusion limitation. This high velocity should be related to the effective increase in the probability to catch more preys and therefore to the heterotrophy metabolism of the algal species.

Movements Other Than Swimming

In some algae movement cannot occur unless the cells are in contact with a solid substratum. This kind of movement, in some cases termed gliding, is present in cyanobacteria, in the red alga *Porphyridium* (Rhodophyta), in diatoms, and in some desmids (Chlorophyta).

The most efficient gliders among the cyanobacteria are found in the filamentous forms such as *Oscillatoria, Spirulina, Phormidium*, and *Anabaena*, which can travel at up to 10 μm sec^{-1}. Some species, such as *Phormidium uncinatum* and *Oscillatoria*, rotate about their long axis while gliding; while others, such as *Anabaena variabilis* translate laterally. Other unicellular coccoid cyanobacteria, such as *Synechocystis*, move by "twitching," a flagella-independent form of translocation over moist surfaces. This type of motility is analogous to social gliding motility (S-motility) in myxobacteria, which involves coordinated movements of cells close to each other (cell–cell interactions) and requires both Type IV pili operating in a manner similar to a grappling hook and fibrils (extracellular matrix material consisting of polysaccharides and protein). While moving, cyanobacterial gliders secrete mucilage, or slime, which plays an active role in gliding. Mucilage is extruded from rows of fine pores clustered circumferentially around the septa. These pores are part of a larger structure called the junctional pore complex (JPC), which span the entire cell wall, peptoglycan layer, and outer membrane. The channels formed by the JPCs are inclined relative to the cell axis, this angle providing directionality to the extruded slime, and are oppositely directed on either side of the septum. Propulsion of the filament results from the adherence of the slime to both the filament surface and the substratum, combined with its extrusion from a row of JPCs on one side of each septum. Switching slime extrusion to the JPCs on the other side of the septum would result in a reversal of the direction of gliding. In *P. uncinatum* the pores are aligned in a single row, whereas in *A. variabilis* several rows of pores line both sides of the septum. The outer surface of gliding cyanobacteria consists of parallelly arranged fibrils of a glycoprotein known as oscillin, a Ca-binding protein required for motility. The surface striations formed by these fibrils would act as channels for the extruded slime to flow along. Therefore, if the fibrils are helically arranged, the cell will rotate as it glides; if the fibrils are aligned radially, the cell will not rotate. In all species studied to date, this correlation is consistent, and provides a structural explanation for why some species rotate as they glide while others do not.

In diatoms, motility is restricted to pennate species possessing a raphe. These diatoms display a characteristic jerky movement forward or backward, with specie-specific path patterns. The general velocity of their movement is $1-25~\mu m~sec^{-1}$, but they can accelerate up to $100-200~\mu m~sec^{-1}$. Raphid diatoms possess an actin-based cytoskeletal system located just beneath the plasma membrane at the raphe. Transmembrane components with an adhesive extracellular domain are connected to these actin bundles, and their interaction is somehow involved in both adhesion and motility mechanisms. Microtubules are also present in this region; in addition secretory vesicles containing polysaccharides often appear near the actin filaments at the raphe, providing the mucilage strands that project from the raphe and adhere to the substratum during the gliding process.

At least two models exist which provide reasonable explanation for diatom locomotion. In the first model, a force applied to the transmembrane protein-actin connectors, parallel to the actin bundles, would result in the movement of trasmembrane proteins through the cell and subsequent movement of the cell in the opposite direction to the force. In the second model, the energy required for motility would be generated by a conformational change of the adhesive mucilage on hydration that occurs when it is secreted from the raphe. In this model, the actin bundles restrict the secretion of mucilage to one end of the raphe, which generates a net force moving the cell over the site of secretion. In both models, the secreted mucilage plays a central role either by providing traction to translate the force into cell movement or by generating the energy through conformational changes on hydration.

A slow gliding movement over solid substrata has been observed in *Porphyridium* sp. (Rhodophyta) and in some desmids (Chlorophyta). In *Porphyridium*, the mucilage produced in mucilage sacs located inside the cell is excreted through the membrane. In desmids mucilage is excreted through the cell wall by flask-shaped pores. As they move, these gliding cells leave behind a fibrillar mucilaginous trail, whose swelling by water pushes the cells forward. Table 2.1 presents swimming and gliding speeds of some planktonic algae.

TABLE 2.1
Swimming and Gliding Speeds of Some Planktonic Algae

Name	Mean Speed
Gymnodinium gracilentum	$500~\mu m~sec^{-1}$
Symbiodinium sp.	$250~\mu m~sec^{-1}$
Tetraselmis suecica	$180~\mu m~sec^{-1}$
Chattonella sp.	$120~\mu m~sec^{-1}$
Cryptomonad	$110~\mu m~sec^{-1}$
Chlorarachnion reptans	$110~\mu m~sec^{-1}$
Euglena gracilis	$100~\mu m~sec^{-1}$
Dunaliella salina	$95~\mu m~sec^{-1}$
Ochromonas danica	$80~\mu m~sec^{-1}$
Bigelowiella natans	$70~\mu m~sec^{-1}$
Pavlova salina	$50~\mu m~sec^{-1}$
Oscillatoria spp.	$10~\mu m~sec^{-1}$
Leptolyngbya spp.	$0.004~\mu m~sec^{-1}$

Buoyancy Control

The alternative to swimming is to float by means of some types of buoyancy device. In some of the attached brown algae of the seashore (*Fucus vesicolosus*, *Ascophyllum nodosum*, and *Sargassum* sp., Heterokontophyta) the fronds gain buoyancy from air bladders (pneumatocysts) within the thallus, which stands erect when submerged. Oxygen and nitrogen, in roughly the same proportion as in air, form the bulk of the gas, but there are also small, variable amount of CO_2 and CO. Oxygen and CO_2 derive partly from the metabolic activities of the cells in the pneumatocyst wall and diurnal changes in the composition and pressure of pneumatocyst gases have been shown. However, equilibration takes place between the gases in the pneumatocyst and in the surrounding water (or air). This is the source of nitrogen in the vesicles and the major source of O_2 and CO_2. In *Enteromorpha* sp. (Chlorophyta), gas bubbles are entrapped in the central area of its tubular hollow thallus, which may aid in keeping the stipe upright by flotation. In other seaweeds such as *Codium fragile* (Chlorophyta), gas trapped among the filaments achieves the same buoyancy effect of pneumatocysts.

Buoyancy regulation in cyanobacteria involves the production of intracellular gas-filled structures (also termed vacuoles), not delimited by membranes, and made up of assemblages of hollow cylinders, whose proteinaceous walls are permeable to gas, but not to water. The density of this structure is about 0.12 g cm^{-3}, about one eighth of that of water, and if sufficient gas-filled structures are present in a cell, it can become positively buoyant. In cyanobacteria buoyancy is regulated by varying gas-filled structure formation and cytoplasmatic composition through synthesis and breakdown of photosynthetic products. The production of gas-filled structures is induced by low-light conditions (e.g., in deep layers with insufficient light). Here, photosynthesis is reduced, osmotic pressure of newly synthesized sugars is small, and ballast materials such as carbohydrates are not produced at a high rate, therefore they will not increase cell density, which in turn would increase sinking. Under these conditions, gas-filled structures can be produced at a high rate, and cells increase their buoyancy. Conversely, if cell osmotic potential is high (high sugar production, increased amount of ballast in the form of secondary photosynthetic products), turgor pressure increases, which may collapse gas-filled structure and cells become negatively buoyant and sink in the water column. The rise in turgor pressure with light irradiance has been found in many cyanobacteria; however, for this rise to result in gas-filled structure regulation, the pressure reached must exceed the lowest pressure of gas-filled structures. This occurs, for example, in *Anabaena flos-aquae*, with a critical collapse pressure distributed about a mean of 6 bars. In *Trichodesmium* sp. gas-filled structures can withstand pressures of $12–37$ bars depending on the species, and turgor pressure collapse is not possible as a buoyancy regulation mechanism in this genus; carbohydrate ballasting is considered the only plausible mechanism for rapid buoyancy shifts in this cyanobacterium.

Other algae obtain buoyancy from liquids of lower specific gravity than seawater or freshwater in a way similar to a bathyscaphe. Liquid-filled floats have the advantage of being virtually incompressible; but because of their higher density they must comprise a much greater proportion of the organism's overall volume than is necessary with gas-filled floats if they are to give equivalent lifts. The large central vacuole of diatoms contains cell sap of reduced density, obtained by the selective accumulation of K^+ and Na^+, which replace the heavier divalent ions, conferring some buoyancy. In young, fast-growing cultures, diatom cells often remain suspended or sink only very slowly, although in older cultures they usually sink more rapidly. Studies of the distribution of diatoms in the sea suggest that some species undergo diurnal changes of depth, usually rising nearer the surface during daylight and sinking lower in darkness, possibly due to slight alterations of their overall density affected by changes in specific gravity of the cell sap, or in some cases by formation or disappearance of gas vacuoles in the cytoplasm. The dinoflagellate *Noctiluca* also gain buoyancy from a high concentration of NH_4^+ ions in its large vacuoles, exclusion of relatively heavy divalent ions, especially sulfate, and a high intracellular content of Na^+ ions relative to K^+. As a result, the density of the cell sap in the vacuoles is less than that of seawater and the cells can therefore be positively buoyant and float.

When buoyancy control is not possible by these mechanisms, algae can keep afloat and regulate their orientation and depth through adaptations reducing sinking rates. The rate at which a small object sinks in water varies with the amount by which its weight exceeds that of the water it displaces, and inversely with the viscous forces between the surface of the object and the water. The viscous forces opposing the motion are approximately proportional to the surface area, and therefore, other things being equal, the greater the surface area, the slower the sinking rate.

There are a number of structural features of planktonic organisms which increase their surface area and must certainly assist in keeping them afloat. The majority of planktonts are of small size, and therefore have a large surface to volume ratio. In many cases, modifications of the body surface increase its area with very little increase in weight. These modifications generally take two forms: a flattening of the body, or an expansion of the body surface into spines, bristles, knobs, wings, or fins.

A great range of flattened or elaborately ornamented shapes occurs in diatoms such as *Chaetoceros* sp. In dinoflagellates also, the cell wall in some cases is prolonged into spines (*Ceratium*) or wings (*Dinophysis*). Among the Chlorophyceae, the wall of the peripheral cells of *Pediastrum* colonies may bear clusters of very long and delicate chitinous bristles regarded as buoyancy devices. In *Scenedesmus* also the cells are clothed by a large number of bristles with a complex structure, which seem to help keep the cells in suspension.

Reduction of the sinking rate is also obtained by an increase in lipid content, which has a density of about $0.86 \, g \, cm^{-3}$. Oil droplets are common inclusions in the cytoplasm of algae; lipids stored in this form are present in the Chrysophyceae and Phaeophyceae (Heterokontophyta), and in the Haptophyta, Cryptophyta, and Dinophyta. The thermal expansion of these compounds may be of some significance in affecting diurnal depth changes, through reduction of cell density, but without producing neutral buoyancy.

How a Flagellum Is Built: The Intraflagellar Transport (IFT)

The mechanisms that determine and preserve the size and function of cellular organelles represent a fundamental question in cell biology up to now only partially understood, and flagella has provided a handy model system to investigate organelles' size-control analysis. It was discovered that flagella are dynamic structures and that flagellar length is regulated by a process called intraflagellar transport. IFT is a motile process within flagella in which large protein complexes move from one end of the flagellum to the other, and flagellar length is regulated by a balance between continuous assembly of tubulin at the tip of the flagellum, counterbalanced by continuous disassembly. According to Iomini et al. (2001), the IFT cycle consists of four phases. In Phase I, which takes place in the basal body region of the flagellum, anterograde particles are assembled from retrograde particles by remodeling or exchange of subunits with the cell body cytoplasm, with a concurrent decrease in number. In this phase, the precursors of the flagellar structures that make up the cargos are also loaded onto the particles. In Phase II, the particles are transported from the base to the distal end of the flagellum by a heterometric kinesin II with a velocity of about $2 \, \mu m \, sec^{-1}$. In Phase III, which occurs at the distal end of the flagellum, anterograde particles are remodeled into retrograde particles with a concurrent increase in number, probably upon or after unloading their cargo. Finally, in Phase IV, retrograde particles are transported by a cytoplasmic flagellar dynein from the distal end back to the basal body region of the flagellum, with a velocity of about $3 \, \mu m \, sec^{-1}$, higher than that of anterograde particles.

How a Flagellar Motor Works

Movement can arise by shape change of permanently linked elements, by reversible interactions causing movement of elements relative to each other, by reversible assembly and disassembly, etc. all of which need energy input. We know that such changes can occur in proteins, the most likely molecules serving these locomotory functions in real movement system. But what drives and controls these changes? In principle, the problem is not difficult. Altering the ionic milieu, changing chemically or electrically its environment can alter the tertiary or quaternary structure of a protein. In most control systems, if not all, a change in the environment brings about a change in the properties of the motor, acting either directly or indirectly on the component of the motor. We need only two proteins to make a motor using the sliding filaments mechanism, that is, a globular protein (such as tubulin) and an anchor protein (such as the dynein–dynactin complex). If the globular protein can polymerize, we can assemble it into a linear polymer that can be attached via the anchor protein to another structure some distance away. The transformation of chemical energy into mechanical work depends on a conformational change of the anchor protein, which uses the hydrolysis of ATP into ADP. Provided the anchor protein repeats the conformational change upon each monomer of the globular protein in turn, the "boat" can be hauled "hand over hand" towards the distant anchorage. Provided some kind of metachrony regulates adjacent motor molecules, we can link our small movements in a temporal series to amplify the amount of movement that can be achieved. Each step costs hydrolysis of one ATP molecule per anchor protein. The simplest and most obvious solution is either to have more than one anchor protein, or to have a dimer, working out-of-phase, being careful not to detach before the new attachment is formed. For instance, most (but not all) microtubular motors (dyneins, kinesins) work as dimers whose subunits walk along microtubule walls just like human legs walk on a surface. Once we have two hands to pull on the rope we can indeed move hand-over-hand; the one-armed man cannot do more than pull once. The flagellum movements are due to the transient interaction between two anchored microtubules, coupled to a linkage control. The generation of sliding of adjacent doublets by flagellar dynein is combined to the resisting forces localized near the active sliding rows of dyneins. During the cycle of binding/release obtained by dynein conformational change coupled to ATP hydrolysis, chemical energy is converted into mechanical energy used for sliding. Owing to their regular spacing every 24 nm along the axoneme, several adjacent dyneins participate to this local sliding and their functioning proceeds by local waves that propagate step by step all the way along the flagella. The postulated regulator has therefore to trigger the functioning of the different dyneins alternatively along the length as well as around the section of the axoneme, but its molecular nature remains unknown.

The model of Lindemann (1994) accounts for wave generation and propagation, regulated by geometrical constraints. This model, the so-called "geometric clutch," is based on the way a cylinder with nine generatrix (the nine outer doublets of the axoneme) changes form when submitted to bending. In the zone of curvature, the doublets located outside the curvature are brought apart while the doublets located inside the curvature come closer to each other. This makes the corresponding dynein molecules efficient for sliding. In contrast, dynein molecules located outside the curvature are too far for binding to adjacent microtubules: no active sliding can occur in this zone. This model is consistent with ultrastructural data of electron microscopy. As above, the functioning of the axonemal mechanics needs transient connections, which could possibly be disrupted by intrinsic proteolytic activities due to the combined activities of a protease/ligase system.

INTERNAL FLAGELLAR STRUCTURE

How a Paraflagellum Rod Works

As the PFR and the axoneme are very tightly connected, both structures have to move together. It can be argued that the PFR contributes to regulate the flagellar movements. This could be done by providing a more rigid structure that can vary its stiffness in time, that is, modifying flagellum beating pattern.

We need only one protein to build a device for information transfer and control, that is, a protein with two conformational, alternate states such as intermediate filament proteins, which can form lattice-like structures. The propagation of conformational changes along these proteins can be used to transport or transduce sensory information. Each protein can be considered as a dipole in one of the two possible states. We can imagine that the conformational change is transmitted by one dipole to the neighbor proteins as a wave. Therefore, a current flow through these lattice-like structures could be generated by the mobile electrons of the proteins that interact with their immediate neighbors via dipole–dipole forces.

The α-helical coiled-coils structural motif in the rod filament is well suited for electron propagation (Figure 2.37b). A current might be propagated distally via PFR_1 and PFR_2 protein–protein charge transfers in the lattice-like rod filaments. The current flows through these lattice-like structures and the sole constraint is that each lattice site should possess a dipole moment proportional to the magnitude of the mobile charge unit and the distance over which it hops, that is, about 1 or 2 nm. This wave produces a contraction in the PFR and a varying internal resistance that modulates the flagellar beats. The contraction occurs by displacement of the goblet appendages of the PFR along the axonemal microtubules, which reduces the distance between the coiled filaments hence generating longitudinal waves of contraction along the paraxial rod. The stiffening should swing the flagellum sideways, damping out some undulatory waves of the axoneme.

PHOTORECEPTOR APPARATA

Life is essentially about information, how information is perceived, how it is stored, passed on and used by organisms as they live and reproduce. In the world of photosynthetic microorganisms, where virtually all life depends on solar energy, light becomes also a source of information, used to orient microorganisms spatially and to guide their movements or growth. Responses using light as a sensory stimulus for orientation towards areas that best match their individual irradiation requirements are thus a virtually universal behavior among algae. The full exploitation of light information necessitates proper perceiving devices, able to change the small signal represented by the light falling upon them in a larger signal and response of an entirely different physical nature, that is, these devices, termed photoreceptors, must perform perception, transduction, amplification, and transmission.

The processing of a photic stimulus and its transformation into an oriented movement can be considered the "vision" phenomenon of motile algae. True vision involves production of a focused image of the external world, and the optical requirements for an eye probably cannot be satisfied by algae, requiring true multicellularity with cell specialization and division of labor. Still, algal "eyes" have many similarities with the complex vision systems of higher organisms, because they do possess optics, photoreceptors, and signal transduction chain components.

The essential elements of these basic visual systems are the shading device(s), for example, the eyespot and the detector, that is, the true photoreceptor(s). When the eyespot is absent its function is performed by the whole algal body.

The eyespot is a sort of roundish shield, inwardly or outwardly concave, made up of one or more layers of lipidic globules closely packed. These globules contain mainly carotenoids that can play the shading role due to their strong absorbance in the 400–500 nm range. The most common type of photoreceptor consists of extensive two-dimensional patches of photosensitive proteins, present in the plasma membrane in close association with the eyespot. Very often the photoreceptor cannot be identified by optical microscopy, while the eyespot can be seen easily because of its size and color, usually orange-red. This is one of the main reasons for the plethora of data present in the literature on the morphology, composition, and ultrastructure of algal eyespots with respect to the few data available on their photoreceptors.

Ultrastructural studies over the past 40 years have shown that photosensory systems in algae have certain characteristics in common, and distinct types of these apparatus can be identified.

On the assumption that photoreception is inseparable from the presence of photoreceptive proteins but not necessarily from the presence of an eyespot, we will group photoreceptive systems into three main types. The description will include also those algae with well documented eyespot but only presumptive photoreceptor localization.

TYPE I

A single layer of photoreceptor molecules is present inside the whole cell membrane, or located in just the patch of membrane that covers the eyespot when present (Figure 2.65). Lacking the eyespot, the whole algal body performs the shading function. This means that this type of photoreceptive system could be not readily visible.

In Cyanophyta, phototactic orientation has been described in *Anabaena variabilis, Pseudoanabaena* sp., and *Phormidium,* although no defined structure for light sensing have been so far detected. The first identification of a complex photoreceptive system leading to the evidence of a photoreceptive protein was performed in *Leptolyngbya* sp. This deep red cyanobacterium lives in Roman hypogea at extremely low light intensity (10^{13} photons m^{-2} sec^{-1}). It possesses an orange eyespot at the tip of the apical cell of the trichome. Electron microscopy revealed that this eyespot is characterized by osmiophilic globules of about 100 nm in diameter arranged in a peripheral cap extending $2-3$ μm from the apex and with a possible layered pattern (Figure 2.2) Microspectrophotometric analysis of the tip of the apical cell of *Leptolyngbya* trichomes revealed a complex absorption spectrum with two main bands. The band centered at 456 nm is due to the absorption of the carotenoid present in the eyespot, whereas the band centered at 504 nm can be assigned to rhodopsin-like molecules packed in the plasma membrane of the tip of the apical cell.

In Heterokontophyta data exist indicating that the photoreceptor molecules are present inside the cell membrane of zygotes of the fucoid brown algae, *Fucus* sp. and *Silvetia compressa.* Experimental work confirmed this localization in *S. compressa* (Figure 2.66a), where a rhodopsin-like

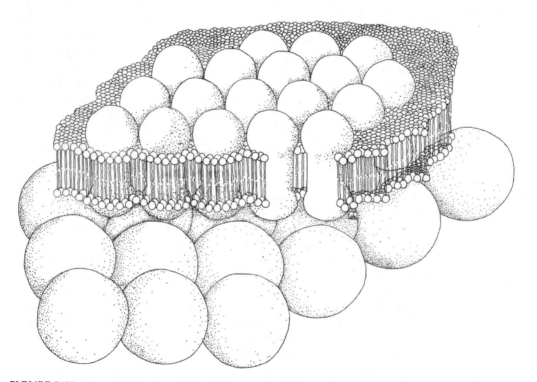

FIGURE 2.65 Schematic drawing of Type I photoreceptor system.

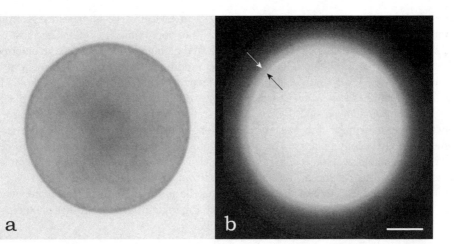

FIGURE 2.66 A *S. compressa* zygote in bright field microscopy (a), and under fluorescence microscopy (b) labeled with anti-rhodopsin antibody. The two arrows point to the cell membrane layer of the cell in which photoreceptive protein are located. (Bar: 2 μm.)

protein was identified in the zygote membrane (Figure 2.66b). Because no eyespot can be detected in these algal stages, the shading function is assigned to the whole cell body.

In the Haptophyta, the eyespot is present only in some species of the order Pavlovales. It consists of a single layer of globules situated at the anterior end of one of the chloroplast, beneath the posteriorly directed flagellum, at the level of its emersion from the cell. In these algae the photoreceptor has not been yet localized, but we can presume it is positioned inside the membrane in close association with the eyespot.

In the Cryptophyta, the presence of an eyespot is limited to a small number of species belonging to the genus *Chroomonas*. The eyespot is situated in the center of the cell, within a conical lobe of the chloroplast. It consists of a single layer of about 35 closely packed globules, attached to the chloroplast envelope and the endoplasmic reticulum. Also in these algae the photoreceptive proteins should be located inside the plasma membrane overlaying the eyespot.

In the Dinophyta, the eyespot is chloroplastic in *Peridiniun* sp., consisting of a layer of globules under the chloroplast envelope, situated behind the longitudinal sulcus, and truly extraplastidic in *Woloszynskia coronata*, where it consists of an irregular cluster of globules located beneath the sulcus, and immediately adjacent to the subthecal microtubules. In the latter, neither connection with the chloroplast nor membranes surrounding the eyespot are present. *Glenodinium foliaceum* and *Peridinium balticum* possess another type of eyespot. It is a roughly triangular body situated behind the sulcus and it is an independent structure bounded by a three-membrane envelope. Basically, there are two layers of pigmented globules, separated by a vesicle of granular material. The eyespot can fold back upon itself, making more layers. In *Amphidinium lacustre* the eyespot is an elongated structure located along the right edge of the sulcus. Its color is a shade of greenish-yellow rather than the red-orange, commonly found in eyespot; it consists of up to six flat rows of brick-like units, each row contained in a vesicle bounded by a unit membrane. No data exist on the photoreceptor location in these dinoflagellates, but the assumption is the same made for the Haptophyta and Cryptophyta, that is, photoreceptive proteins must be located inside the plasma membrane close to the eyespot. A separate case is that of *Alexandrium hiranoi* and *Gymnodinium mikimotoi*. Both dinoflagellates show phototactic responses but lack a detectable eyespot, hence the shading function is performed by the cell body.

In Chlorophyta such as *Haematococcus* sp., *Spermatozopsis* sp., and *Dunaliella* sp., the eyespot is situated on one side of the cell, sometimes slightly protruding beyond the cell surface, as in the gametes of *Ulva*; its area can range from about 0.3 to 10 μm^2. The globules range from 80 to 190 nm in diameter and their number varies from 30 to approximately 2000. The most common organization consists of a single layer of closely packed globules lying between the outermost thylakoid and the two-layered chloroplast membrane. Additional layers of globules can be present underneath the first layer, individual layers subtended or not by a single thylakoid. In most species the globules show a hexagonal packing, which enable the highest possible packing density. In both *Pandorina* and *Volvox* colonies, the eyespot of cells in the anterior of the colony are larger than those of the posterior, consisting of up to nine layers, marking the occurrence of some degree of colony polarity. The photoreceptor of Chlamydomonas can be considered the model of Type I photoreceptor. It consists of an extensive two-dimensional patch of photosensitive proteins, identified as rhodopsin-like protein, localized in the plasma membrane overlying the eyespot. The layered structure of the shading organelle in this type of photoreceptor works as a quarter wave interference reflector that reflects the impinging light toward the photoreceptor, in order to increase detectability of the light signal.

Type II

The photoreceptor consists of a multilayered membrane structure of photoreceptive protein. The eyespot is outwardly concave and is located close to this structure.

In Heterokontophyta, complex photoreceptors, consisting of layered electron-dense material organized in a rounded, wedge-shaped, or T-shaped organelle are present inside the smooth flagellum of the motile stages of Xanthophyceae, Eustigmatophyceae, and Phaeophyceae. In the Xanthophyceae, the eyespot consists of a single layer of about 40 globules located at one side of the anterior end of the chloroplast. It is contained by the outermost thylakoid of the chloroplast, bounded by the chloroplast envelope and its associated endoplasmic reticulum. The cell membrane above the eyespot forms a depression through which the posterior smooth flagellum passes. The eyespot depression accommodates the photoreceptor. In the Eustigmatophyceae, the prominent eyespot occupies nearly the whole anterior part of the cell, adjacent to the flagellar insertion. It consists of a somewhat irregular collection of globules situated in a slight bulge of the zoospore, but not enclosed by a membrane. The anterior hairy flagellum bears a photoreceptor swelling which fits alongside the eyespot. In the Phaeophyceae, the eyespot is situated in the posterior part of the cell, inside a strongly reduced chloroplast, and behind a depression of the cell surface through which the posterior flagellum runs. The eyespot appears concave in shape and prominent, containing a single layer of about 60 globules. The photoreceptor swelling is localized at level of the eyespot.

The flagellate species of the Chrysophyceae possess eyespot within a chloroplast and closely associated with a flagellum. The eyespot anatomy is similar to the Xanthophyceae, while the photoreceptor consisting of a 3D assemblage of rhodopsin-like proteins, with a presumptive regular organization, is found in association with the smooth flagellum directly above the eyespot depression (Figure 2.67).

In the members of the Euglenophyta the eyespot consists of a loose collection of globules situated on the dorsal side of the reservoir, the anterior invagination characteristic of these organisms. The globules vary in size (from 240 to 1200 nm) and number, can lie in a single layer, or be bunched together. Individual globules may be membrane-bound, but there is never a membrane surrounding the whole complex, and no association with any chloroplast component is present. The position of the eyespot within the reservoir region can vary among the species, but it is always in front of the photoreceptor situated on the long or emergent flagellum (Figure 2.68). This organelle is a three-dimensional assemblage consisting of a stack of more than 100 membrane protein layers with a regular organization. First-order crystallographic analysis suggests a crystalline structure, with a monoclinic unit cell. Each layer has a height of about 70 Å, which is the height of the model cell membrane (Figure 2.69). Figure 2.70 shows the isolated *Euglena* photoreceptor–PFR complex.

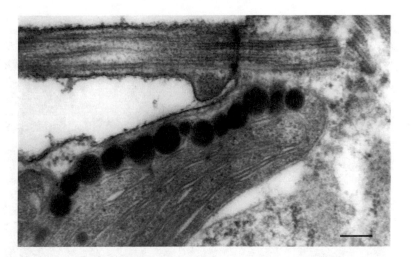

FIGURE 2.67 Transmission electron microscopy image of the photoreceptive system of *Ochromonas danica* in longitudinal section, showing the photoreceptor inside the trailing flagellum. (Bar: 0.30 μm.)

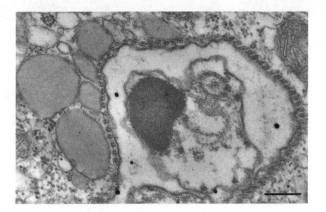

FIGURE 2.68 Transmission electron microscopy image of the photoreceptive system of *Euglena gracilis* in transverse section, showing the photoreceptor and the PFR inside the trailing flagellum, and the eyespot in front of them inside the cell. (Bar: 0.50 μm.)

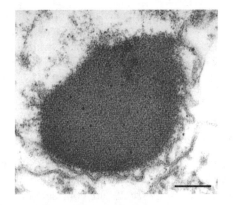

FIGURE 2.69 Transmission electron microscopy image of the photoreceptor of *Euglena gracilis* in skew section, showing the regular layer organization of this structure. (Bar: 0.20 μm.)

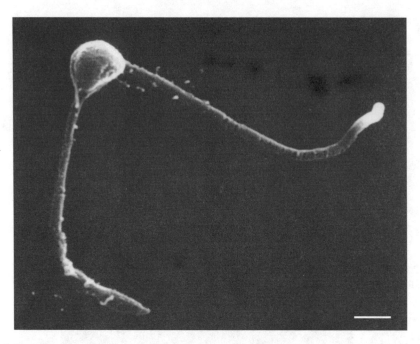

FIGURE 2.70 Scanning electron microscopy image of the isolated photoreceptor–PFR complex of *Euglena gracilis.* (Bar: 0.50 μm.)

Type III

This organization is present only in some dinoflagellates of the order Warnowiales, such as *Nematopsides* sp. and *Erytropsidinium* sp. The photoreceptor system is very specialized and it is termed ocellus. It is situated towards the left side of the ventral surface of the cell. It consists of a refractile structure termed hyalosome, thought to act as a lens, subtended by a domed pigmented part, divided into two sections, a retinoid and a pigmented cup. Between the lens and the retinoid is a chamber representing an invagination of the cell covering, which is lined by the cell membrane, and allows the contact of the ocellus with the external medium. The pigmented cup wrapping the retinoid represents the eyespot and is made up of pigment containing droplets enclosed in a vesicular layer. Small droplets contain carotenoid pigments, large droplets contain melanoid pigments. The retinoid is an extremely complex membranous construction made up of numerous regularly arranged layers giving an almost paracrystalline appearance.

No data are available on the structure and localization of the photoreceptive system in the divisions and classes of algae other than those listed above. It seems unlikely that Type II and Type III systems have not been identified so far in other algal groups, while it is more reasonable to assume that this lack of information is mainly due to the difficulty to reveal photoreceptor systems belonging to Type I. We can conclude that those algae should possess photoreceptor systems that can be taken back to Type I.

Photosensory Proteins and Methods for Their Investigation

Nature evolved a very limited number of photoreceptive molecules in the different evolutionary branches of the tree of life because of the closely similar needs of organisms for the detection of the external world. Two photoreceptor molecules are considered as underlying "vision" in algae: rhodopsin-like proteins and flavoproteins. Photoreceptor proteins should occur only in small amounts in algal structures, because they are used to detect light, not to collect it. They are concentrated in photoreceptive structures and do not contribute to the color of the organism. This is an important consideration because a primary, historical criterion for algal taxonomy and phylogeny has been their bulk pigment composition. Sensory pigments have played no role at all in these determinations, because it is generally believed that different algal groups have different photosensory pigments; furthermore, the association of particular sensory pigments with particular algal groups has been based largely upon the examination of only one or a few species.

For a long time the behavior of algae in response to light stimuli has attracted the attention of scientists, who hoped that the simplicity of the material would make this phenomenon easy to investigate. On the contrary, the undertaking turned out to be more difficult than expected and to date a clear understanding of the phenomenon has still to be attained, while comprehension of the visual process in humans is extensive. Initial ideas on the chemical nature of receptors for photobehavioral responses were gained by measurements of their spectral sensitivity. Action spectroscopy results were then screened and supported by other methods such as biochemistry, absorption spectroscopy, electrophysiology, and molecular biology, which when integrated with different investigative approaches shed more and more light on nature and functioning of photoreceptive pigments.

In the following we will briefly describe the characteristics of rhodopsin-like proteins and flavoproteins and compare spectroscopic and biochemical techniques that have been used to define the light-absorbing properties of photoreceptive proteins either *in vivo* (single cell or cell population) or *in vitro* (extracted material). To provide a clear perspective on the efficacy of different techniques as tools for the study of photoreceptive structures, a discussion of the pros and cons, the advantages and limitations of each technique are included.

Rhodopsin-Like Proteins

Rhodopsins are photoreceptor proteins, universally used from archeabacteria to humans, consisting of a proteic part, the opsin, organized in seven transmembrane α-helices, and a light absorbing group, the retinal (i.e., the chromophore). The retinal is located inside a pocket of the opsin, approximately in its center.

Why these proteins are so special? First, retinal–opsin complex has an intense absorption band whose maximum can be shifted into the visible region of the spectrum, over the entire range from 380 to 640 nm. Second, light isomerizes the retinal inside the protein very efficiently and rapidly. This isomerization, that is, the event initiating the vision reaction cascade, can be triggered almost exclusively by light; in the dark it occurs only about once in a thousand years. Third, remarkable structural changes (movements of single α-helix) are produced by isomerization of retinal. Light is converted into atomic motion of sufficient magnitude to trigger a signal reliably and reproducibly. Fourth, the photocycle (the photoreceptive protein upon light excitation undergoes a series of conformational changes which can be driven back to the original conformational state) is very fast, hence the intracellular photoreceptive machinery is immediately reset for a new response. Fifth, retinal is derived from β-carotene, a precursor with a widespread biological distribution.

About 1000 rhodopsin-like protein genes have been so far detected in the microbial world. Genes were detected in *Anabaena* (aka *Nostoc*) sp. (Cyanophyta), *Guillarda theta* (Cryptophyta), *Pyrocystis lunula* (Dynophyta), and *Chlamydomonas* sp. (Chlorophyta). Biochemical and spectroscopical evidences for a retinal-based photoreceptor were reported in one lineage of prokaryotes, that is, in *Leptolyngbya* sp., (Cyanophyta), and at least in four lineages of eukaryotic algae, that is, *Ochromonas* sp. and *Silvetia*, sp. (Heterokonthophyta); *Euglena* sp. (Euglenophyta); *Gymnodinium* sp. (Dinophyta), and *Dunaliella* sp., *Spermatozopsis* sp., and *Volvox* sp. (Chlorophyta).

Flavoproteins

Flavoproteins, or yellow enzymes, are a diverse group of more than 70 oxidoreductases found in animal, plants, and microorganisms, which have a flavin as a prosthetic group covalently attached to the protein. The proposition that these proteins could function as a near-UV-visible-light detector dates back more than 40 years. Despite the ubiquity and ancient origins of flavoproteins, their role in acquiring information from the radiation environment still remains a complex area of study. Apart from difficulties in their identification, much of the reason for the lack of understanding lies in their diversity of function. Typical absorption spectra of flavoproteins show a dominant protein peak at 280 nm, and major peaks at 380 and 460 nm. The overall similarity of many blue-light action spectra with flavoprotein absorption spectra is one of the main reasons for the belief that flavoproteins can function as blue-light photoreceptors. So far, the only biochemical identification of flavin-based photoreceptor has been carried out in *Euglena gracilis*. Despite this paucity of evidence, the hypothesis of a flavin-based photoreceptor in algae has withstood time and still remains an accepted working hypothesis.

Action Spectroscopy

Action spectroscopy still represents the classic way to investigate photopigments. By means of action spectroscopy the photosensitivity of a cell at different wavelengths can be measured, thus providing information on the nature of the pigments involved in photoreception. It is still a common belief that this approach represents the only feasible way to study the photosensory pigments of a large number of species. However, direct measure of an action spectrum is much more difficult than that of an absorption spectrum. Moreover, action spectra may not be directly correlated with the absorption peaks of the pigments involved in photoreception, as light scattering can cause several errors. When many pigments with similar absorption characteristics in the same visible range are present, action spectroscopy often fails to discriminate between them. Even when there is only one predominant pigment, it is not always possible to identify it. To obtain more reliable results, threshold action spectra should be preferred to eliminate adaptation phenomena and screening modulation, limiting the utilization of action spectroscopy to the study of changes or increases in the photosensitivity of a mutant cell, after the exogenous addition of a presumptive photoreceptor pigment which the cell lacks. It may be hasty to indicate the nature of a photoreceptor only on the basis of data obtained from action spectroscopy. This is especially true in the case of photoreceptors such as rhodopsins, which have retinal as the chromophoric group. Retinal absorption can be fine-tuned by amino acid charges of the retinal pocket, which allows the entire spectrum between 380 and 640 nm to be covered. Moreover, the presence or formation of photo-intermediates may shift the absorption maxima, and make the interpretation of the action spectrum difficult.

Absorption and Fluorescence Microspectroscopy

These techniques do not disturb the integrity of the organism or subcellular components, and allow the examination of an uninjured system with its physiological functions intact. The spectroscopic overshadowing of one pigment by another is avoided because each pigment is packaged in a different structure. Thus, cellular structure can be easily correlated with pigment type by direct observation. It is possible to make exact quantitative determinations of various reactions at the time of their occurrence in the sample, the progressive changes in these reactions, and their relationships to different conditions in the external medium. Because of the fundamental connection between optical parameters and properties of molecular structures, microspectroscopy allows assessments of minute changes in the state of the molecules of various substances in the organism, the degree of their aggregation, and the interconversions of various forms of pigments and other important biochemical compounds with characteristic spectra. In many cases the lability and reversibility of such changes make microspectroscopy the only possible method of investigation. There is virtually no light scattering problem associated with microspectroscopic measurements, even if the analyzed structure has a dimension of 1 μm. Obviously, the absorption spectrum cannot provide adequate information about the photochemical action of photons as a function of their fundamental energy; however, the identification of the chromophores in the photoreceptive structures does provide information about possible mechanisms of energy transfer. The measurements are very difficult when small changes in absorption have to be measured in the presence of a strong total signal (luminous background), as photon noise is proportional to the square root of the intensity of the incident light. Then, fluorescence spectroscopy is recommended: it can achieve more reliable results compared with absorption spectroscopy, because the background emission is much reduced. In this case the sensitivity of detection is not limited by the signal-to-noise ratio, but rather by the presence, virtually unavoidable, of fluorescent contaminants.

Biochemical and Spectroscopic Study of Extracted Visual Pigments

Extraction of visual pigments (chromophore or protein), either by means of detergents such as digitonin and TRITON or by organic solvents, could be the best method for providing large quantities of photoreceptive pigments in an accessible *in vitro* form for subsequent detailed biochemical analysis. Such samples allow a very accurate determination of spectroscopic parameters. Spectroscopy of solubilized pigment may be complicated, however, by the simultaneous extraction of several other pigments in the cell, which cause distortion of absolute spectra and necessitate special procedures of purification. These problems can be solved by separating the different pigments after extraction, for example, by high performance liquid chromatography (HPLC) and final identification with gas chromatography – mass spectrography (GC–MS) for chromophores or affinity chromatography for proteins. As pigment extraction permanently removes the identifying link to a particular cell structure and may change the spectroscopic properties of a receptor because native interactions are disrupted, these detrimental factors mandate a careful evaluation of the results obtained by this method.

Electrophysiology

Light excitation of the photoreceptor generates a cascade of electrical events. Electrophysiology was the elective method in the study of vertebrate photoreceptors. However, this technique has been applied with less success in the algae because it is very difficult to locate the photoreceptors in the cell body and, when this is possible, to produce a good sealing between the glass pipette and the cell membrane.

Flash-induced transient depolarizing potentials using intracellular glass microelectrode were first identified in *Acetabularia crenulata*. However, the first detailed analysis of photocurrents were possible by employing a suction pipette technique (patch clamp technique) in *Haematococcus pluvialis* and in the wall-free mutants of the unicellular green alga *Chlamydomonas* sp. In these experiments whole cells were gently sucked into fire-polished pipettes, forming seals with resistances up to 250 MW, allowing cell attached recordings from a relatively large membrane area, though higher resistance seals were not achieved. Recently, Negel et al. (2002) demonstrated by means of electrophysiology that the rhodopsin-like protein of Chlamydomonas, expressed in *Xenopus laevis* oocytes in the presence of all-trans retinal produces a light gated conductance that shows characteristics of a channel selectively permeable for protons.

Molecular Biology Investigations

DNA hybridization is useful in attempting to determine phylogenetic interrelationship between species. The rationale is that similarities between DNA structures correlate to interrelatedness. It is used to detect and isolate specific sequences and to measure the extent of homology between nucleic acids. It represents an alternative to the study of visual pigments at the protein level, as the genes encoding these proteins can be identified, their sequences determined, and the comparative genetic information assessed. Genomic Southern blot hybridization is used to probe the genomes of a variety of species in a manner analogous to that reported for other protein families. The potential for using bovine rhodopsin opsin complementary DNA (cDNA) probe to identify homologous genes in other species was demonstrated by Martin et al. (1986). These authors identified coding regions of bovine opsin that are homologous with visual pigment genes of vertebrate, invertebrate, and phototactic unicellular species. Successful application of this method requires closely homologous genes, and in general additional criteria, such as protein sequence information, is desirable for eliminating false positives on Southern blots. A molecular biology approach has been used also by Sineshchekov et al. (2002) in Chlamydomonas. These authors identified gene fragments with homology to the archaeal rhodopsin apoprotein genes in the expressed sequence-tag data bank of *Chlamydomonas reinhardtii*. Two quite similar genes were identified having almost all the residues of bacteriorhodopsin in the retinal binding site. The authors suspected that these genes were related to the putative retinal-based pigments already suggested for Chlamydomonas.

However, to show that the pigments are a part of the genuine signaling system, ideally one would like to delete each gene by using homologous recombination, but it is not easy to do such gene knockouts in any algal species. The problem can be overcome partially by using RNA interference (RNAi) technology to preferentially suppress the synthesis of the pigments to convincingly show that the pigment is a genuine segment of the algal phototactic response.

Understanding the molecular mechanism used by algal cells to "see the light," as we have tried to explain, is a very difficult task. At least a century has been wasted without any success. It is discouraging to think that even if the algae are not as intelligent as men are, they have "understood" very well how to orientate themselves in their light environment, and do it very efficiently. Maybe the compass mechanism they use is too simple for our complex brain.

HOW ALGAE USE LIGHT INFORMATION

No physical quantity regulates and stimulates the developments of algae as strongly as light. Light is an electromagnetic radiation characterized by its quality (different wavelengths) and intensity. To detect light and to measure both parameters and react to them, algae photoreceptor systems have to satisfy five main requirements:

- They should possess a photocycling protein
- They should possess high sensitivity
- They should be characterized by a low noise level
- They should detect either spatial or temporal patterns of light
- They should transmit the detected signal in order to modify the cell behavior

Photocycling Proteins

Upon absorption of a photon, the photocycling protein undergoes a series of conformational changes generating intermediate state(s); one of these states is the "active" state that will start signal transmission. The last intermediate state is driven back to the original state of the protein, by either a thermal process, or a second absorbed photon of different wavelength. The primary event in the photoreceptive process is the structural change of the chromophore (isomerization) to which the protein adapts. It occurs within a few picoseconds after the absorption of a photon, and this is one of the fastest biological processes in nature. The whole photocycle is very fast (order of microseconds or less), hence the intracellular response is immediately reset so that the system is prepared for a new light signal, and algae must respond rapidly on a time scale of milliseconds to seconds as environmental conditions change or as they change position relative to their static surroundings. A photoreceptor protein capable of photocycling is mandatory for algae whose photoreceptive systems are an integral part of the cell body.

This localization would not allow the continuous recovery of the exhausted photoreceptive proteins without interfering with a continuous and immediate response of the alga cell to the light.

Sensitivity

The ability to perceive and adapt to changing light conditions is critical to the life and growth of photosynthetic microorganisms. Light quality and quantity varies diurnally, seasonally, and with latitude, and is influenced by cloud conditions and atmospheric absorption (e.g., pollution). Competition for light in aquatic environment may be particularly fierce because of shading among the different organisms and the rapid absorption of light in the water column. Illumination of the surface layers varies with place, time, and conditions depending on the intensity of light penetrating the surface and upon the transparency of the water. Hence, detecting light as low as possible (i.e., a single photon) becomes an adaptive advantage, because a photosynthetic organism in dim light can obtain more metabolic energy if it is able to move to more lighted and suitable areas.

For detecting the direction of light of a specific spectral range, a photoreceptor demands a high packing density of chromophore molecules organized in a lattice structure, with high absorption cross-section of the chromophore, that is, high probability of photon absorption by the chromophore, and very low dark noise (see later). For detecting patterns of light, the number and location of photoreceptors having fixed size and exposure time must be viewed according to the pattern of motion of algae. For transmitting the detected signal a photoreceptor must generate a potential difference, or a current.

The most investigated photoreception system is that of Chlamydomonas. It consists of a patch of rhodopsin-like proteins in the plasma membrane (Type I). The packing density of these molecules appears to be about $20,000-30,000$ μm^{-2} of membrane, with a molar absorption coefficient ε of $40,000-60,000$ M^{-1} cm^{-1} and a dark noise (see later) approximately equal to zero. The number of embedded molecules per square micrometer of membrane, the absorption cross-section, and the dark noise are at the best of theoretical limits. Nevertheless, the fraction of photon absorbed from a single layer of these molecules is less than 0.05% (each layer contributes approximately to 0.005 OD).

Let us calculate how many photons this simplest but real photoreceptive system can absorb (Type I). In a sunny day about 10^{18} photons per square meter per second per nanometer are emitted by the sun (in a cloudy day the number of photons lowers to 10^{17}). As algae dwell in an aquatic environment we have to consider absorption and reflection effects of the water, and we lower this figure to 10^{17} photons per second per nanometer (in a cloudy day the number of photons lowers to 10^{16}). This means that a photoreceptor of 1 μm^2 can catch at most 10^5 photons per second per nanometer, that is, 10^7 photons in its 100 nm absorbance window in the sunniest day. As only 0.05% of the incident radiation will be effectively absorbed by a single-layer photoreceptor, the amount of photons lowers to about 10^3 photons per second. The true signal that the algae should discriminate results from the difference between the number of photons absorbed by the photoreceptor when it is illuminated (for about 400 msec) and the number of photons absorbed by the photoreceptor when it is shaded by a screen such as the eyespot (for about 600 msec).

In the case of the sunniest day the highest signal is lower than 100 photons per second, which lowers to about 10 photons per second in a cloudy day. Even in the sunniest day the number of photons absorbed for detecting light direction is very low. As this photoreceptor does not form any image, but detects only light intensity, this amount of photons is enough; however, this photoreceptor must posses a very low threshold and a very low or negligible dark noise (see later).

It has been demonstrated that not only single photons induce transient direction changes but also fluence rates as low as 1 photon cell^{-1} sec^{-1} can actually lead to a persistent orientation in Chlamydomonas (Chlorophyceae).

Strategies have been evolved to increase the sensitivity of a photoreceptor, as stacking-up many pigment-containing membranes in the direction of the light path (*Euglena gracilis* is a wonderful example of this solution) or exploiting the reflecting properties of the eyespot as in Chlamydomonas and in different species of dinoflagellates. The effectiveness of the multilayer strategy has been experimentally tested by Robert R. Birge of the W.M. Keck Center of the Syracuse University (personal communication). The absorption value recorded on a preparation of precisely ordered rhodopsin-like proteins multilayers was about 95% of the incident light, very close to the absorption value recorded on the photoreceptor crystal of *Euglena*, which consists of more than 100 layers of proteins.

Noise

A signal consists of a *true* signal component and a noise component. Hence the ability to detect a signal is limited by noise. Organisms must deal with noise of various kinds:

- The dark noise is the noise inherent in a receptor, constant and independent of light level. It arises from the random thermal motions of the molecules.
- The photon noise (also known as shot noise) is due to the quantic nature of the light stimulus and to the resulting statistical fluctuation in the capture of a photon. This noise can be accurately estimated as the square root of the number of photons captured.
- The response noise leads to random variations in the locomotory response of the microorganism. Response noise can be further divided into motor noise, in which variations occur in locomotion from one time to another in an individual; and developmental noise, in which variations occur between individuals, as a bias to turn toward one side.
- The environmental noise consists in extraneous signal arising from sources other than those of the light signal.

The general problem is to decide whether a *true* signal is indeed present when a *given* signal is observed. In the simplest case the problem becomes that of determining the threshold intensity upon which to make the decision, that is, if the signal intensity is higher than the threshold intensity, a *true* signal is detected; if the observed intensity is below the threshold it can be concluded that a *true* signal is not present. Usually, if a situation is not well defined optimal threshold intensity can be chosen, as the *true* signal intensity and noise intensity distributions overlap. The degree of overlapping is measured by the signal-to-noise (S/N) ratio.

In the absence of true signal (i.e., the signal is only the noise component and $S/N = 1$) the only strategy available is pure guessing, but the detector could guess in same way, for example, the signal is present or the signal is absent, or choose some other combinations. If a unequivocal *true* signal is available ($S/N > 1000$), a correct decision can always be made. For ambiguous signals ($S/N \approx 10$), right decisions are approximately 70%.

After all this theoretical stuff, let us calculate if the numbers of photons that reach the algal photoreceptors are enough to be detected or can generate false alarms. These photoreceptors have only to discriminate between two intensity levels (dark and light).

The light stimulus (S) that maximizes information transmission is intrinsically random and has a root-mean-square deviation of standard deviation $S^{1/2}$. As we have seen before, 100 photons are the true light signal in the sunniest day for the photoreceptor and 10 photons are the standard deviation of this signal (ten and three, respectively, in a cloudy day). So the light intensity level can oscillate from 110 to 90 photons (13 and 7 in a cloudy day). We can assume that the dark noise level (N, with a standard deviation of $N^{1/2}$) is about zero when a rhodopsin-like protein, with the chemical–physical properties described earlier, is used as photoreceptive molecule. So the dark intensity level is about zero. Detection theory states that the two intensity levels must differ by $2N^{1/2}$ to be distinguishable with the 95% reliability.

On the basis of what we have said and, once calculated, the number of photons in the sunniest and cloudy days, we can state that the simplest photoreceptor system of most algae could recognize light from dark even in a cloudy day. Theory allows us to understand how one photon cell^{-1} sec^{-1} can actually lead to a persistent orientation in Chlamydomonas photon and is able to elicit an algae response.

Direction

In order to determine the direction of a light source, an alga must possess both a light detector and a light screen to provide directional selectivity. Such a screen can be either an absorbing element that prevents light coming from certain direction from reaching the detector or a refractive element that focuses light onto the detector only from specific directions.

Femtoplankton ($0.02-0.2$ μm) are too small when compared with the wavelength of light to create differential light intensity; hence, they cannot determine the direction of a light source. Still, these microorganisms can use light, but can only measure its intensity and move in a light-intensity gradient. In contrast, algae are large enough to determine light direction and move along a beam of light by scanning the environment by means of their directionally sensitive receptor.

Guiding

To determine the orientation of a stimulus field, an organism has to measure the stimulus intensity at different positions. The two fundamental alternative strategies for obtaining information are:

- Parallel sampling of the stimulus by multiple separated receptors positioned on different parts of the organism surface. In this case the organism measures directly the spatial gradient by comparing the intensities at the different positions (one instant mechanism) (Figure 2.71).
- Sequential sampling of the stimulus by a single receptor that moves from one place to another. In this situation the organism measures directly a temporal gradient, and then infers the spatial gradient from the information on the movements of the receptor, (two instants mechanism) (Figure 2.72).

The simultaneous comparison of signals requires widely spaced receptors to detect intensity gradient, which makes large body size advantageous. On the other hand, sequential sampling requires a coherent pattern of movement such as locomotion in a helical path. Sequential sampling also requires some form of memory to allow the comparison with previously recorded intensities.

Another fundamental distinction is based on whether an organism is able to make turns in its motion path, which will direct it toward its destination. Depending on the characteristic of the stimulus and the abilities of the microorganism, guiding may be either direct in the sense of taking a straight-line path to the destination or indirect, as in the case of a biased random walk, to reach the vicinity of the destination.

Phototaxis is the behavior that involves orientation in response to light stimuli that carry the information. This behavior consists in a migration oriented with respect to the stimulus direction or gradient, which is established and maintained by direct turns (phototaxis is different from phototropism. This term is frequently used for the behavior in which the organisms respond to the light response with an oriented growth). Photokinesis is the behavior that comprises undirected responses dependent on either the intensity or temporal changes in intensity of the stimulus.

Trajectory Control

Control characteristics, and thus behavioral peculiarities, are connected with the functioning of the propelling structure of the algae, that is, the flagellum. If the cell is asymmetric, it advances spinning along its axis; it can correct its trajectory by either sudden steering obtained by changing the insertion angle of flagella or stiffening of the flagellum via accessory structure of the axoneme. This behavior can be attributed to all heterokont or uniflagellate algae. In the case of a symmetric cell, it can accomplish a gradual smooth correction of its trajectory going forward without spinning (or rotating with a very long period) and displacing the barycenter of the motor couple. This behavior can be attributed to all isokont cells.

FIGURE 2.71 Schematic drawing of one instant mechanism used by *Silvetia* sp. zygote to orientate thallus-rhizoid growth direction. Light 2 is turned on after turning off Light 1.

Signal Transmission

Algae are aneural organisms, lacking any system for the transmission of the stimuli received from the outside. The light signal will be first transduced in an electric signal by means of electron or ion flux, and this electric signal will be transmitted to the algae motor apparatus; in this case the flagellum using non-conventional routes are mostly unknown. PFR could represent a good candidate for such a role.

An Example: Photoreceptor and Photoreception in Euglena

We can utilize all the earlier mentioned theoretical information as a frame and put in this frame the particular experimental knowledge that we possess on a typical algae photoreceptor system. Let us examine in detail the case of *Euglena* that possesses a Type II photoreceptor system. This system consists of the eyespot and the photoreceptor located inside the locomotory flagellum. As the cell

FIGURE 2.72 Schematic drawing of two instants mechanism used by *Euglena gracilis* to orientate. The asterisk indicates the extension of the flagellum and the tumbling of the cell.

rotates while swimming, the eyespot comes between the light source and the photoreceptor, thus modulating the light that reaches it, and regulating the steering of the locomotory flagellum. Different experimental data exist providing evidence that the photoreceptor of *Euglena* could use either rhodopsin-like proteins or flavoproteins. In 2000 Barsanti et al. isolated the photoreceptors of Euglena in very high yield and performed the purification of a rhodopsin-like protein from this sample. Just recently Iseki et al. (2002) biochemically characterized a new type of blue-light receptor flavoprotein, a photoactivated adenyl cyclase, in the photoreceptor organelle of *Euglena*. Though is not yet certain which is the primary light detecting protein in the photoreceptor, still a model of how *Euglena* orients itself in a luminous field can be provided.

The photoreceptor possesses optical bistability, that is, upon photoexcitation the ground state generates a stable excited state, which can be photochemically driven back to the ground state. *Euglena* photoreceptor undergoes a very simple photocycle: UV light (365 nm) of the non-fluorescent ground state leads to the photogeneration of the fluorescent excited state, which in turn is driven back to the ground state by blue light (436 nm). Figure 2.73 shows the absorption spectrum of the ground and excited states of the photoreceptor, together with the absorption spectrum of the eyespot. The latter spectrum clearly overlaps the excited state spectrum of the photoreceptor. The photocycle of the photoreceptor and shading effect of the eyespot are shown in the bottom part of the figure.

As the cell rotates while swimming (rotation frequency: 2 Hz, that is, two complete revolutions per second), the photoreceptor experiences periodic shading by the eyespot, which comes between the light source and the organelle. As the absorption spectra of the eyespot and that of the excited state of the photoreceptor are superimposable, the screening effect of the eyespot leads to the interruption of the photocycle, preventing the transition of the excited state of the photoreceptor to its ground state, hence a signal can arise. Cell reorientation is brought about by a selective screening of the absorption window of the excited state of the photoreceptor by the eyespot.

During the entire shading period, the photoreceptor is in the excited state, which could generate a photoelectric signal due to the change in dipole orientations between the excited and parent conformers of the photoreceptive protein. Under the influence of this electric field, a displacement

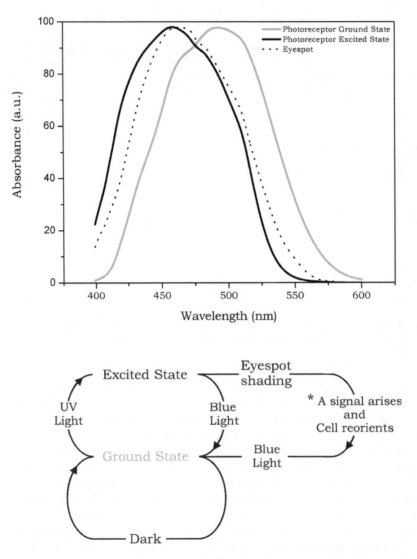

FIGURE 2.73 Absorption spectra of the photoreceptor ground state (continuous gray line) and excited state (continuous black line) and absorption spectrum of eyespot (dotted line) of *Euglena gracilis*. The photocycle of the photoreceptor and shading effect of the eyespot are shown in the bottom part of the figure.

photocurrent could be propagated through paraxial rod filaments via charge transfer between rod proteins, which in turn generates longitudinal waves of contraction along the paraxial rod. The contractility of the rod should regulate the flagellum movement, turning the cell body towards the light source. The ground state of the photoreceptor is restored when the photoreceptor again achieves the alignment with the light.

The dynamics of the light alignment of a model *Euglena gracilis* that obeys the earlier mentioned constraints is shown in Figure 2.74. The initial orientation of the cell is perpendicular to the light direction, but in few seconds, *E. gracilis* is completely aligned with the light. *The CRC Handbook on Organic Photochemistry and Photobiology* (2004) is an excellent text that covers all the aspects of algae photobiology.

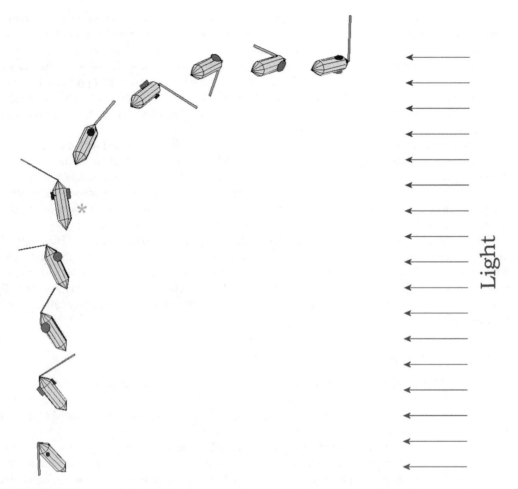

FIGURE 2.74 Dynamics of the light alignment of a model *Euglena gracilis*. The cell is represented by an ellipsoid, the eyespot by a big spot, the photoreceptor by a small spot. The asterisk indicates the extension of the flagellum and the tumbling of the cell.

CHLOROPLASTS

The photosynthetic compartment contains the pigments for absorbing light and channeling the energy of the excited pigment molecules into a series of photochemical and enzymatic reactions. These pigments are organized in proteic complexes embedded in the membrane of sac-like flat compressed vesicles known as the thylakoids. These vesicles are about 24 nm thick, and enclose a space, termed lumen, 10 nm wide.

In prokaryotes the thylakoids are free within the cytoplasm, while in eukaryotes they are enclosed within bounding membranes to form the chloroplast. The colorless matrix of the chloroplast is known as the stroma. Inside the chloroplast, thylakoids are organized into two different compartments, granal thylakoids, stacked into hollow disks termed grana; and stromal thylakoids forming multiple connections between the grana. All thylakoids surfaces run parallel to the plane of the maximum chloroplast cross-section. Chloroplasts contain nucleic acids and ribosomes. DNA is naked, that is, not associated with proteins, and occurs in two configurations: scattered, but not connected, small nucleoids or as a peripheral ring. Chloroplasts are semiautonomous organelles that replicate their own DNA, and this replication is not linked to the division of the organelle. Synthesis of RNA and proteins is possible inside the chloroplast, though they are not strictly autonomous from the nuclear genome. The plastid genes are transcribed and translated within the plastids. The machinery of protein synthesis is, in any case, partially composed of imported nuclear gene products. In this respect, as in others, the chloroplast is no longer independent. Nevertheless, about half of the plastid genome consists of genes that contribute to the machine of gene expression, for example, genes for rRNA, tRNAs, RNA polymerase subunits, and ribosomal proteins. Plastids code for, and synthesize, some proteins that are components for photosystems and particular subunits of photosynthetic enzymes. The missing subunits of these complexes are coded in the nucleus and must be imported from the cytoplasm.

Chloroplast development and division (self-replication) may be coordinated with that of the cell or may proceed independently. They divide in mother cells and they are inherited by daughter cells during vegetative division and usually only from the maternal side in sexual reproduction. The shape and number of chloroplasts are extremely variable from the single cup-shaped chloroplast of *Dunaliella salina*, the ribbon-like chloroplast of *Spirogyra* or stellate one of *Zygnema* to the numerous (about 10^8) ellipsoidal chloroplasts of *Acetabularia* giant cell.

The fact that algae of different divisions have different colors, due to the presence of a variety of pigments in the photosynthetic membrane system, might lead to the supposition that the photosynthetic membrane structures are variable. This is true because the features of photosynthetic membrane system represent a diagnostic important element at the class level. In this chapter we will consider the structure, composition and location of the photosynthetic membrane system in each algal division.

Cyanophyta and Prochlorophyta

The photosynthetic apparatus of these algae is localized on intracytoplasmic membranes termed thylakoids. The thylakoids membranes show considerable variations in structure and arrangements depending on the species. Also the amount of thylakoid membranes per cell is variable due to growth conditions and taxonomic specificity.

In most species the thylakoids are arranged peripherally in three to six layers running parallel to the cell membrane, forming an anastomosing network of concentric shells. This peripheral region has been called the chromatoplast, which is separated from the inner nucleoplasmic region called the centroplasm. This type of thylakoid arrangement is characteristic of many unicellular and filamentous cyanobacteria, such as *Synechococcus planctibus* and *Anabaena* sp. In some other organisms such as *Oscillatoria* and *Arthrospira* the thylakoids are orientated perpendicular to the longitudinal cell wall. Radial arrangement is present in *Phormidium retzi*. Thylakoids are not always restricted to the cell periphery but can be found scattered throughout the cell as in *Gloeotrichia* sp. However, the arrangement of thylakoids can change from cell to cell and cell types (vegetative cells, heterocyst, akinete) within the same culture from parallel to a convoluted appearance. Thylakoids can fuse with each other and form an anastomosing network; unlike the chloroplasts of eukaryotic algae, the thylakoids of cyanobacteria form stacking regions only to a very limited extent. During cell division the thylakoids have to be separated and divided into the daughter cells; generally this division is an active process. Cell division starts by a centripetal growth of cytoplasmic membrane and peptidoglycan layer, to form the septum; the thylakoid cylinder is narrowed at the level of the cross wall by invagination before the septum is formed. Proteins, lipids, carotenoids, and chlorophyll *a* are major components of thylakoids. On the outer surface of the thylakoids regular rows of electron dense granular structures are closely attached. These granules termed phycobilisomes contain the light harvesting phycobiliproteins, that is, allophycocyanin, phycocyanin, and phycoerythrin (Figure 2.75). The phycobilisome structure consists of a three-cylinder core of four stacked molecules of allophycocyanin, closest to the thylakoid membrane, on which converge rod-shaped assemblies of coaxially stacked hexameric molecules of only phycocyanin or both phycocyanin and phycoerythrin (Figure 2.76). Phycobiliproteins are accessory pigments for the operation of photosystem II also in Glaucophyta, Cryptophyta, and Rhodophyta.

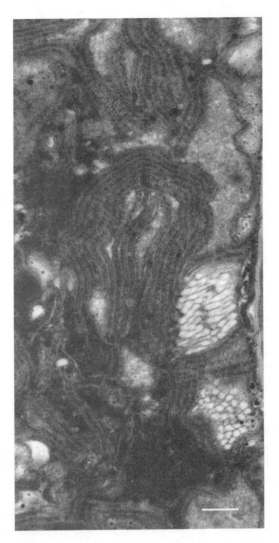

FIGURE 2.75 Transmission electron microscopy image of a cyanobacterium in longitudinal section, showing the thylakoid membranes with phycobilisomes. (Bar: 0.20 μm.) (Courtesy of Dr. Luisa Tomaselli.)

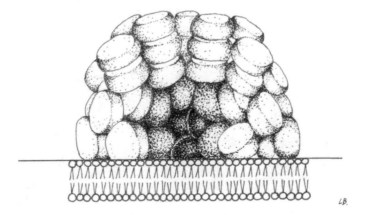

FIGURE 2.76 Schematic drawing of a phycobilisome.

The thylakoid membranes of *Prochloron* (Prochlorophyta) are organized into stacked and unstacked regions reminiscent of the grana and stromal lamellae of higher plant chloroplasts. They differ from cyanobacteria in that they contain chlorophyll a/b light harvesting systems rather than phycobiliproteins organized in the phycobilisomes.

Glaucophyta

As already described in Chapter 1 these algae possess inclusion termed cyanelles that are probably symbiotic cyanobacteria functioning as chloroplasts. Each cyanelle, surrounded by a reduced peptoglycan cell wall (except in *Glaucosphaera* sp.), is enclosed in a vesicle of the host cytoplasm. Cyanelles do not fix molecular nitrogen in contrast with cyanobacteria; they contain polyphospate granules and a conspicuous central carboxisome similar to the pyrenoids of other algae. The thylakoid are not stacked but they are single and equidistant with a concentric arrangement. Cyanelle pigments are chlorophyll a and β-carotene, which represents the main carotenoid. Interthylakoidal phycobilisomes contain allophycocyanin and phycocianin. Phycoerytrin is absent from Glaucophyta but phycoerythrocyanin can be found in some species.

Rhodophyta

Ultrastructurally, red algal chloroplasts are composed of a double-membrane envelope inside of which are one or more parallel, thylakoidal photosynthetic lamellae. These chloroplasts are not associated with the endoplasmic reticulum, a feature shared with Glaucophyta and Chlorophyta. Encircling thylakoids are present in all Florideophyceae and in some taxa of Bangiophyceae while the thylakoids in the other red algae lie equidistant and are single, that is, not stacked, unlike any other group of eukaryotic algae (except Glaucophyta), and typically oriented parallel to each other. All thylakoids have phycobilisomes attached to their stromal surface, which contain the accessory phycobiliprotein pigments, that is, allophycocyanin, phycocyanin, and five forms of phycoerythrin (Figure 2.77). Chlorophyll a is the only chlorophyll present in the thylakoid membrane, together with carotenoids such as β-carotene and lutein. Plastid number, shape, and position (many, discoid, and parietal) is rather uniform throughout the Florideophyceae, and pyrenoids may or may not be present. A single stellate plastid with a central pyrenoid is commonly associated with bangiophycidean red algae, such as *Phorphyridium*. DNA is organized into numerous nucleoids scattered throughout the chloroplast.

Heterokontophyta

Some ultrastructural features of the chloroplast compartments of these algae are common to all the seven classes of the division, with few exceptions. Four membranes surround the chloroplasts, the outer two being the chloroplast endoplasmic reticulum and the inner two being the chloroplast

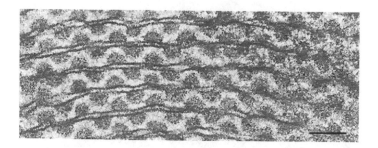

FIGURE 2.77 Transmission electron microscopy image of rhodophyte thylakoid showing with regularly arranged phycobilisomes. (Bar: 0.05 μm.)

envelope. When the chloroplasts are located close to the nucleus, the chloroplast endoplasmic reticulum is continuous with the nuclear envelope. In the Xanthophyceae, this connection is not the rule. Thylakoids are grouped into lamellae of three, which are two in some Raphidophyceae, with varying degrees of coherence depending on the species. Thylakoids from adjacent lamellae frequently interconnect across the stroma. In all the classes, with the exception of Eustigmatophyceae and *Chattonella* (Raphidophyceae), one lamella runs around the periphery of the chloroplast beneath the chloroplast envelope, enclosing all the other lamellae. The lamella is called girdle lamella.

One or more plate-like (Chrysophyceae, Figure 2.78a and 2.78b; Eustigmatophyceae, Figure 2.79a and b), discoid (Xanthophyceae, Raphidophyceae, and Dictyochophyceae) or ribbon-like (Phaeophyceae), plastids are typically present, often lobed, parietal or located in close connection with the nucleus. In the Bacillariophyceae, chloroplasts are the most conspicuous feature, and their number and shape are consistent features of taxonomic importance. They may be rounded or lobed discs or large plate-like with or without lobed margins, and may range from one to two, four or more. A typical centric diatom has many disc-shaped plastids, arranged close to the periphery surrounding a large central vacuole or scattered throughout the cell. The raphid diatoms tend to have large chloroplasts (one to four) lying along the girdle with central nucleus flanked by two large vacuoles.

The chloroplast DNA is ring-shaped and located in the region between the girdle lamella and the others in all the classes, with the exception of the Eustigmatophyceae, where it is organized into many dot-like nucleoids, which may be united in a sort of reticulum. The main photosynthetic pigment is chlorophyll a, which is the only chlorophyll present in the Eustigmatophyceae. In addition, chlorophylls of the c group occur, both c_1 and c_2 (Chrysophyceae, Rhaphidophyceae, Phaeophyceae, Dictyocophyceae, and in only extremely low concentrations in Xanthophyceae) or only c_2 (Bacillariophyceae). The most important accessory pigment is fucoxanthin in Chrysophyceae, Bacillariophyceae, Dictyocophyceae, and Phaeophyceae, violaxanthin in Eustigmatophyceae, and vaucheriaxanthin in Xanthophyceae. Other accessory pigments are β-carotene and xanthophylls such as diadinoxanthin, heteroxanthin, vaucheriaxanthin, antheraxanthin, and lutein. In the Raphidophyceae marine and freshwater species differ in their accessory pigments, marine species possessing mainly β-carotene, fucoxanthin, and violaxanthin, and freshwater species having β-carotene, diadinoxanthin, heteroxanthin, and vaucheriaxanthin.

Pyrenoids are present in all the classes, except in the zoospores of the Eustigmatophyceae and in the freshwater species of Raphidophyceae. They are of a semi-immersed type, attached to the

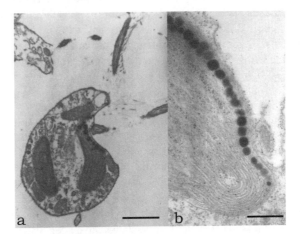

FIGURE 2.78 Transmission electron microscopy image of *O. danica* in longitudinal section, showing the chloroplast (a) (Bar: 3 μm). Transmission electron microscopy image of a chloroplast at higher magnification showing the thylakoid membrane and the eyespot globules (b). (Bar: 1 μm.)

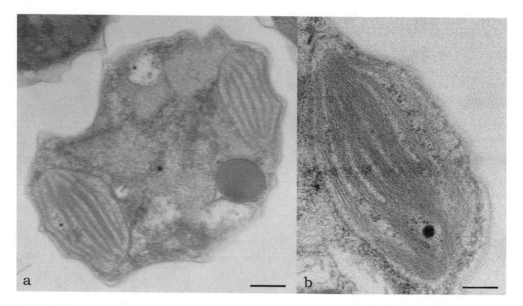

FIGURE 2.79 Transmission electron microscopy image of *Nannochloropsis* sp. in transverse section, showing the chloroplast (a) (Bar: 0.50 μm); chloroplast at higher magnification (b) (Bar: 0.10 μm).

inner face of the chloroplast, pear-shaped in the Phaeophyceae, or stalked in the Eustigmatophyceae. They can be one or many (Bacillariophyceae and Phaeophyceae). No storage material or capping vesicles have been found to be associated with pyrenoids, but lipid or oil droplets normally distributed randomly in the chloroplast matrix are often concentrated at the periphery of the pyrenoid.

Haptophyta

Cells of Haptophyceae species normally possess one or two chloroplasts containing thylakoids stacked in three to form lamellae. There is no girdle lamella. Spindle-shaped pyrenoids are commonly immersed within the chloroplast, penetrated by one or a few pairs of thylakoids, but in some genera they bulge from the inner face of the organelle. Both the chloroplast and the pyrenoid are surrounded by endoplasmic reticulum confluent with the nuclear envelope, the nucleus itself lying close to the chloroplast. Chlorophylls a, c_1, and c_2 and in some genera also c_3 are found in the thylakoidal membranes together with carotenes, such as β-carotene, and xanthophylls, the most commonly occurring being fucoxanthin. DNA is organized into numerous nucleoids scattered throughout the chloroplast.

Cryptophyta

The chloroplasts of these algae, one or two per cell, are unusual in both their pigment composition and ultrastructure. Four membranes enclose these organelles: the inner pair forms the plastid envelope and the outer pair forms the plastid endoplasmic reticulum. This four-membrane configuration is common in chlorophyll c-containing algae. An expanded space is present between the plastid endoplasmic reticulum and the plastid envelope on its inward face. This compartment contains 80S ribosomes, starch grains, and the nucleomorph. Thylakoids inside the chloroplast are typically in pairs, although single thylakoids as well as large stacks have been also observed, with no girdle lamella. A pyrenoid is present, projecting from the inner side of the chloroplast. Cryptomonads are characterized by the presence of chlorophylls a and c_2; phycoerythrin, phycocyanin, and allophycocyanin; carotenes, and several xanthophylls.

The phycobiliproteins differ from those in the red algae and cyanobacteria, because they are of lower molecular weight, and do not aggregate to form discrete phycobilisomes, being localized within the lumen of the thylakoids. DNA is organized into numerous nucleoids scattered throughout the chloroplast.

Dinophyta

Chloroplasts may be present or absent in these algae, depending on the nutritional regimen. When present, they are characterized by triple-membrane envelopes not connected with the endoplasmic reticulum; thylakoids are usually in group of three, unappressed, and girdle lamellae are generally absent. Pyrenoids show various types, stalked or embedded within the chloroplast. The pigment assortment includes chlorophylls a and c_2, β-carotene, and several xanthophylls, among which is piridinin. DNA is organized into numerous nodules scattered throughout the chloroplast. This description corresponds to the most important dinoflagellates plastid, the peridinin-type plastid.

About 50% of dinoflagellates with plastids acquired them from a variety of photosynthetic eukaryotes by endosymbiosis (cf. Chapter 1). Those containing three-membrane, peridinin-containing plastids, such as *Amphidinium carterae*, probably derived from the red lineage by secondary endosymbiosis, that is, through the uptake of primary plastid-containing endosymbiont. Other groups of dinoflagellates have plastids derived from a tertiary endosymbiotic event, that is, the uptake of a secondary plastid-containing endosymbiont. Tertiary plastids are present in the toxic genus *Dinophysis*, characterized by two-membrane, cryptophyte-derived plastids, in other important species (*Karenia* sp., *Gyrodinium aureolum*, and *Gymnodinium galatheanum*) with fucoxanthin as accessory pigment, which possess haptophyte-derived plastids surrounded by two or four membranes, and in *Kryptoperidinium* sp. and close relatives, which have a five-membrane, heterokontophyte-derived (diatom) plastid that includes a diatom nucleus of unknown complexity. Another case is that of *Gymnodinium acidotum*, characterized by a cryptophyte endosymbiont, which may or may not represent a permanent endosymbiosis, because the endosymbiont may be acquired as prey and retained for long periods of time, but not kept permanently.

A serial secondary endosymbiosis (the uptake of a new primary plastid-containing endosymbiont) occurred in *Lepidodinium* sp. and its close relatives, in which the peridinin plastid has apparently been replaced by a secondary plastid derived probably from a prasinophycean alga (Chlorophyta) and surrounded by two membranes. This dinoflagellate also has external scales atypical for its division, but closely resembling those of Prasinophyceae, suggesting that *Lepidodinium* expresses genes for scale formation acquired from its endosimbiont. Also in *Noctiluca* sp. a Prasinophyceae endosymbiont has been observed.

Euglenophyta

As in the Dinophyta the chloroplast envelope consists of three membranes. Chloroplasts are typically many per cell and show considerable diversity of size, shape, and morphology (Figure 2.80a and 2.80b).

Five main types can be recognized:

- Discoid chloroplasts with no pyrenoids (*Phacus*)
- Elongated or shield shaped chloroplasts with a central naked pyrenoid (*Trachelomonas*)
- Large plated chloroplasts with the so-called double sheeted pyrenoid, that is, a pyrenoid which carries on both plastid surfaces a watch glass shaped cup of paramylon (*Euglena obtusa*)
- Plate chloroplasts having a large pyrenoid projecting from the inner surface, which is covered by a cylindrical or spherical cup of paramylon (*Colacium*)
- Chloroplast ribbons radiating from one or three paramylon centers (*Eutreptia*)

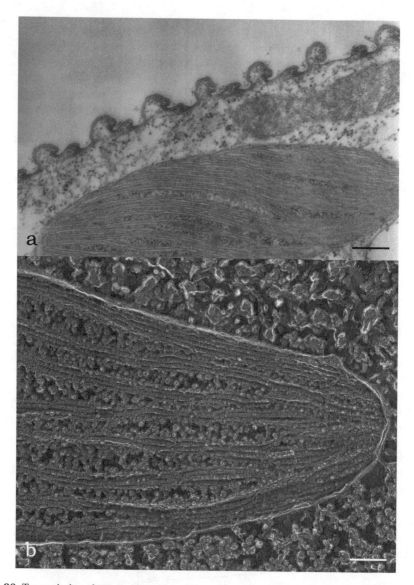

FIGURE 2.80 Transmission electron microscopy image of a chloroplast of *Euglena gracilis* in longitudinal section (a) (Bar: 0.10 μm). Deep-etching image of a chloroplast of *E. gracilis* (b) (Bar: 0.05 μm).

The chloroplasts are never connected to the nucleus by the endoplasmic reticulum, and thylakoids are usually grouped into threes forming lamellae as in Heterokontophyta and Dynophyta. Girdle lamellae are never found in this group. The photosynthetic pigments are chlorophylls *a* and *b* with carotenes and xanthophylls as accessory pigments. Chloroplast DNA occurs as tiny granules throughout the whole stroma.

Chlorarachniophyta

Five to seven bi-lobed chloroplasts are present inside these algae in a peripheral position. Each chloroplast is bounded by a system of membranes that may appear either as four separate membranes, as a pair of membranes with a sort of flattened vesicles between them, or as three membranes. Four separate membranes are always found near the proximal end of the pyrenoid and over the pyrenoid itself. The outer pair of membranes, when four are present, is referred to as a type of chloroplast endoplasmic reticulum; however, the inner pair is interpreted as the chloroplast envelope. The thylakoids are often loosely stacked in three, with no girdle lamella. Each chloroplast bears a central, pear-shaped pyrenoid projecting inward. Around the pyrenoid, often tightly associated with it, there are vesicles containing β-1,3-glucan, which is the principal storage carbohydrate. A nucleomorph, that contains DNA and a nucleolus-like body, is present between the second and third envelope of each chloroplast, in a pocket located in the pyrenoid surface. Chloroplasts contain chlorophylls *a* and *b* and xanthophylls.

Chlorophyta

The Chlorophyta are not uniform in the ultrastructure of chloroplast, still some generalization can be made. The chloroplasts of these algae are enclosed only by the double membrane of the chloroplast envelope; there is no additional envelope of endoplasmic reticulum or nuclear membrane. In this respect they resemble Rhodophyta and Glaucophyta in the cell compartmentalization. The chloroplasts vary greatly in shape and size. In unicellular forms it is often cup-shaped with a thick base (*Dunaliella*); in filamentous forms it is often ring-like or net-like shaped and lies against the cell wall (*Oedogonium*). More massive and elaborate plastids, lying along the longitudinal axis of the cell are particularly characteristic of members of the Zygnematophyceae. Thylakoids are arranged in stacks of two to six or more; their multilayered arrangement may take on the appearance of grana with membrane interconnections, as in higher plants. Girdle lamellae are absent. One to several pyrenoids occur in most of the algae of this division embedded within the chloroplast, and are often penetrated by thylakoids. The DNA organized in small nucleoids is distributed throughout the chloroplast matrix. Both chlorophylls *a* and *b* are present; accessory pigments include different xanthophylls such as lutein, zeaxantin, and violaxantin; β-carotene is always present together with other carotenoids.

NUCLEUS, NUCLEAR DIVISION, AND CYTOKINESIS

An organized nucleus is absent in both Cyanophyta and Prochlorophyta, where DNA molecules are free in the cytoplasm. The central region of those algae features the naked (without histone proteins) circular DNA genome, which is not contained within a double membrane, and consists of a single unbranched molecule. Transcription and translation processes are accomplished by the assistance of abundant 70S ribosomes located in the centroplasm. These ribosomes are smaller than their counterparts in the eukaryotic cytoplasm, but are typical of all prokaryotes, mitocondria, and chloroplasts. Cyanobacteria can only reproduce asexually, but appear to have some forms of genetic recombination possible, which are divided into two categories: transformation and conjugation. Transformation occurs when DNA is shed by one cell into the environment and is taken up by another cell and incorporated into its genome replacing homologous sections of DNA. The ability to be transformed by external DNA is specie-specific and typically requires special environmental conditions. Conjugation is a "parasexual" process in which one of the partners develops a conjugation tube that connects to the recipient cell. Typically a plasmid (a small circle of DNA) from the donor cell passes through the conjugation tube. It is thought that genes for gas vacuolation, antibiotic resistance, and toxin production are carried on plasmids in cyanobacteria.

In eukaryotic algae genetic information in the form of DNA, together with the controlling services for its selective expression, occurs in plastids (plastome), mitochondria (chondrome), and in the cell nucleus (genome). In the previous section, we have described organization and behavior of chloroplast DNA, which are similar to those of mitochondria. It is worthwhile to recall that plastid and mithocondrial genes and the way in which they are expressed have much in common with the system of gene expression of prokaryotes, and there are many plastid genes with extended introns that are very rare in prokaryotes (genes are mosaics of introns and exons. Introns are the DNA sequences of unknown function that are removed in the primary mRNA transcript. Exons are the DNA sequences that code for amino acids).

In the nucleus of eukaryotic algae, long, linear, and unbranched molecules of DNA are associated with proteins and small amount of RNA. Two types of proteins are found: the relatively uniform histones (about 20), involved in the structural organization of the DNA, and the very variable proteins involved in gene activity regulation, such as DNA and RNA polymerases. The DNA–protein complex, made up of repeating units termed nucleosomes, is known as chromatin, which is usually highly dispersed in the interphase nucleus. During mitosis and meiosis metaphases the DNA–protein complexes are more helically condensed around the nucleosome and form the chromosomes, each chromosome consisting of a single DNA molecule.

The nucleus is surrounded by a two-membrane envelope continuous with the endoplasmic reticulum. Between the two membranes there is a narrow perinuclear space about 20 nm wide. The nuclear envelope is perforated by numerous pores 60–100 nm in diameter. Nuclear pores have a complicated superstructure; they are not simply free openings, but are gateways in nuclear envelope through which the controlled transport of macromolecules (RNA, proteins) takes place. Pores can be arranged in straight lines as in *Bumilleria* (Xanthophyceae, Heterokontophyta), in closely hexagonal groups as in *Prorocentrum* (Dinophyta), or can be randomly distributed as in *Glenodinium* (Dinophyta).

All algal nuclei possess nucleoli that vary in shape, size, and number in the different algal divisions. Nucleoli are dense concentrations of ribonucleoprotein-rich material, which are intimately associated with the specific region of the chromosomal DNA coding for ribosomal RNA. Nucleoli can be single and central as in *Cryptomonas* (Cryptophyta) or be more than 20 scattered in the nucleus as in *Euglena acus* (Euglenophyta).

In eukaryotic algae, the features of the interphase nucleus, DNA replication, and processes of transcription and translation in the expression of genetic information and cytokinesis are comparable with those of all the other eukaryotes. Exceptions to this rule will be described.

Rhodophyta

The basic features of mitosis and cytokinesis are the same throughout the division, though some variation may occur. Unlike other groups of eukaryotic algae, where the mitotic spindle is formed between two pairs of centrioles, one at each spindle pole, in red algae the poles of the spindle are marked by ring-shaped structures named polar rings. The absence of centrioles reflects the complete absence of flagella in the red algae. The chromosomal and interzonal microtubules do not converge towards the polar rings, so that the spindle poles are very broad. The nuclear envelope does not break down, though it is perforated by large holes, hence the mitosis is closed. The spindle persists for some time at early telophase, holding daughter nuclei apart; it collapses only at late telophase, when daughter nuclei are kept separated by a vacuole. Cytokinesis is by furrowing, but is typically incomplete; the furrow impinges upon the vacuole but leaves open a cytoplasmic connection between sibling cells. This connection is then filled and blocked by a proteinaceous stopper, named pit plug. The simplest type of pit plug consists only of a proteinaceous core; two- or one-layered caps partly composed of polysaccharides can border the core of both sides with different thickness. A cap membrane can be present between the two layers of the cap or bounding the plug core.

Cryptophyta

These algae possess a peculiar organelle termed nucleomorph located in the space between the chloroplast endoplasmic reticulum and the chloroplast envelope. This organelle has a double membrane around it, pierced by pores and contains both DNA and a nucleolus-like structure where rRNA genes are transcribed. The DNA is organized in three small chromosomes encoding genes for its own maintenance. It possesses the ability to self-replicate, without forming a spindle during mitosis. It is considered a vestigial nucleus belonging to a photosynthetic eukaryotic symbiont.

Dinophyta

The dinoflagellate nucleus, known as dinokaryon, is bounded in the usual eukaryotic fashion by a nuclear envelope penetrated by pores. However, it possesses a number of unusual features, including high amounts of DNA per cell (five to ten times the most common eukaryotic levels, up to a maximum of 200 pg in *Gonyaulax polyedra*). It is relatively large, often occupying about one half of the volume of the cell. Nuclear shapes are variable, ranging from spheroid to U-, V-, or Y-shaped configurations. In most dinoflagellates the chromosomes remain continuously condensed and visible, by both light and electron microscopy, during interphase and mitosis. Chromosome counts range from 12 to around 400, but may be variable within a species. A prominent nucleolus is also persistent. Each chromosome consists of a tightly super-coiled structure of DNA double helix. The permanent condensed chromosome shows a swirled, fibrillar appearance due to the naked DNA double helices (i.e., no histones are present). The 3–6 nm fibrils are packed in a highly ordered state, up to six level of coiling. A small amount of basic protein is present in a few species, such as *Oxyrrhis marina*, but none of it corresponds in amino acid content to the histones normally present in eukaryotic chromosomes. A further peculiarity of dinoflagellate chromosomes is the high amount of calcium and other divalent metals, such as iron, nickel, or copper, which may play a role in chromosomal organization. In some dinoflagellate, such as *Noctiluca* and *Blastodinium*, the chromosomes undergo an expanded, decondensed change during interphase and are termed noctikaryotic.

Dinoflagellate mitosis is also unusual. At the time of division, chromosomes divide longitudinally, the split starting at one end of the chromosome and moving along the entire length. The nucleus is invaded by cytoplasmic channels that pass from one pole to the other. Microtubules are present in these channels. The nuclear envelope and the nucleolus are persistent throughout nuclear division. The chromosomes upon dividing assume a V- or Y-configuration, and the apices of such chromosomes are closely associated with the nuclear envelope surrounding a cytoplasmic channel. This association suggests that the cytoplasmic channels serve as mechanism for the movement of chromosomes.

The nuclear envelope persists during mitosis (closed mitosis), as it does in other algae, for example, Euglenophyta and Raphidophyceae. However, with the exception of *O. marina*, where mitotic spindle is intranuclear, the chromosomes do not appear fibrillar, and the mitosis strongly resembles that of *Euglena*, the mitotic spindle is always extranuclear. The spindle microtubules pass through furrows and tunnels that form in the nucleus at prophase. No obvious spindle pole bodies other than concentric aggregation of Golgi bodies are present. Some microtubules contact the nuclear envelope, lining the tunnels at points where the chromosomes also contact. The chromosomes have differentiated dense regions inserted into the envelope.

Dinoflagellate cells undergo binary fission, and each daughter cell can retain half the parent cell wall, which splits along a predetermined fission line. This mode of reproduction is called "desmoschisis," and examples are *Ceratium* and many *Gonyaulax* species. In other dinoflagellates, daughter cells do not share the parent cell wall; this mode of division is called "eleutheroschisis." Binary fission may take place either inside the mother cell (many freshwater *Peridinium* species), or the protoplast may leave the mother cell wall before division through a hole or slit. In thecate species (many marine *Protoperidinium* species) the protoplast escapes, after special thecal plates are dislocated, through a hole. In several species the protoplast has no flagella and deforms its cell wall during the escape from the mother theca; however, this is not a typical ameboid stage. After cell division, each daughter cell produces a new cell wall, or a new theca in thecate cells.

Euglenophyta

In these algae the interphase nucleus lies in the central or posterior region of the cell; it is spherical in most spindle-shaped species, and ovoid or long and narrow in elongate cells. In the largest species the nucleus is at least 30×15 μm in dimensions, but in small species the spherical nucleus may be less than 2 μm in diameter. The chromosomes retain their condensed condition throughout interphase, appearing as granular or filamentous threads (Figure 2.81a and 2.81b). In some nuclei the chromosomes radiate from the central endosome, while in others (even in the same species) they coil haphazardly throughout the nucleoplasm. Mitosis begins with a forward migration of the nucleus so that it comes to lie immediately posterior to the reservoir. In species with several endosomes in the interphase nucleus, these usually fuse to form a single body. The endosome then elongates along the division axis, perpendicular to the long axis of the cell, and the chromosomes orient into the metaphase position, following three main types of orientation:

- Pairs of chromatids from late prophase orient into a circlet of single chromatids, separation and segregation having occurred during orientation (*E. gracilis*)
- Pairs of chromatids from interphase or prophase come to lie along the division axis still as pairs, parallel to one another and to the elongated endosomes (*Euglena communis, Euglena viridis*)
- Single chromosomes from prophase line up along the division axis and there undergo duplication into the pairs of chromatids of that mitosis (*E. acus, Euglena spirogyra*)

These different types overlap to a certain extent, species differ mainly in the time at which the double structure of the chromosomes first becomes microscopically visible. In all cases, the endosome continues to elongate and the chromatids segregate towards the ends of the endosome into two daughter groups. This stage, with most but not quite all of the daughter chromosomes separated, must be called metaphase in *Euglena*. During this early-to-late metaphase succession, the locomotor apparatus (flagella, photoreceptor, and eyespot) replicates and the reservoir divides. The daughter reservoirs open into the still single canal, but each now has its own contractile vacuole, eyespot, and flagella. Separation, segregation, and anaphasic movements of the chromatids are irregular, and this, coupled with a very low chromatid velocity, results in an extremely long anaphase.

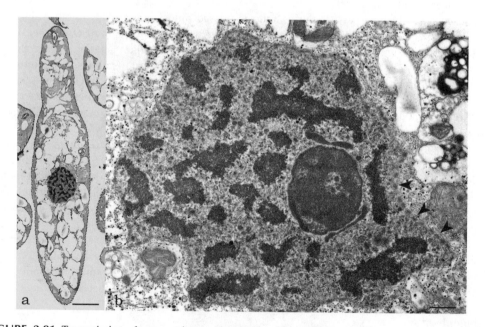

FIGURE 2.81 Transmission electron microscopy image of the WZSL mutant of *Euglena gracilis* in longitudinal section, showing the central nucleous with the condensed chromosomes (a) (Bar: 3 μm). Transmission electron microscopy image of the nucleus of the WZSL mutant of *E. gracilis*, showing the nucleolus, the satellite nucleoli and the nuclear membrane pores (arrowhead) (b). (Bar: 0.3 μm.) (Courtesy of Dr. Giovanna Rosati.)

The end of the anaphase is marked by a sudden flow to the poles of the central region of the elongated endosome, and the persistent nuclear envelope seals around the groups of chromatids and the daughter endosomes to form the telophase nuclei. Once telophase is established, with separate daughter nuclei, one of the two flagella in each daughter reservoir grows to emerge as a locomotory flagellum. A cleavage line is initiated between the now distinct daughter canals and progress helically backward to separate the daughter cells. High chromosome numbers are the rule for species of Euglenophyta, indicating a possible polyploidy.

Chlorophyta

Nuclear and cell division has been intensively and extensively studied in the green algae at light-microscopic and ultrastructural level, and several different patterns have been recognized. Two basic patterns have merged from these studies regarding nuclear and cell division in green algae:

- Intranuclear mitosis, with the nuclear envelope closed at metaphase or open at the poles, interrupted by the microtubules of the spindle; other microtubules, transverse to the longitudinal axis of the spindle, are present at telophase and are called the phycoplast; the latter functions somehow in cytokinesis, by furrowing or by cell-plate formation; the daughter nuclei at telophase are in close proximity
- The spindle and nuclear envelope are open and a phycoplast is absent; at telophase the microtubules of the intranuclear spindle are persistent and a phragmoplast-like structure is organized as cytokinesis by furrowing proceeds

Cylindrocapsa is an example of the first type; the parietal chloroplast divides before mitosis. Two pairs of centrioles are already present at the beginning of the interphase. In early prophase, the nuclear envelope is surrounded by one or two layers of endoplasmic reticulum, which is rough, that is, covered by ribosomes. Perinuclear microtubules appear around the nucleus and in late prophase they proliferate within the nucleus to form a tilted mitotic spindle between the pairs of centrioles lying at the spindle poles. At metaphase the nuclear envelope is still intact and surrounded by the endoplasmic reticulum, so that mitosis is closed. The fully condensed chromosomes become aligned to form a distinct metaphase plate and have plate-like layered kinetochores. The non-persistent telophase spindle soon degenerates, though a few microtubules can still be found around the reformed nuclear enevlopes. The pairs of centrioles migrate around the telophase nuclei, away from the former spindle poles and towards the center of the equatorial plane, where they remain until after cytokinesis.

Cisternae of endoplasmic reticulum proliferate in the narrow zone of cytoplasm present between the two daughter nuclei at the center of the cell. They bleb off smooth endoplasmic reticulum vesicles, which become aligned in the equatorial plane to form a cell plate of smooth vesicles. These vesicles coalesce to form a transverse system separating the daughter cells. Here the vesicles accumulate within a phycoplast, that is, a plate of microtubules lying in the future plan of division. After completion of the transverse septum and the resultant separation of the daughter cells, a new cell wall is secreted around each daughter protoplast by exocytosis of Golgi-derived vesicles containing wall material. Each daughter cell thus gains a complete new wall; in the case of *Cylindrocapsa*, daughter cells remain united to form filaments, because the parental walls are persistent. In the case of other green algae with this type of mitosis and cytokinesis, daughter cells are liberated from the parent cell wall as non-flagellate autospores, or as zoospores with centrioles ready to form flagella.

Coleochaete possesses the second type of nuclear and cell division; the chloroplast begins to cleave at prophase; the single centriolar pair present during the interphase replicates at prophase and each of the two pairs takes up a position at one pole of the future spindle. Microtubules then form between the centriolar pairs, outside the envelope of the elongate prophase nucleus. The mitosis is open, because the nuclear envelope breaks down during metaphase; endoplasmic reticulum vesicles are present among the spindle microtubules. The chromosomes align in a distinct metaphase plate but the chromosomal microtubules do not attach to defined kinetochores. At early telophase daughter nuclei become separated through elongation of the spindle, which is persistent during telophase and holds the nuclei far apart.

By the end of the telophase, new microtubules proliferate and surround the spindle microtubules, forming the phragmoplast, which includes also actin filaments. Golgi-derived vesicles appear within the phragmoplast, guided by the microtubules and the filaments to the future plane of division, where they become arranged to form a cell plate. The vesicles contain cell wall material and their coalescence produces a transverse septum consisting of two cell membranes with the new transverse wall between them. As the coalescence is not complete, connections leading to plasmodesmata are left between the daughter cells. This type of mitosis and cytokinesis is rare in the green algae, but is the common mode of division in plants such as bryophytes and tracheophytes.

EJECTILE ORGANELLES AND FEEDING APPARATA

In this section we will describe organelles that upon stimulation by contact, heat, or chemicals discharge a structure such as a thread or a tube from the surface of the cell. These organelles may serve for defense purpose or as feeding adjuvant.

Heterokontophyta

Many species of Raphidophyceae have extrusome organelles that explode on strong stimulation throwing out up to 200 μm long slime threads. The material produced may surround a motile individual with mucilage so that it becomes palmelloid.

Haptophyta

During its motile stage, *Phaeocystis* can eject ribbon-like filaments a few tens of nanometers wide and several tens of micrometers long. Their interconnections form a five-branch star-like configuration (Figure 2.82). The winding of the filaments inside vesicles, as well as their axial twist, are probably the consequences of their biosynthesis within a confined space. As a result, the filaments behave like spring-coils whose stored energy is released once the vesicles are broken and the filaments ejected. Using electron diffraction techniques, the filaments have been unambiguously characterized as being made of α-chitin crystals, the polymer chain axis lying along the filament direction. These chains are arranged antiparallel (allomorph) and this arrangement has never been reported before in the algal world.

Cryptophyta

These algae possess large refractile ejectosomes, lining the oral groove, and small ejectosomes scattered around the cell surface at the anterior corners of the periplast plates. An undischarged ejectosome is a tightly coiled, tapered ribbon that is wound with the wider end towards the outside; a smaller coil is attached to it and lies in the depression of the larger one. Prior to release, ejectosomes are enclosed within vesicles. The large ejectosomes are explosive organelles. If cryptophyte cells are irritated by mechanical or chemical stress, they escape the potentially lethal influences by discharging the ejectosomes: the cells jump backwards in fast zig-zag movements through the water. When discharged, the ribbon unfurls, with the shorter segment forming a beaklike tip on the longer. The edges of the ribbon tend to curl inwards, producing circular and e-shaped profiles in cross-section (Figure 2.83).

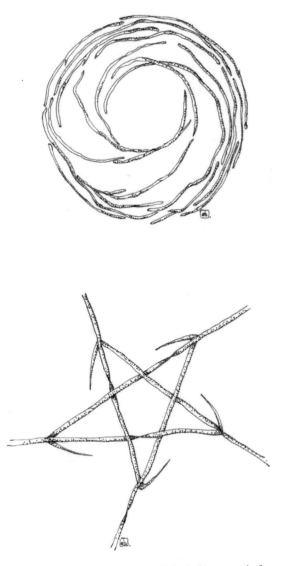

FIGURE 2.82 Ejectile organelles of *Phaeocystis* sp. Coiled filaments before release (top); star-like configuration after release (bottom).

Dinophyta

The most common type of extrusomes, of almost universal occurrence in the motile phase of these algae, are trichocysts, that is, rod-shaped bodies that, when mature, usually lie in the amphiesma perpendicular to the cell membrane (Figure 2.84). The shaft is a paracrystalline, proteinaceous rod a few micrometers long and rectangular in cross-section. At its distal end it extends as a group of twisted fibers. The whole is enclosed within a membranous sac, and there is a sheathing material between the rod and the membrane. The tip of the sac is in contact with the cell membrane, passing through the amphiesmal vesicles, and the thecal plates, if present. Thrichocysts are formed in the vicinity of the Golgi body and move to the cell periphery. They are discharged apparently by a rapid hydration process, the discharged structures measuring up to 200 μm in length. Their function is unknown, but it is assumed to be defensive, excretory, or both.

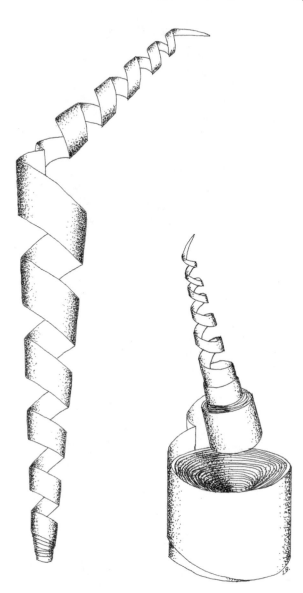

FIGURE 2.83 Ejectile organelles of a cryptophyte. Discharging ejectosome (left) and reel of an undischarged ejectosome just begining to be pulled out (right).

A less ordered type of extrusomes in dinoflagellates is the mucocyst, a simple sac with granular content, associated with the release of mucoid material. They are located just beneath the cell membrane, and are often aggregated in the region of the sulcus; their release in some species is correlated with the psamnophilous existence of these algae, facilitating the attachment of the cells to sand grains along the seashore.

More elaborate extrusomes termed nematocysts are found in genera such as *Polykrikos kofoidi* and *Nematodinium*; usually only about eight to ten nematocysts are present per cell. These organelles are larger than trichocysts and can reach 20 μm in length. They are conical, fluid-filled sacs with a capitate blunt end. Most of the body consists of a large posterior chamber, from which a smaller anterior chamber is isolated; the whole structure is capped by a lid-like operculum. A sharp stylet in the anterior chamber is connected to a tubular filament in the posterior chamber (Figure 2.85).

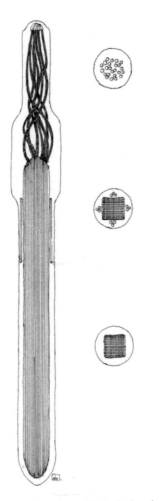

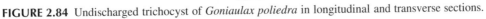

FIGURE 2.84 Undischarged trichocyst of *Goniaulax poliedra* in longitudinal and transverse sections.

Feeding structures are present in dinoflagellates, depending on the type of feeding mechanisms present: feeding tube (peduncle, or tentacle) and feeding veil (pallium). Tube feeding is commonly found among both naked and thecate species of dinoflagellates (e.g., *Dinophysis*, *Amphidinium*, *Gyrodinum*, and *Peridinopsis*). Cell membrane is extruded from the cell to form a tube, which can engulf whole cells or penetrate prey cell walls and suck in prey cytoplasma.

Pallium feeding has been described only for heterotrophic thecate species such as *Protoperidinium* and *Diplopsalis*. The prey is captured via a primary attachment filament; an extension of the cytoplasm then emerges from the region of the sulcus-cingulum, which encloses the prey as a veil (pallium). Enzymatical digestion of the prey cytoplasm is brought about inside the veil and the products are then transported to the predator.

Direct engulfment is mainly found among naked species (e.g., *Gyrodinium*, *Gymnodinium*, and *Noctiluca*); recently, however, some thecate species have been shown to use this feeding mechanism as well. Feeding behavior in dinoflagellates involves several steps prior to actual ingestion, including precapture, capture, and prey manipulation. As feeding mechanisms allow the ingestion of relatively large preys, or parts thereof, dinoflagellates are regarded as raptorial feeders. While prey size plays an important role for the ability of dinoflagellates to ingest food, this cannot alone explain prey preferences actually found. Some dinoflagellate species can be very selective in their choice of prey, while others show a remarkable versatility.

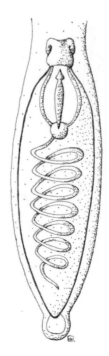

FIGURE 2.85 Nematocysts of *Polykrikos kofoidi*.

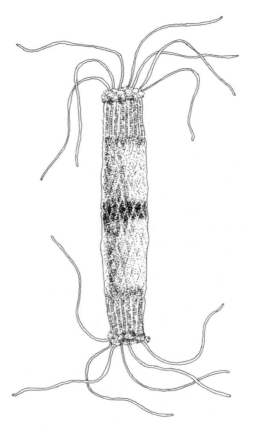

FIGURE 2.86 Ejected mucocyst of *Peranema trichophorum*.

Euglenophyta

Trichocysts are present only in some phagotrophic species, such as *Entosiphon* and *Peranema tri-chophorum*. In *Peranema* ejectile organelles or mucocysts are located beneath the cell surface, often subtending the pellicular articulations. In transverse sections they appear as hollow tubes of amorphous material, with low electron density, enclosed within a vesicle bounded by single membrane, often found near the Golgi apparatus; or they show a content of greater density, and may subtend the pellicle or protrude through it. Mucocysts subtending the pellicular articulations are ejected from the cell through the grooves. They are characterized by three distinct regions: an inner tube 1.20 mm long and 1.18 mm wide with helical striations; an outer tube 0.77 mm long and 0.21 mm wide with diamond-shaped pattern; and a dense middle band 0.07 mm in width. Often a crown of dense fibrillar material occurs at the tips of the mucocyst (Figure 2.86). Small slime-extruding mucus bodies are the trichocyst counterparts commonly found in the non-phagotrophs.

A feeding apparatus is present in many Euglenophyceae, such as *Diplonema*, *Ploeotia*, or *Entosiphon*. In the latter (Figure 2.87a), it consists of a cytostome surrounded by four centrally located vanes (series of plicate folds or curved ribbons) and supported by rods composed of closely packed microtubules. Near the base of the feeding apparatus, one of the rods splits giving rise to a total of three stout rods that extend nearly the length of the cell. About one third the distance down the apparatus (from the anterior end) the number of microtubules per rod increases dramatically and then decreases in number towards the base giving the apparatus the overall appearance of a cone with one side open (Figure 2.87b). This feeding apparatus can be moved in and out, extending toward the anterior of the cell and then withdrawing down into the cell for a distance of 3–5 mm. This motion allows food to be drawn into the cell.

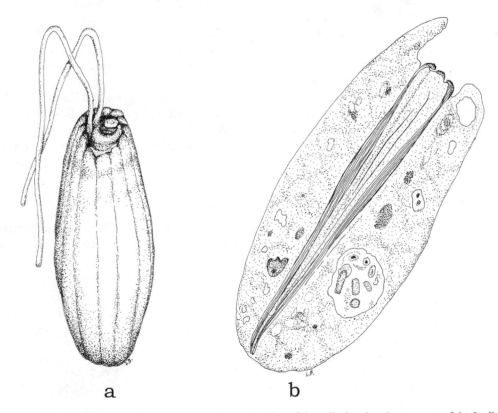

a   b

FIGURE 2.87 Cell of *Entosiphon* sp. (a); longitudinal section of the cell, showing the structure of the feeding apparatus (b).

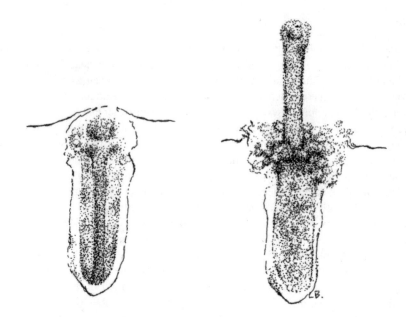

FIGURE 2.88 Ejectile organelle of *Bigelowiella natans* before (left) and during discharge (right).

Chlorarachniophyta

Structures interpreted as ejectile organelles (extrusomes) are commonly seen in *Bigelowiella natans*. Each organelle resides in a vesicle and consists of two parts. The anterior part is very short, measuring ca. 0.15 μm in length, whereas the posterior part is ca. 0.55 μm long. The posterior part has a very distinct substructure, comprising a core surrounded by a less opaque cylinder. The cylinder is lined by a thin membrane-like structure, whereas the core comprises a thinner membranous structure surrounded by a more opaque one. The structure of the anterior part of the ejectile organelle is less well defined. Although it is clear that the membrane of the surrounding vesicle forms an indentation ring that separates the two parts of the ejectile organelle, the path of the membrane in this area is less clear. It appears that only the core of the organelle is discharged, the core apparently turning inside out during discharge. The surrounding cylinder seems to remain in place. The fate of the anterior part of the ejectile organelle after discharge has not been ascertained (Figure 2.88).

Chlorophyta

Some *Pyramimonas* species (Prasinophyceae) contain ejectile structures similar to the ejectosomes of Cryptophyta. In the undischarged state, they consist of a coiled ribbon, which upon discharge forms a hollow tube tapered at both ends.

SUGGESTED READING

Albertano, P., Barsanti, L., Passarelli, V., and Gualtieri, P., A complex photoreceptive structure in the cyano-bacterium *Leptolyngbya* sp., *Micron*, 31, 27, 2000.

Andersen, R. A., Biology and systematic of heterokont and haptophyte algae, *American Journal of Botany*, 91, 1508–1522, 2004.

Barlow, H. B., The physical limits of visual discrimination, in *Photophysiology*, Giese, A., Ed., Academic Press, New York, 1964.

Barsanti, V., Passarelli, V., Walne, P.L., and Gualtieri, P. *In vivo* photocycle of the *Euglena gracilis* photo-receptor, *Biophysical Journal*, 72, 545, 1997.

Barsanti, L., Passarelli, V., Walne, P.L., and Gualtieri, P. The photoreceptor protein of *Euglena gracilis, FEBS Letters* 482, 247–251, 2000.

Beek, P.L., The long and the short, of flagellar length control, *Journal of Phycology*, 39, 837–839, 2003.

Bouck, G. B., Extracellular microtubules, *Journal of Cell Biology*, 40, 446–460, 1969.

Bouck, G. B., The structure, origin, isolation and composition of the tubular mastigonemes of *Euglena gracilis, Journal of Cell Biology*, 50, 362–384, 1971.

Bouck, G. B. and Ngo, H., Cortical structure and function in euglenoids with reference to trypanosomes, ciliates, and dinoflagellates, *International Review of Cytology*, 169, 267–318, 1996.

Brett, S. J., Perasso, L., and Wetherbee, R., Structure and development of the cryptomonad periplast: a review, *Protoplasma*, 181, 106–122, 1994.

Buetow, D. E., Ed., *The Biology of Euglena. General Biology and Ultrastructure*, Academic Press, New York, 1968a.

Buetow, D. E., Ed., *The Biology of Euglena. Biochemistry*, Academic Press, New York, 1968b.

Buetow, D. E., Ed., *The Biology of Euglena. Physiology*, Academic Press, New York, 1982.

Buetow, D. E., Ed., *The Biology of Euglena. Subcellular Biochemistry and Molecular Biology*, Academic Press, New York, 1989.

Daugbjerg, N., *Pyramimonas tychotreta*, a new marine species from Antarctica: light and electron microscopy of the motile stage and notes on growth rates, *Journal of Phycology*, 36, 160–171, 2000.

Deininger, W., Kroger, P., Hegemann, U., Lottspeich, F., and Hegemann, P., Chlamyrhodopsin represents a new type of sensory photoreceptor, *The EMBO Journal*, 14, 5849–5854, 1995.

Dodge, J. H., *The Fine Structure of Algal Cells*, Academic Press, London, 1973.

Dodge, J., Photosensory systems in eukaryotic algae, in *Evolution of the Eye and Visual System*, Cronly-Dillon, J. R. and Gregory, R. L., Eds., Macmillan Press, London, 1991.

Dodge, J. D. and Crawford, R. M., A survey of thecal fine structure in the Dinophyceae, *Botanical Journal of Linnean Society*, 63, 53–67, 1970.

Dusenbery, D. B., *Sensory Ecology*, Freeman, New York, 1992.

Ebnet, E., Fischer, M., Deininger, W., and Hegemann, P., Volvoxrhodopsin, a light-regulated sensory photo-receptor of the spheroidal green alga *Volvox carteri, The Plant Cell*, 11, 1473, 1999.

Fenchel, T., How Dinoflagellate swim, *Protist*, 152, 329–338, 2001.

Fredrick, J. F., Ed., *Origins and Evolution of Eukaryotic Intracellular Organelles*, The New York Academy of Sciences, New York, 1981.

Gibbs, S. P., Cheng, D., and Slankis, T., The chloroplast nucleoid in *Ochromonas danica*, I and II, *Journal of Cell Science*, 557–591, 1974.

Gilson, P. R. and McFadden, G. I., Molecular, morphological and phylogenetic characterization of six chlor-arachniophytes strains, *Phycological Research*, 47, 7–19, 1999.

Grain, J., Mignot, J. P., and de Puytorac, P., Ultrastructures and evolutionary modalities of flagellar and cyliary system in protists, *Biology of the Cell*, 63, 219–237, 1988.

Gregson, A. J., Green, J. C., and Leadbeater, B. S. C., Structure and physiology of Haptonema in *Chrisochromulina*. Fine structure of the flagellar/haptonematal root system in *C. acanthi* and *C. simplex, Journal of Phycology*, 29, 674–686, 1993a.

Gregson, A. J., Green, J. C., and Leadbeater, B. S. C., Structure and physiology of Haptonema in *Chrisochromulina*. Mechanism of haptonematal coiling and the regeneration process, *Journal of Phycology*, 29, 686–700, 1993b.

Gualtieri, P., Microspectroscopy of photoreceptor pigments in flagellated algae, *Critical Review of Plant Science*, 9 (6), 474–495, 1991.

Gualtieri, P., Morphology of photoreceptor systems in microalgae, *Micron*, 32, 411, 2001.

Gualtieri, P. and Robinson, K. R., A rhodopsin-like protein in the plasma membrane of *Silvetia compressa* eggs, *Photochemistry Photobiology*, 75, 76, 2002.

Gualtieri, P., Barsanti, L., and Passarelli, V., Absorption spectrum of a single isolated paraflagellar swelling of *Euglena gracilis, Biochemical Biophysical Acta*, 993, 293, 1989.

Gualtieri, P., Pelosi, P., Passarelli, V., and Barsanti, L., Identification of a rhodopsin photoreceptor in *Euglena gracilis, Biochemical Biophysical Acta*, 1117, 55, 1992.

Hackett, J. D., Anderson, D. M., Erdner, D. L., and Bhattacharya, D., Dinoflagellates: a remarkable evolutionary experiment, *American Journal of Botany*, 91, 1523–1534, 2004.

Harrison, F. W. and Corliss, J. O., *Microscopy Anatomy of Invertebrate. Protozoa*. Wiley-Liss, New York, 1991.

Hegemann, P., Algal sensory photoreceptors, *Journal of Phycology*, 37, 668, 2001.

Hibbed, D. J. and Norris, R. E., Cytology and ultrastructure of *Chlorarachnion reptans* (Chlorarachniophyta *divisio nova*, Chlorarachniophyceaea *classis nova*), *Journal of Phycology*, 20, 310–330, 1984.

Hilenski, L. L. and Walne, P. L., Ultrastructure of mucocysts in *Peranema trichophorum, Journal of Protozoology*, 30, 491–496, 1983.

Hoiczyk, E. and Hansel, A., Cyanobacterial cell walls: news from and unusual prokaryotic envelope, *Journal of Bacteriology*, 182, 1191–1199, 2000.

Honda, D. and Inouye, I., Ultrastructure and taxonomy of a marine photosynthetic stramenopile *Phaemonas parva* with emphasis on the flagellar apparatus archittetture, *Phycological Research*, 50, 75–89, 2002.

Horiguchi, T. and Hoppenrath, M., *Haramonas viridis*, a new sand-dwelling raphidophyte from cold temperate waters, *Phycological Research*, 51, 61–67, 2003.

Horiguchi, T., Kawai, H., Kubota, M., Takahashi, T., and Watanabe, M., Phototactic response of four marine dinoflagellates with different types of eyespot and chloroplast, *Phycological Research*, 47, 101–107, 1999.

Horspool, W. and Lenci, F., Eds., *Organic Photochemistry and Photobiology*, CRC Press, Boca Raton, 2004.

Huitorel, P., Audebert, S., White, D., Cosson, J., and Gagnon, C., Role of tubulin epitopes in the regulation of flagellar motility, in *The Male Gamete: From Basic Science to Clinical Applications*, Gagnon, C., Ed., Cache River Press, Vienna, 1999, pp. 475–491.

Hyams, J. S., The *Euglena* paraflagellar rod: structure, relationship to the other flagellar components and preliminary biochemical characterization, *Journal of Cell Science*, 55, 199–210, 1982.

Iomini, C. V., Babaev-Khaimov, M., Sassaroli, D., and Piperno, G., Protein particles in *Chlamydomonas* flagella undergo a transport cycle consisting of four phases, *Journal of Cell Biology*, 153 (1), 13–24, 2001.

Inuhe, I. and Hori, T., High-speed video analysis of the flagellar beat and swimming patterns of algae: possible evolutionary trends in green algae, *Protoplasma*, 164, 54–69, 1991.

Iseki, M., Matsunaga, S., Murakami, A., Ohno, K., Shiga, K., Yoshida, K., Sugai, M., Takahashi, T., Hori, T., and Watanabe, M., A blue-light-activated adenylyl cyclase mediates photoavoidance in *Euglena gracilis, Nature*, 415, 1047, 2002.

James, T. W., Crescitelli, F., Loew, E. R., and McFarland, W. N., The eyespot of *Euglena gracilis*: a microspectrophotometric study, *Vision Research*, 32, 1583, 1992.

Kreimer, G., Cell biology of phototaxis in flagellated algae, *International Review of Cytology*, 148, 229–310, 1994.

Kreimer, G., Reflective properties of different eyespot types in dinoflagellates, *Protist*, 150, 311, 1999.

Kreimer, G. and Melkonian, M., Reflection confocal laser scanning microscope of eyespot in flagellated green algae, *European Journal of Cell Biology*, 53, 101, 1990.

Kreimer, G., Marner, F. J., Brohsonn, U., and Melkonian, M., Identification of 11-cis and all-trans-retinal in the photoreceptive organelle of a flagellate green alga, *FEBS Letters*, 293, 49, 1991.

Kugrens, P., Clay, B. L., and Lee, R. E., Ultrastructure and systematic of two new freshwater red cryptomonads, *Storeatula rhinosa*, and *Pyrenomonaas ovalis, Journal of Phycology*, 35, 1079–1089, 1999a.

Kugrens, P., Clay, B. L., Meyer, C. J., and Lee, R. E., Ultrastructure and description of *Cyanophora biloba* with additional observations on *C. paradoxa, Journal of Phycology*, 35, 844–854, 1999b.

Lakie, J. M., *Cell Movement and Cell Behaviour*, Allen and Unwin, London, 1986.

Lavau, S. and Wetherbee, R., Structure and development of the scale case of *Mallomonas adamas, Protoplasma*, 181, 259–268, 1994.

Leadbeater, B. S. C. and Green, J. C., Eds., *The Flagellates. Unit, Diversity and Evolution*, Taylor and Francis, London, 2000.

Leander, B. S. and Farmer, M. A., Comparative morphology of euglenid pellicle. Patterns of stripes and pores, *Journal of Eukariotic Microbiology*, 46, 469–479, 2000.

Leander, B. S. and Farmer, M. A., Comparative morphology of euglenid pellicle. Diversity of strip substructures, *Journal of Eukariotic Microbiology*, 48, 202–217, 2001.

Leander, B. S., Witek, R. P., and Farmer, M. A., Trends in the evolution of the Euglenoid pellicle, *Evolution*, 55, 2215–2235, 2001.

Lima-de-Faria, A., Ed., *Handbook of Molecular Cytology*, North-Holland Publishing Co., Amsterdam 1969.

Lindemann, C. B., A "Geometric Clutch" hypothesis to explain oscillations of the axoneme of cilia and flagella, *Journal of Theoretical Biology*, 168, 175–189, 1994.

Lobban, C. S. and Harrison, P. J., *Seaweed Ecology and Physiology*, Cambridge University Press, Cambridge, 1997.

Maga, J. A. and LeBowitz, J. H., Unravelling the kinetoplastid paraflagellar rod, *Trends Cell Biology*, 9, 409–413, 1999.

Mann, K. H. and Lazier, J. R. N., *Dinamics of Marine Ecosystems*, Blackwell Science, Cambridge, 1996.

Marian, B. and Melkonian, M., Flagellar hairs in Prasinophytes: ultrastructure and distribution on the flagellar surface, *Journal of Phycology*, 30, 659–678, 1994.

Martin, R. L., Wood, C., Baehr, W., and Applebury, M., Visual pigment homologies revealed by DNA hybridization, *Science* 232, 1266–1269, 1986.

Maruyama, T., Fine structure of the longitudinal flagellum in *Ceratium tripos*, a marine dinoflagellate, *Journal of Cell Science*, 58, 109–123, 1982.

Melkonian, M., Ed., *Algal Cell Motility*, Chapman Hall, New York, 1992.

Melkonian, M. and Robenek, H., The eyespot apparatus of flagellated green algae: a critical review, in *Progress in Phycological Research*, Vol. 3, Round, F. E. and Chapman, D. J., Eds., Biopress Ltd., Bristol, 1984, pp. 193–268.

Menzel, D., Ed., *The Cytoskeleton of the Algae*, CRC Press Inc., Boca Raton, 1992.

Miyasaka, I., Nanba, K., Furuya, K., Nimura, Y., and Azuma, A., Functional roles of the transverse and longitudinal flagella in the swimming motility of *Prorocentrum minimum, Journal of Experimental Biology*, 207, 3055–3066, 2004.

Moenstrup, Ø., Flagellar structure in algae: a review, with new observations particularly on the Chrysophyceae, Phaeophyceae, Euglenophyceae, and *Reckertia, Phycologia*, 21 (4), 427–528, 1982.

Moenstrup, Ø. and Sengco, M., Ultrastructural studies on *Bigelowiella natans*, a Chlorarachniophyte flagellate, *Journal of Phycology*, 37, 624–646, 2001.

Nagel, G., Ollig, D., Fuhrmann, M., Kateriya, S., Musti, A. M., Bamberg, E., and Hegemann, P., Channel-rhodopsin-1: a light-gated proton channel in green algae, *Science* 296, 2395–2398, 2002.

Niklas, K. J., The cell walls the bind the tree of life, *Bioscience*, 54, 831–841, 2004.

Norton, T. A., Andersen R. A., and Melkonian, M., Algal biodiversity, *Phycologia*, 35, 308, 1996.

Novarino, G., A companion to the identification of cyptomonad flagellates, *Hydrobiologia*, 502, 225–270, 2003.

Parkinson, P. R., Ontogeny v. Phylogeny. The strange case of the silicoflagellates, *Constancea*, 83, 1–29, 2002.

Passarelli, V., Barsanti, L., Evangelista, V., Frassanito, A. M., and Gualtieri, P., *Euglena gracilis* photoreception interpreted by microspectroscopy, *European Journal of Protistology*, 39, 404–408, 2003.

Patterson, D. J. and Larsen, J., *The Biology of Free-living Heterotrophic Flagellates*, Claredon Press, Oxford, 1991.

Pickett-Heaps, J. D., Carpenter, J., and Koutoulis, A., Valve and sseta (spine) morphogenesis in the centric diatom *Chaetoceros peruvianus, Protoplasma*, 181, 269–282, 1994.

Preisig, H. R., Siliceous structure and silicification in flagellate protists, *Protoplasma*, 181, 29–42, 1994.

Preisig, H. R., Anderson, O. R., Corliss, J. O., Moestrup, Ø., Powell, M. J., Roberson, R. W., and Wetherbee, R., Terminology and nomenclature of protest cell surface structures, *Protoplasma*, 181,1–28, 1994.

Purcell, E. M., Life at low Reynold number, *American Journal of Physics*, 45 (1), 3–11, 1977.

Rave, J. A. and Brownlee, C., Understanding membrane function, *Journal of Phycology*, 37, 960–967, 2001.

Rizzo, P. J., Those amazing dinoflagellate chromosomes, *Cell Research*, 13, 215–217, 2002.

Rosati, G., Verni, F., Barsanti, L., Passarelli, V., and Gualtieri, P., Ultrastructure of the apical zone of *Euglena gracilis*: photoreceptor and motor apparatus, *Electron Microscopy Review*, 4, 319–342, 1991.

Saunders, G. W. and Hommersand, M. H., Assessing red algae supraordinary diversity and taxonomy in the context of contemporary systematic data, *American Journal of Botany*, 91, 1494–1507, 2004.

Sato, H., Greuet, C., Cachon, M., and Cosson, J., Analysis of the contraction of an organelle using its birefringency: the R-fiber of the *Ceratium* flagellum, *Cell Biology International*, 28, 387–396, 2004.

Schaller, K. and Uhl, R., A microspectrophotometric study of the shielding proprieties of eyespot and cell body in *Chlamydomonas, Biophysic Journal*, 73, 1573, 1997.

Schaller, K., David, R., and Uhl, R., How *Chlamydomonas* keeps track of the light once it has reached the right phototactic orientation, *Biophysical Journal*, 73, 1562–1567, 1997.

Sineshchekov, O. A., Jung, K.-H., and Spudich, J. L., Two rhodopsins mediate phototaxis to low and high-intensity light in *Chlamydomonas reinhardtii, Proceedings of the National Academy of Science*, 99, 8689–8693, 2002.

Solomon, C. M., Evelyn, J., Lessard, R. G., Keil, M. S., Foy, Characterization of extracellular polymers of *Phaeocystis globosa* and *P. antarctica, Marine Ecology Progress Series*, 250, 81–89, 2003.

Suneel, K., Nagel, G., Bamber, E., and Hegemann, P., Vision in single-celled algae, *News Physiological Science*, 19, 133–137, 2004.

Stoebe, B. and Maier, U.G., One, two, and three: nature's tool box for building plastids, *Protoplasma*, 219, 123–130, 2002.

Taylor Max, F. J. R., Charles Atwood Kafoid and his dinoflagellate tabulation system: an appraisal and evaluation of the phylogenitic value of tabulation, *Protist*, 150, 213–220, 1999.

Verspagen, J. M. H., Snelder, E. O. F. M., Visser, P. M., Huisman, J., Mur, L. R., and Ibelings, B. W., Recruitment of benthic *Microcystis* to the water column: internal bouyance changes or rresuspention? *Journal of Phycology*, 40, 260–270, 2004.

Vogeley, L., Sineshchekov, O. A., Trivedi, V. D., Sasaki, J., Spudich, J. L., and Luecke, H., *Anabaena* sensory rhodopsin: a photochromic color sensor at 2.0 Å *Science*, 306, 1390–1393, 2004.

Walne, P. L. and Gualtieri, P., Algal visual proteins: an evolutionary point of view, *Critical Review of Plant Science*, 13, 185, 1994.

Walne, P. L., Passarelli, V., Barsanti, L., and Gualtieri, P., Rhodopsin: a photopigment for phototaxis in *Euglena gracilis, Critical Review of Plant Science*, 17, 559, 1998.

Westfall, J. A., Bradbury, P. C., and Townsend, J. W., Ultrastructure of the dynoflagellate *Polykrikos, Journal of Cell Science*, 63, 245–261, 1983.

Young, J. R., Davis, S. A., Brown, P. R., and Mann, S., Coccolith ultrastructure and biomineralization, *Journal of Structural Biology*, 126, 195–215, 1999.

Zingone, A., Chretiennot-Dinet, M. J., Lange, M., and Medlin, L., Morphological and genetic characterization *Phaeocystis cordata* and *P. jahnii*, two new species from the Mediterranean Sea, *Journal of Phycology* 35, 1322–1337, 1999.

3 Photosynthesis

LIGHT

The Sun is the universal source of energy in the biosphere. During the nuclear fusion processes occurring in the Sun, matter is changed into energy, which is emitted into space in the form of electromagnetic radiation, having both wave and particle properties. The electromagnetic radiation has a spectrum or wavelength distribution from short wavelength (10^{-6} nm, γ- and x-rays) to long wavelength (10^{15} nm, long radio waves). About 99% of the Sun radiation is in the wavelength region from 300 to 4000 nm and it is called the broadband or total solar radiation. Within this broadband, different forms of energy exist, which can be associated with specific phenomena such as harmful and potentially mutagen ultraviolet radiation (UV 100–400 nm), sight (visible light 400–700 nm), and heat (infrared radiation 700–4000 nm). The particles producing the electromagnetic waves are called photons or quanta. The energy of a photon or quantum can be expressed as $h\nu$, where h is the Planck's constant (6.626×10^{-34} J sec) and ν is the frequency of the photon. The frequency is in turn equal to $c\lambda^{-1}$, where c is the speed of light (3×10^8 m sec^{-1}) and λ is the wavelength of the photon in nanometres (nm). According to this formula the shorter the photon wavelength, the higher its energy; for example, the energy of one photon of 300 nm light is 6.63×10^{-19} J, the energy of one photon of 400 nm light is 4.97×10^{-19} J, the energy of one photon of 700 nm light is 2.84×10^{-19} J, and the energy of one photon of 4000 nm light is 0.49×10^{-19} J.

The energy of photons can also be expressed in terms of electron volts (eV). Absorption of a photon can lead to excitation of an electron and hence of a molecule. This excited electron acquires potential energy (capacity of producing chemical work) measured in eV. An electron volt is the potential energy of 1 V gained by the excited electron, which is equal to 1.60×10^{-19} J. Thus the energy of one photon of 300 nm light is equal to 4.14 eV, the energy of one photon of 400 nm light is equal to 3.11 eV, the energy of one photon of 700 nm light is equal to 1.77 eV, and the energy of one photon of 4000 nm light is equal to 0.30 eV.

The average intensity of the total solar radiation reaching the upper atmosphere is about 1.4 kW m^{-2} (UV 8%, visible light 41%, and infrared radiation 51%). The amount of this energy that reaches any one "spot" on the Earth's surface will vary according to atmospheric and meteorological (weather) conditions, the latitude and altitude of the spot, and local landscape features that may block the Sun at different times of the day. In fact, as sunlight passes through the atmosphere, some of it is absorbed, scattered, and reflected by air molecules, water vapor, clouds, dust, and pollutants from power plants, forest fires, and volcanoes. Atmospheric conditions can reduce solar radiation by 10% on clear, dry days, and by 100% during periods of thick clouds. At sea level, in an ordinary clear day, the average intensity of solar radiation is less than 1.0 kW m^{-2}, (UV 3%, visible light 42%, infrared radiation 55%). Penetrating water, much of the incident light is reflected from the water surface, more light being reflected from a ruffled surface than a calm one and reflection increases as the Sun descends in the sky (Table 3.1). As light travels through the water column, it undergoes a decrease in its intensity (attenuation) and a narrowing of the radiation band is caused by the combined absorption and scattering of everything in the

TABLE 3.1
Sun Light Reflected by Sea Surface

Angle between Sun rays and zenith	0°	10°	20°	30°	40°	50°	60°	70°	80°	90°
Percentage of reflected light	2	2	2.1	2.1	2.5	3.4	6	13.4	34.8	100

water column including water. In fact, different wavelengths of light do not penetrate equally, infrared light (700–4000 nm) penetrates least, being almost entirely absorbed within the top 2 m, and ultraviolet light (300–400 nm) is also rapidly absorbed. Within the visible spectrum (400–700 nm), red light is absorbed first, much of it within the first 5 m. In clear water the greatest penetration is by the blue-green region of the spectrum (450–550 nm), while under more turbid conditions the penetration of blue rays is often reduced to a greater extent than that of the yellow-red wavelengths (550–700 nm). Depending on the conditions about 3–50% of incident light is usually reflected, and Beer's law can describe mathematically the way the light decreases as function of depth,

$$I_z = I_0 * e^{-kz} \tag{3.1}$$

where I_z is the intensity of light at depth z, I_0 is the intensity of light at depth 0, that is, at the surface, and k is the attenuation coefficient, which describes how quickly light attenuates in a particular body of water. Algae use the light eventually available in two main ways:

- As information in sensing processes, supported by the photoreceptors systems, which has been already explained in Chapter 2
- As energy in transduction processes, supported by chloroplasts in photosynthesis

Both types of processes depend on the absorption of photons by electrons of chromophore molecules with extensive systems of conjugated double bonds. These conjugated double bonds create a distribution of delocalized *pi* electrons over the plane of the molecule. *Pi* electrons are characterized by an available electronic "excited state" (an unoccupied orbital of higher energy, higher meaning the electron is less tightly bound) to which they can be driven upon absorption of a photon in the range of 400–700 nm, that is, the photosynthetic active radiation (PAR). Only absorption of a photon in this range can lead to excitation of the electron and hence of the molecule, because the lower energy of an infrared photon could be confused with the energy derived by molecular collisions, eventually increasing the noise of the system and not its information. The higher energy of an UV photon could dislodge the electron from the electronic cloud and destroy the molecular bonds of the chromophore. Charge separation is produced in the chromophore molecule elevated to the excited state by the absorption of a photon, which increases the capability of the molecule to perform work. In sensing processes, charge separation is produced by the photoisomerization of the chromophore around a double bond, thus storing electrostatic energy, which triggers a chain of conformational changes in the protein that induces the signal transduction cascade. In photosynthesis, a charge separation is produced between a photo-excited molecule of a special chlorophyll (electron donor) and an electron-deficient molecule (electron acceptor) located within van der Waals distance, that is, a few Å. The electron acceptor in turn becomes a donor for a second acceptor and so on; this chain ends in an electron-deficient trap. In this way, the free energy of the photon absorbed by the chlorophyll can thereby be used to carry out useful electrochemical work, avoiding its dissipation as heat or fluorescence. The ability to perform electrochemical work for each electron that is transferred is termed redox potential; a negative redox potential indicates a reducing capability of the system (the system possesses available electrons), while a positive redox potential indicates an oxidizing capability of the system (the system lacks available electrons).

Photosynthetic activity of algae, which roughly accounts for more than 50% of global photosynthesis, make it possible to convert the energy of PAR into biologically usable energy, by means of reduction and oxidation reactions; hence, photosynthesis and respiration must be regarded as complex redox processes.

As shown in Equation (3.2), during photosynthesis, carbon is converted from its maximally oxidized state ($+4$ in CO_2) to strongly reduced compounds (0 in carbohydrates, $[CH_2O]_n$) using the light energy.

$$nCO_2 + nH_2O + light \xrightarrow{\text{Chlorophyll } a} (CH_2O)_n + nO_2 \qquad (3.2)$$

In this equation, light is specified as a substrate, chlorophyll a is a requisite catalytic agent, and $(CH_2O)_n$ represents organic matter reduced to the level of carbohydrate. These reduced compounds may be reoxidized to CO_2 during respiration, liberating energy. The process of photosynthetic electron transport takes place between $+0.82$ eV (redox potential of the H_2O/O_2 couple) and -0.42 eV (redox potential of the CO_2/CH_2O couple).

Approximately half of the incident light intensity impinging on the Earth's surface (0.42 kW m^{-2}) belongs to PAR. In the water, as explained earlier, the useful energy for photobiochemical processes is even lower and distributed within a narrower wavelength range. About 95% of the PAR impinging on algal cell is mainly lost due to the absorption by components other than chloroplasts and the ineffectiveness of the transduction of light energy into chemical energy. Only 5% of the PAR is used by photosynthetic processes. Despite this high energy waste, photosynthetic energy transformation is the basic energy-supplying process for algae.

PHOTOSYNTHESIS

Photosynthesis encompasses two major groups of reactions. Those in the first group, the "light-dependent reactions," involve the capture of the light energy and its conversion to energy currency as NADPH and ATP. These reactions are absorption and transfer of photon energy, trapping of this energy, and generation of a chemical potential. The latter reaction follows two routes: the first one generates NADPH due to the falling of the high energy excited electron along an electron transport system; the second one generates ATP by means of a proton gradient across the thylakoid membrane. Water splitting is the source of both electrons and protons. Oxygen is released as a by-product of the water splitting. The reactions of the second group are the "light-independent reactions," and involve the sequence of reactions by which this chemical potential is used to fix and reduce inorganic carbon in triose phosphates (Figure 3.1).

Light Dependent Reactions

Photosynthetic light reactions take place in thylakoid membranes where chromophore–protein complexes and membrane-bound enzymes are situated. The thylakoid membrane cannot be considered as a rigid, immutable structure. It is rather a highly dynamic system, the molecular compositions and conformation of which, including the spatial pattern of its components, can change very rapidly. This flexibility, is, however, combined with a high degree of order necessary for the energy-transforming processes.

Quantitative analysis established that the 7 nm thick thylakoid membrane consists of approximately 50% lipids and 50% proteins. Galactolipids, a constituent that is specific of thylakoid membranes, make up approximately 40% of the lipid fraction. Chlorophylls a, b, c_1 and c_2, phospholipids, sulfolipids, carotenoids, xanthophylls, quinones, and sterols, all components occurring in

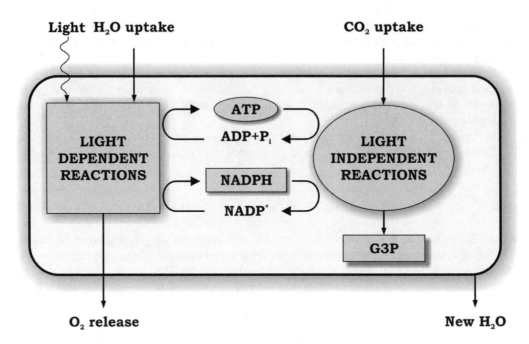

FIGURE 3.1 Schematic drawing of the photosynthetic machinery.

a bound form, represent the remainder 10%. Chlorophyll a consists of a hydrophilic porphyrin head formed by four linked pyrrole rings with a magnesium atom chelated (Mg^{2+}) at the center and a hydrophobic phytol tail. Chlorophyll b possesses the same structure as chlorophyll a but a keto group ($-CH{=}O$) is present in the second pyrrole ring instead of a methyl group ($-CH_3$). Chlorophyll c possesses only the hydrophilic porphyrin head without the phytol tail; chlorophyll c_2 differs from chlorophyll c_1 by possessing two vinyl groups ($-CH{=}CH_2$) instead of one. In the phycobiliproteins the four pyrrolic rings are linearly arranged, and unlike the chlorophylls they are strongly covalently bound to a protein. Carotenoids are C_{40} hydrocarbon chains, strongly hydrophobic, with one or two terminal ionone rings. The xanthophylls are carotenoid derivates with a hydroxyl group in the ring (Figure 3.2).

The protein complex content consists mainly of the highly organized energy transforming units, enzymes for the electron transport, and ATP-synthesis, more or less integrated into the thylakoid membrane. The energy transforming units are two large protein complexes termed photosystems I (PSI) and II (PSII), surrounded by light harvesting complexes (LHCs). Photons absorbed by PSI and PSII induce excitation of special chlorophylls, P_{700} and P_{680} (P stands for pigment and 700/680 stand for the wavelength in nanometer of maximal absorption), initiating translocation of an electron across the thylakoid membrane along organic and inorganic redox couples forming the electron transfer chains (ETCs). The main components of these chains are plastoquinones, cytochromes, and ferredoxin. This electron translocation process eventually leads to a reduction of $NADP^+$ to NADPH and to a transmembrane difference in the electrical potential and H^+ concentration, which drives ATP-synthesis by means of an ATP-synthase.

Thylakoid membranes are differentiated into stacked and unstacked regions. Stacking increases the amount of thylakoid membrane in a given volume. Both regions surround a common internal thylakoid space, but only unstacked regions make direct contact with the chloroplast stroma. The two regions differ in their content of photosynthetic assemblies; PSI and ATP-synthase are located almost exclusively in unstacked regions, whereas PSII and LCHII are present mostly in stacked regions. This topology derived from protein–protein interactions rather than lipid bilayers interactions. A common internal thylakoid space enables protons liberated by PSII in

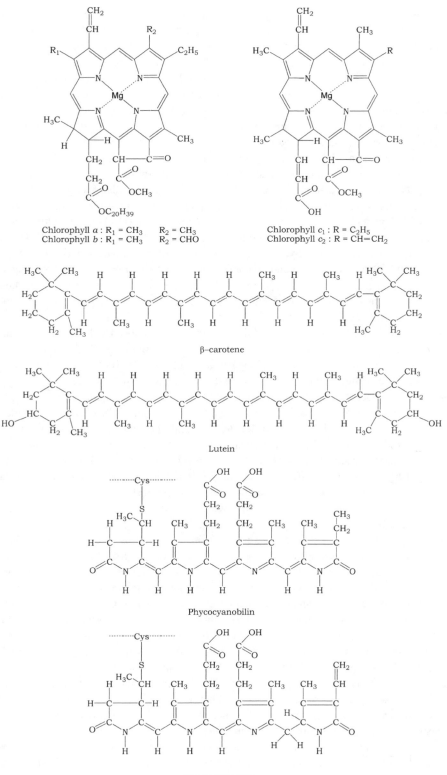

β-carotene

Lutein

Phycocyanobilin

Phycoerythrobilin

Chlorophyll a : R_1 = CH_3 R_2 = CH_3
Chlorophyll b : R_1 = CH_3 R_2 = CHO

Chlorophyll c_1 : R = C_2H_5
Chlorophyll c_2 : R = $CH=CH_2$

FIGURE 3.2 Structure of the main pigments of the thylakoid membrane.

stacked membranes to be utilized by ATP-synthase molecules that are located far away in unstacked membranes. What is the functional significance of this lateral differentiation of the thylakoid membrane system? If both photosystems were present at high density in the same membrane region, a high proportion of photons absorbed by PSII would be transferred to PSI because the energy level of the excited state P_{680}^* relative to its ground state P_{680} is higher than that of P_{700}^* relative to P_{700}. A lateral separation of photosystems solves this problem by placing P_{680}^* more than 100 Å away from P_{700}. The positioning of PSI in the unstacked membranes gives it a direct access to the stroma for the reduction of $NADP^+$. In fact the stroma-exposed surface of PSI, which contains the iron-sulfur proteins that carry electron to ferredoxin and ultimately to $NADP^+$, protrudes about 50 Å beyond the membrane surface and could not possibly be accommodated within the stacks, where adjacent thylakoids are separated by no more than 40 Å. It seems likely that ATP-synthase is also located in unstacked regions to provide space for its large protruding portion and access to ADP. In contrast, the tight quarters of the appressed regions do not pose a problem for PSII, which interacts with a small polar electron donor (H_2O) and a highly lipid-soluble electron carrier (plastoquinone). According to the model of Allen and Forsberg (2001), the close appression of grana (stacks of thylakoids) membranes arises because the flat stroma-exposed surfaces of LHCII form recognition and contact surfaces for each other, causing opposing surfaces of thylakoids to interact. There is not steric hindrance to this close opposition of stacked grana membranes, because similar to LHCII PSII presents a flat surface that protrudes not more than 10–20 Å beyond the membrane surface.

The functional significance of thylakoid stacking is presumably to allow a large, connected, light-harvesting antenna to form both within and between membranes. Within this antenna both the excitation energies can pass between chlorophylls located in LHCII complexes that are adjacent to each other, both within a single membrane and between appressed membranes.

The degree of stacking and the proportion of different photosynthetic assemblies are regulated in response to environmental variables such as the intensity and spectral characters of incident light. The lateral distribution of LHC is controlled by reversible phosphorylation. At low light levels, LHC is bound to PSII. At high light levels, a specific kinase is activated by plastoquinol, and phosphorylation of threonine side chains of LHC leads to its release from PSII. The phosphorylated form of these light harvesting units diffused freely in the thylakoid membrane and may become associated with PSI to increase its absorbance coefficient (Figure 3.3).

Central to the photosynthetic process is PSII, which catalyzes one of the most thermodynamically demanding reactions in biology: the photo-induced oxidation of water ($2H_2O \rightarrow 4e^- + 4H^+ + O_2$). PSII has the power to split water and use its electrons and protons to drive photosynthesis. The first ancestor bacteria carrying on anoxygenic photosynthesis probably synthesized ATP by oxidation of H_2S and FeS compounds, abundant in the environment. The released energy could have been harnessed via production of a proton gradient, stimulating evolution of electron transport chains, and the reducing equivalents (electrons) generated used in CO_2 fixation and hence biosynthesis. This was the precursor of the PSI. About 2800 million years ago the evolutionary pressure to use less strongly reducing (and therefore more abundant) source of electrons appears to have culminated in the development of the singularly useful trick of supplying the electrons to the oxidized reaction center from a tyrosine side chain, generating tyrosine cation radicals that are capable of sequential abstraction of electrons from water. Oxygenic photosynthesis, which requires coupling in series of two distinct types of reaction centers (PSI and PSII) must have depended on later transfer of genes between the evolutionary precursors of the modern sulfur bacteria (whose single reaction center resembles PSI) and those of purple bacteria (whose single reaction center resembles PSII). Thus the cyanobacteria appeared. They were the first dominant organisms to use photosynthesis. As a by-product of photosynthesis, oxygen gas (O_2) was produced for the first time in abundance. Initially, oxygen released by photosynthesis was absorbed by iron(II), then abundant in the sea, thus oxidizing it to insoluble iron(III) oxide (rust). Red "banded iron deposits" of iron(III) oxide are marked in marine sediments of ca. 2500

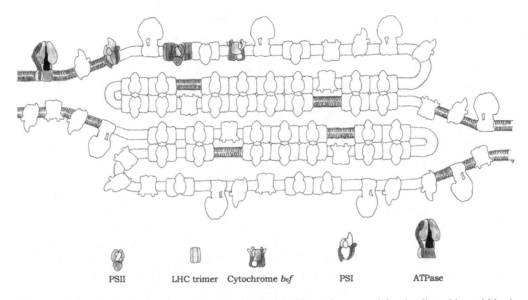

PSII   LHC trimer   Cytochrome b_6f PSI   ATPase

FIGURE 3.3 Model for the topology of chloroplast thylakoid membrane, and for the disposition within the chloroplast of the major intrinsic protein complexes, PSI, PSII, LHCII trimer, Cytochrome b_6f dimer and ATPase. (Redrawn after Allen and Forsberg, 2001.)

million years ago. Once most/all iron(II) had been oxidized to iron(III), then oxygen appeared in, and began to increase in the atmosphere, gradually building up from zero ca. 2500 million years ago to approximately present levels ca. 500 million years ago. This was the "oxygen revolution." Oxygen is corrosive, so prokaryotic life then either became extinct, survived in anaerobic (oxygen free) environments (and do so to this day), or evolved antioxidant protective mechanisms. The latter could begin to use oxygen to pull electrons from organic molecules, leading to aerobic respiration. The respiratory ETC probably evolved from established photosynthetic electron transport, and the citric acid cycle probably evolved using steps from several biosynthetic pathways.

Hence cyanobacteria marked the planet in a very permanent way and paved the way for the subsequent evolution of oxidative respiratory biochemistry. This change marks the end of the Archaean Era of the Precambrian Time.

PSII and PSI: Structure, Function and Organization

The PSII and PSI photosynthetic complexes are very similar in eukaryotic algae (and plants) and cyanobacteria, as are many elements of the light capture, electron transport, and carbon dioxide (CO_2) fixation systems. The PSI and PSII complexes contain an internal antenna-domain carrying light harvesting chlorophylls and carotenoids, both non-covalently bound to a protein moiety, and a central core domain where biochemical reactions occur. In the internal antenna complexes, chlorophylls do most of the light harvesting, whereas carotenoids and xanthophylls mainly protect against excess light energy, and possibly transfer the absorbed radiation. In all photosynthetic eukaryotes, PSI and PSII form a supercomplex because they are associated with an external antenna termed LHC. The main function of LHCs is the absorption of solar radiation and the efficient transmittance of excitation energy towards reaction center chlorophylls. LHCs are composed of a protein moiety to which chlorophylls and carotenoids are non-covalently bound. In eukaryotic algae, ten distinct light harvesting apoproteins (Lhc) can be distinguished. Four of them are exclusively associated with PSI (Lhca1–4), another four with PSII (Lhcb3–6), and two (Lhcb1 and Lhcb2) are preferentially but not exclusively associated with PSII, that is they can shuttle between the two photosystems. The apoproteins are three membrane-spanning α-helices and are nuclear-encoded. LHCs are arranged externally with respect to the photosystems. In Cyanophyta, Glaucophyta, Rhodophyta, and Cryptophyta, no LHCs are present and the light-harvesting function is performed by phycobiliproteins organized in phycobilisomes peripheral to the thylakoid membranes in the first three divisions, and localized within the lumen of thylakoids in the latter division. The phycobilisome structure consists of a three-cylinder core of four stacked molecules of allophycocyanin, close to the thylakoid membrane, on which converge rod-shaped assemblies of coaxially stacked hexameric molecules of only phycocyanin or both phycocyanin and phycoerythrin, (cf. Chapter 2, Figure 2.76). Phycobilisomes are linked to PSII but they can diffuse along the surface of the thylakoids, at a rate sufficient to allow movements from PSII to PSI within 100 ms. Among prokaryotes, Prochlorophyta (*Prochlorococcus* sp., *Prochlorothrix* sp. and *Prochloron* sp.), differ from cyanobacteria in possessing an external chlorophyll *a* and *b* antenna, like eukaryotic algae, instead of the large extrinsic phycobilisomes.

PSII complex can be divided into two main protein superfamilies differing in the number of membrane-spanning α-helices, that is, the six-helix protein superfamily, which includes the internal antennae CP43 and CP47 (CP stands for Chlorophyll–Protein complex), and the five-helix proteins of the reaction center core D1 and D2 (so-called because they were first identified as two *diffuse* bands by gel electrophoresis and staining) where ETC components are located. External antenna proteins of Prochlorophyta belong to the six-helix CP43 and CP47 superfamily and not to the three-helix LHCs superfamily.

PSII is a homodimer, where the two monomers in the dimers are almost identical. The monomer consists of over 20 subunits. All the redox active cofactors involved in the activity of PSII are bound to the reaction center proteins D1 and D2. Closely associated with these two proteins are the chlorophyll *a* binding proteins CP43 and CP47 and the extrinsic luminally bound proteins of the oxygen evolving complex. Each monomer also includes one heme *b*, one heme *c*, two plastoquinones, two pheophytins (a chlorophyll *a* without Mg^{2+}), and one non-heme Fe and contains 36 chlorophylls *a* and 7 *all-trans* carotenoids assumed to be β-carotene molecules. Eukaryotic and cyanobacterial PSII are structurally very similar at the level of both their oligomeric states and organization of the transmembrane helices of their major subunits. The eukaryotic PSII dimer is flanked by two clusters of Lhcb proteins. Each cluster contains two trimers of Lhcb1, Lhcb2, and Lhcb3 and the other three monomers, Lhcb4, Lhcb5, and Lhcb6.

The reactions of PSII are powered by light-driven primary and secondary electron transfer processes across the reaction center (D1 and D2 subunits). Upon illumination, an electron is dislodged from the excited primary electron donor P_{680}, a chlorophyll *a* molecule located towards the luminal surface. The electron is quickly transferred towards the stromal surface to the final electron acceptor, a plastoquinone, via a pheophytin. After accepting two electrons and undergoing protonation, plastoquinone is reduced to plastoquinol, and it is then released from PSII into the membrane matrix. The cation P_{680}^{+} is reduced by a redox active tyrosine, which in turn is reduced by a Mn ion within a cluster of four. When the $(Mn)_4$ cluster accumulates four oxidizing equivalents (electrons), two water molecules are oxidized to yield one molecule of O_2 and four proton. All the redox active cofactors involved in the electron transfer processes are located on the D1 side of the reaction center.

PSI complex possesses only eleven-helix PsaA and PsaB protein superfamilies. Each 11 transmembrane helices subunit has six N terminal transmembrane helices that bind light-harvesting chlorophylls and carotenoids and act as internal antennae and five C terminal transmembrane helices that bind Fe_4S_4 clusters as terminal electron acceptors. The N terminal part of the PsaA and PsaB proteins are structurally and functionally homologues to CP43 and CP47 proteins of PSII; the C terminal part of the PsaA and PsaB proteins are structurally and functionally homologues to D1 and D2 proteins of PSII. Eukaryotic PSI is a monomer that is loosely associated with the Lhca moiety, with a deep cleft between them. The four antenna proteins assemble into two heterodimers composed of Lhca1 and Lhca4 and homodimers composed of Lhca2 and Lhca3. Those dimers create a half-moon-shaped belt that docks to PsaA and PsaB and to other 12 proteic subunits of PSI, termed PsaC to PsaN that contribute to the coordination of antenna chromophores. On the whole PSI binds approximately 200 chromophore molecules. The cyanobacterial PSI exists as a trimer. One monomer consists of at least 12 different protein subunits, (PsaA, PsaB, PsaC, PsaD, PsaE, PsaF, PsaI, PsaJ, PsaK, PsaL, PsaM, and PsaX) coordinating more than 100 chromophores.

After primary charge separation initiated by excitation of the chlorophyll a pair P_{700}, the electron passes along the ETC consisting of another chlorophyll a molecule, a phylloquinone, and the Fe_4S_4 clusters. At the stromal side, the electron is donated by Fe_4S_4 to ferredoxin and then transferred to $NADP^+$ reductase. The reaction cycle is completed by re-reduction of P_{700}^+ by plastocyanin (or the interchangeable cytochrome c_6) at the inner (lumenal) side of the membrane. The electron carried by plastocyanin is provided by PSII by the way of a pool of plastoquinones and the cytochrome b_6f complex.

Photosynthetic eukaryotes such as Chlorophyta, Rhodophyta, and Glaucophyta have evolved by primary endosymbiosis involving a eukaryotic host and a prokaryotic endosymbiont. All other algae groups have evolved by secondary (or higher order) endosymbiosis between a simple eukaryotic alga and a non-photosynthetic eukaryotic host. Although the basic photosynthetic machinery is conserved in all these organisms, it should be emphasized that PSI does not necessarily have the same composition and fine-tuning in all of them. The subunits that have only been found in eukaryotes, that is, PsaG, PsaH, and PsaN, have actually only been found in plants and in Chlorophyta. Other groups of algae appear to have a more cyanobacteria-like PSI. PsaM is also peculiar because it has been found in several groups of algae including green algae, in mosses, and in gymnosperms. Thus, the PsaM subunit appears to be absent only in angiosperms. With respect to the peripheral antenna proteins, algae are in fact very divergent. All photosynthetic eukaryotes have Lhcs that belong to the same class of proteins. However, the Lhca associated with PSI appear to have diverged relatively early and the stoichiometry and interaction with PSI may well differ significantly between species. Even the green algae do not possess the same set of four Lhca subunits that is found in plants.

Are all those light harvesting complexes necessary? They substantially increase the light harvesting capacity of both photosystems by increasing the photon collecting surface with an associated resonance energy transfer to reaction centers, facilitated by specific pigment–pigment interactions. This process is related to the transition dipole–dipole interactions between the involved donor and acceptor antenna molecules that can be weakly or strongly coupled depending on the distance between and relative orientation of these dipoles. The energy migrates along a spreading wave because the energy of the photon can be found at a given moment in one or the other of the many resonating antenna molecules. This wave describes merely the spread of the probability of finding the photon in different chlorophyll antenna molecules. Energy resonance occurs in the chromophores of the antenna molecules at the lowest electronic excited state available for an electron, because only this state has a life time (10^{-8} sec) long enough to allow energy migration (10^{-12} sec). The radiationless process of energy transfer occurs towards pigments with lower excitation energy (longer wavelength absorption bands). Within the bulk of pigment–protein complexes forming the external and internal antenna system, the energy transfer is directed to chlorophyll a with an absorption peak at longest wavelengths. Special chlorophylls (P_{680} at PSII and P_{700} at PSI) located in the reaction center cores represent the final step of the photon trip, because once excited ($P_{680} + h\nu \rightarrow P_{680}^*$; $P_{700} + h\nu \rightarrow P_{700}^*$) they become redox active species ($P_{680}^* \rightarrow P_{680}^+ + e^-$; $P_{700}^* \rightarrow P_{700}^+ + e^-$), that is, each donor releases one electron per excitation and activates different ETCs.

For an image gallery of the three-dimensional models of the two photosystems and LHCs in prokaryotic and eukaryotic algae refer to the websites of Jon Nield and James Barber at the Imperial College of London (U.K.).

ATP-Synthase

ATP production was probably one of the earliest cellular processes to evolve, and the synthesis of ATP from two precursor molecules is the most prevalent chemical reaction in the world. The enzyme that catalyzes the synthesis of ATP is the ATP-synthase or F_0F_1-ATPase, one of the most ubiquitous proteins on Earth. The F_1F_0-ATPases comprise a huge family of enzymes with members found not only in the thylakoid membrane of chloroplasts but also in the bacterial cytoplasmic membrane and in the inner membrane of mitochondria. The source of energy for the functioning of ATP-synthase is provided by photosynthetic metabolism in the form of a proton gradient across the thylakoid membrane, that is, a higher concentration of positively charged protons in the thylakoid lumen than in the stroma.

The F_0F_1-ATPase molecule is divided into two portions termed F_1 and F_0. The F_0 portion is embedded in the thylakoid membrane, while the F_1 portion projects into the lumen. Each portion is in turn made up of several different proteins or subunits. In F_0, the subunits are named a, b, and c. There is one a subunit, two b subunits, and 9–12 c subunits. The large a subunit provides the channel through which H^+ ions flow back into the stroma. Rotation of the c subunits, which form a ring in the membrane, is chemically coupled to this flow of H^+ ions. The b subunits are believed to help stabilize the F_0F_1 complex by acting as a tether between the two portions. The subunits of F_1 are called α, β, γ, δ, and ε. F_1 has three copies each of α and β subunits which are arranged in an alternating configuration to form the catalytic "head" of F_1. The γ and ε subunits form an axis that links the catalytic head of F_1 to the ring of c subunits in F_0. When proton translocation in F_0 causes the ring of c subunits to spin, the γ–ε axis also spins because it is bound to the ring. The opposite end of the γ subunit rotates within the complex of α and β subunits. This rotation causes important conformational changes in the β subunits resulting in the synthesis of ATP from ADP and P_i (inorganic phosphate) and to its release.

For an image gallery of the three-dimensional models of the ATPase refer to the website of Michael Börsch at the Stuttgart University (www.atpase.de).

ETC Components

Components of the electron transport system in order are plastoquinone, cytochrome b_6f complex, plastocyanin, and ferredoxin. Each of the components of the ETC has the ability to transfer an electron from a donor to an acceptor, though plastoquinone also transfers a proton. Each of these components undergoes successive rounds of oxidation and reduction, receiving an electron from the PSII and donating the electron to PSI.

Plastoquinone refers to a family of lipid-soluble benzoquinone derivatives with an isoprenoid side chain. In chloroplasts, the common form of plastoquinone contains nine repeating isoprenoid units. Plastoquinone possesses varied redox states, which together with its ability to bind protons and its small size enables it to act as a mobile electron carrier shuttling hydrogen atoms from PSII to the cytochrome b_6f complex.

Plastoquinone is present in the thylakoid membrane as a pool of 6–8 molecules per PSII. Plastoquinone exists as quinone A (Q_A) and quinone B (Q_B); Q_A is tightly bound to the reaction center complex of PSII and it is immovable. It is the primary stable electron acceptor of PSII, and it accepts and transfers one electron at time. Q_B is a loosely bound molecule, which accepts two electrons and then takes on two protons before it detaches and becomes Q_BH_2, the mobile reduced form of plastoquinone (plastoquinol). Q_BH_2 is mobile within the thylakoid membrane, allowing a single PSII reaction center to interact with a number of cytochrome b_6f complexes.

Plastoquinone plays an additional role in the cytochrome b_6f complex, operating in a complicated reaction sequence known as a Q-cycle. When Q_B is reduced in PSII, it not only receives two electrons from Q_A but it also picks up two protons from the stroma matrix and becomes Q_BH_2. It is able to carry both electrons and protons (e^- and H^+ carrier). At the cytochrome b_6f complex level it is then oxidized, but FeS and cytochrome b_6 can accept only electrons and not protons. So the two protons are released into the lumen. The Q-cycle of the cytochrome b_6f complex is great because it provides extra protons into the lumen. Here two electrons travel through the two hemes of cytochrome b_6 and then reduce Q_B on the stroma side of the membrane. The reduced Q_B takes on two protons from the stroma, becoming Q_BH_2, which migrates to the lumen side of the cytochrome b_6f complex where it is again oxidized, releasing two more protons into the lumen. Thus the Q-cycle allows the formation of more ATP. This Q-cycle links the oxidation of plastoquinol (Q_BH_2) at one site on the cytochrome b_6f complex to the reduction of plastoquinone at a second site on the complex in a process that contributes additional free energy to the electrochemical proton potential.

The cytochrome b_6f complex is the intermediate protein complex in linear photosynthetic electron transport. The cytochrome b_6f complex essentially couples PSII and PSI and also provides the means of proton gradient formation by using cytochrome groups as redox centers in the ETC thereby separating the electron/hydrogen equivalent into its electron and proton components. The electrons are transferred to PSI via plastocyanin and the protons are released into the thylakoid lumen of the chloroplast. The electron transport from PSII to PSI via cytochrome b_6f complex occurs in about 7 ms, representing the rate limiting step of the photosynthetic process.

The cytochrome b_6f exists as a dimer of 217 kDa. The monomeric complex contains four large subunits (18–32 kDa), including cytochrome f, cytochrome b_6, the Rieske FeS iron-sulfur protein (ISP), and subunit IV, as well as four small hydrophobic subunits, PetG, PetL, PetM, and PetN. The monomeric unit contains 13 transmembrane helices: four in cytochrome b_6 (helices A to D); three in subunit IV (helices E to G); and one each in cytochrome f, the ISP, and the four small hydrophobic subunits PetG, PetL, PetM, and PetN. The monomer includes four hemes, one [2Fe-2S] cluster, one chlorophyll a, one β-carotene, and one plastoquinone. The extrinsic domains of cytochrome f and the ISP are on the luminal side of the membrane and are ordered in the crystal structure. Loops and chain termini on the stromal side are less well ordered. The ISP contributes to dimer stability by domain swapping, its transmembrane helix obliquely spans the membrane in one monomer, and its extrinsic domain is part of the other monomer. The two monomers form a protein-free central cavity on each side of the transmembrane interface.

Cytochrome c_6 is a small soluble electron carrier. It is a highly α-helical heme-containing protein. It is located on the luminal side of the thylakoid membrane where it catalyzes the electron transport from the membrane-bound cytochrome b_6f complex to PSI. It is the sole electron carrier in some cyanobacteria.

Plastocyanin operates in the inner aqueous phase of the photosynthetic vesicle, transferring electrons from cytochrome f to PSI. It is a small protein (10 kDa) composed of a single polypeptide that is coded for in the nuclear genome. Plastocyanin is a β-sheet protein with copper as the central ion that is ligated to four residues of the polypeptide. The copper ion serves as a one-electron carrier with a midpoint redox potential (0.37 eV) near that of cytochrome f. Plastocyanin shuttles electrons from the cytochrome b_6f complex to PSI by diffusion. Plastocyanin is more common in green algae and completely substitutes for cytochrome c_6 in the chloroplasts of higher plants. In cyanobacteria and green algae where both cytochrome c_6 and plastocyanin are encoded, the alternative expression of the homologous protein is regulated by the availability of copper.

Ferredoxin is a small protein (11 kDa), and has the distinction of being one of the strongest soluble reductants found in cells (midpoint redox potential = −0.42 eV). The amino acid sequence of ferredoxin and the three-dimensional structure are known in different species. Plants contain different forms of ferredoxin, all of which are encoded in the nuclear genome. In some algae and cyanobacteria, ferredoxin can be replaced by a flavoprotein. Ferredoxin operates in the stromal aqueous phase of the chloroplast, transferring electrons from PSI to a membrane associated flavoprotein, known as FNR. A 2Fe2S cluster, ligated by four cysteine residues, serves as one-electron carrier.

Once an electron reaches ferredoxin, however, the electron pathway branches, enabling redox free energy to enter other metabolic pathways in the chloroplast. For example, ferredoxin can transfer electrons to nitrite reductase, glutamate synthase, and thioredoxin reductase.

Electron Transport: The Z-Scheme

The fate of the released electrons is determined by the sequential arrangement of all the components of PSII and PSI, which are connected by a pool of plastoquinones, the cytochrome b_6f complex, and the soluble proteins cytochrome c_6 and plastocyanin cooperating in series. The electrons from PSII are finally transferred to the stromal side of PSI and used to reduce $NADP^+$ to NADPH, which is catalyzed by ferredoxin-$NADP^+$ oxidoreductase (FNR). In this process, water acts as electron donor to the oxidized P_{680} in PSII, and dioxygen (O_2) evolves as a by-product.

Photosystem II uses light energy to drive two chemical reactions: the oxidation of water and the reduction of plastoquinone. Photochemistry in PSII is initiated by charge separation between P_{680} and pheophytin, creating the redox couple $P_{680}^+/Pheo^-$. The primary charge separation reaction takes only a few picoseconds. Subsequent electron transfer steps prevent the separated charges from recombining by transferring the electron from pheophytin to a plastoquinone molecule within 200 ps. The electron on Q_A^- is then transferred to Q_B-site. As already stated, plastoquinone at the Q_B-site differs from plastoquinone at the Q_A-site in that it works as a two-electron acceptor and becomes fully reduced and protonated after two photochemical turnovers of the reaction center. The full reduction of plastoquinone at the Q_B-site requires the addition of two electrons and two protons. The reduced plastoquinone (plastoquinol, Q_BH_2) then unbinds from the reaction center and diffuses in the hydrophobic core of the membrane, after which an oxidized plastoquinone molecule finds its way to the Q_B-binding site and the process is repeated. Because the Q_B-site is near the outer aqueous phase, the protons added to plastoquinone during its reduction are taken from the outside of the membrane. Electrons are passed from Q_BH_2 to a membrane-bound cytochrome b_6f, concomitant with the release of two protons to the luminal side of the membrane. The cytochrome b_6f then transfers one electron to a mobile carrier in the thylakoid lumen, either plastocyanin or cytochrome c_6. This mobile carrier serves an electron donor to PSI reaction center, the P_{700}. Upon photon absorption by PSI a charge separation occurs with the electron fed into a bound chain of redox sites; a chlorophyll a (A_0), a quinone acceptor (A_1) and then a bound Fe–S cluster, and then two Fe–S cluster in ferredoxin, a soluble mobile carrier on the stromal side. Two ferredoxin molecules can reduce $NADP^+$ to NADPH, via the flavoprotein ferredoxin-$NADP^+$ oxidoreductase. NADPH is used as redox currency for many biosynthesis reactions such as CO_2 fixation. The energy conserved in a mole of NADPH is about 52.5 kcal/mol, whereas in an ATP hole is 7.3 kcal/mol.

The photochemical reaction triggered by P_{700} is a redox process. In its ground state, P_{700} has a redox potential of 0.45 eV and can take up an electron from a suitable donor, hence it can perform an oxidizing action. In its excited state it possesses a redox potential of more than -1.0 eV and can perform a reducing action donating an electron to an acceptor, and becoming P_{700}^+. The couple P_{700}/P_{700}^+ is thus a light-dependent redox enzyme and possesses the capability to reduce the most electron-negative redox system of the chloroplast, the ferredoxin-$NADP^+$ oxidoreductase (redox potential $= -0.42$ eV). In contrast, P_{700} in its ground state (redox potential $= 0.45$ eV) is not able to oxidize, that is, to take electrons from water that has a higher redox potential (0.82 eV). The transfer of electrons from water is driven by the P_{680} at PSII, which in its ground state has a sufficiently positive redox potential (1.22 eV) to oxidize water. On its excited state, P_{680} at PSII reaches a redox potential of about -0.60 eV that is enough to donate electron to a plastoquinone (redox potential $= 0$ eV) and then via cytochrome b_6f complex to P_{700}^+ at PSI so that it can return to P_{700} and be excited once again. This reaction pathway is called the "Z-scheme of photosynthesis," because the redox diagram from P_{680} to P_{700} looks like a big "Z" (Figure 3.4).

From this scheme it is evident that only approximately one third of the energy absorbed by the two primary electron donors P_{680} and P_{700} is turned into chemical form. A 680 nm photon has an energy of 1.82 eV, a 700 nm photon has an energy of 1.77 eV (total $= 3.59$ eV) that is three times more than sufficient to change the potential of an electron by 1.24 eV, from the redox potential of the water (0.82 eV) to that of ferredoxin-$NADP^+$ oxidoreductase (-0.42 eV).

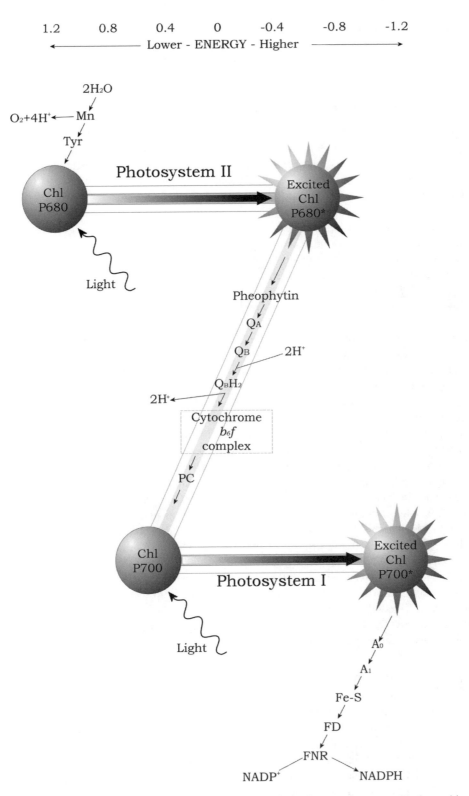

FIGURE 3.4 Schematic drawing of the Z-scheme of photosynthetic electron transport, with the positions of the participants on the oxido-reduction scale.

It is worthwhile to emphasize that any photon that is absorbed by any chlorophyll molecule is energetically equivalent to a red photon because the extra energy of an absorbed photon of shorter wavelength (<680 nm) is lost during the quick fall to the red energy level that represents the lowest excited level.

Proton Transport: Mechanism of Photosynthetic Phosphorylation

The interplay of PSI and PSII leads to the transfer of electrons from H_2O to NADPH, and the concomitant generation of proton gradient across the thylakoid membrane for ATP synthesis. The thylakoid space becomes markedly acidic with pH approaching 4. The light-induced trans-membrane proton gradient is about 3.5 pH units. The generation of these protons follows two routes:

- For the splitting of two water molecules and the release of one oxygen molecule four proton are released in the thylakoid space.
- The transport of four electrons through the cytochrome $b_6 f$ complex leads to the trans-location of eight protons from the stroma to the thylakoid space.

Therefore about 12 protons for each O_2 molecule released are translocated. The proton-motive force Δp, that is, the force created by the accumulation of hydrogen ions on one side of the thyla-koid membrane, consists of a pH gradient contribution and a membrane-potential contribution. In chloroplasts, nearly all of Δp arises from the pH gradient, whereas in the mitochondria the contri-bution from the membrane potential is larger. This difference is due to the thylakoid membrane per-meability to Cl^- and Mg^{2+}. The light-induced transfer of H^+ into the thylakoid space is accompanied by the transfer of either Cl^- in the same direction or Mg^{2+} in the opposite direction (1 per $2H^+$). Consequently, electrical neutrality is maintained and no membrane potential is gen-erated. A pH of 3.5 units across the thylakoid membrane corresponds to Δp of 0.22 V or a ΔG (change in Gibbs free energy) of -4.8 kcal/mol. The change in Gibbs free energy associated with a chemical reaction is a useful indicator of whether the reaction will proceed spontaneously. This energy is called free energy because it is the energy that will be released or freed up to do work. As the change in free energy is equal to the maximum useful work which can be accomplished by the reaction, then a negative ΔG associated with a reaction indicates that it can happen spon-taneously. About three protons flow through the F_0F_1-ATPase complex per ATP synthesized, which corresponds to a free energy input of 14.4 kcal/mol of ATP, but in which only 7.3 kcal are stored in the ATP molecule, a yield of about 50%. No ATP is synthesized if the pH gradient is less than two units because the gradient force is then too small. The newly synthesized ATP is released into the stromal space. Likewise, NADPH formed by PSI is released into the stromal space. Thus ATP and NADPH, the products of light reactions of photosynthesis, are appro-priately positioned for the subsequent light-independent reactions, in which CO_2 is converted into carbohydrates. The overall reaction can be expressed as:

$$4NADP^+ + 2H_2O + 4ADP + 4P_i \xrightarrow{\text{8 photons and 4e}^-} 4NADPH + 4ATP + O_2 \qquad (3.3)$$

This equation implies that each H_2O is split in the thylakoids under the influence of the light to give off $\frac{1}{2}O_2$ molecule, and that the two electrons so freed are then transferred to two molecules of $NADP^+$, along with H^+s, to produce the strong reducing agent NADPH. Two molecules of ATP can be simultaneously formed from two ADP and two inorganic phosphates (P_i) so that the energy is stored in high energy compounds. NADPH and ATP are the assimilatory power required to reduce CO_2 to carbohydrates in the light-independent phase. The generation of ATP following this route is termed non-cyclic phosphorylation because electrons are just transported from water to $NADP^+$ and do not come back.

An alternative pathway for ATP production is cyclic phosphorylation, in which electrons from PSI cycle in a closed system through the phosphorylation sites and ATP is the only product formed. Electron arising from P_{700} are transferred to ferredoxin and then to the cytochrome $b_6 f$ complex. Protons are pumped by this complex as electrons return to the oxidized form of reaction center P_{700} through plastocyanine. This cyclic phosphorylation takes place when NADP is unavailable to accept electrons from reduced ferredoxin because of a very high ratio of NADPH to $NADP^+$.

The electrochemical potential of the proton gradient drives the synthesis of ATP through an ATP-synthase situated, as we have seen, anisotropically in the thylakoid membrane.

Pigment Distribution in PSII and PSI Super-Complexes of Algal Division

Absorption spectra can give us information about the spectral range in which pigment molecules organized in the thylakoid membranes capture photons. Absorption spectra in the visible range, from 400 to 700 nm, have been measured *in vivo* on photosynthetic compartments (thylakoid membranes and chloroplasts) of single cells belonging to each algal division. Each spectrum represents the envelope of the real absorption data, and is coupled to the plot of the fourth-derivative absorption spectrum. This mathematical tool allows the resolution of absorption maxima relative to the different components of the pigment moiety characterizing the division, which cannot be detected in the envelope spectrum because of the overlapping of their multiple spectra. These components have been grouped and related to the following pigment classes: chlorophylls a, b, c_1, and c_2, carotenoids, cytochromes, and phycobiliproteins. Each pigment possesses its own distinctive absorption spectrum in the visible range, which have been decomposed in its Gaussian bands and relative absorption maxima. By relating the absorption maxima resolved by means of fourth-derivative analysis with the absorption maxima of the Gaussian bands of the different pigments, it is possible to both predict the presence of a specific pigment in an alga and give an unknown alga a plausible taxonomic framing.

Chlorophylls and cytochromes always show the same absorption maxima independently of the algal division, though the intensity of their absorption bands may change; phycobiliproteins, carotenoids, and xanthophylls show a variable distribution of their single components, which is characteristic of each algal division. Figure 3.5–Figure 3.8 show the absorption spectra of the photosynthetic compartments in all the different algal divisions. The absorption spectrum of the tissue of a higher plant (*Hedysarum coronarium*) is shown for highlighting the uniformity of the pigment distribution throughout the algal green lineage and plants.

It should be stressed that all the pigments other than chlorophyll a perform two main functions: protection of photosynthetic assemblies from photosensitization processes (mainly carotenoids); absorption of light at wavelengths other than those absorbed by chlorophyll a and transfer of its energy to P_{680} and P_{700}.

LIGHT-INDEPENDENT REACTIONS

We have seen that NADPH and ATP are produced in the light phase of photosynthesis. The next phase of photosynthesis involves the fixation of CO_2 into carbohydrates. Although many textbooks state that glucose ($C_6H_{12}O_6$) is the major product of photosynthesis, the actual carbohydrate end-products are those listed in Table 1.4 in Chapter 1 (sucrose, paramylon, starch, etc.). The fixation of CO_2 takes place during the light independent phase using the assimilatory power of NADPH and ATP in the chloroplast stroma (eukaryotic algae) or in the cytoplasm (prokaryotic algae). The light-independent reactions do not occur in the dark; rather they occur simultaneously with the light reactions. However, light is not directly involved. The light-independent reactions are commonly referred to as the Calvin Benson Bassham cycle (CBB cycle) after the pioneering work of its discoverers.

The first metabolite was a 3-carbon organic acid known as 3-phosphoglycerate (3-PG). For this reason, the pathway of carbon fixation in algae and most plants is referred to as C3 photosynthesis.

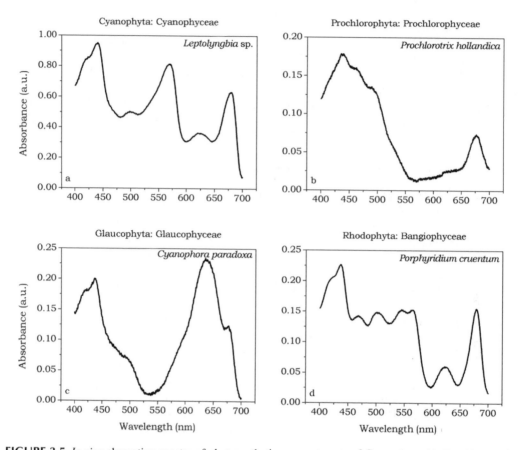

FIGURE 3.5 *In vivo* absorption spectra of photosynthetic compartments of Cyanophyta (a), Prochlorophyta (b), Glaucophyta (c), and Rhodophyta (d).

As the first product was a C3 acid, Calvin hypothesized that the CO_2 acceptor would be a C2 compound. However, no such C2 substrate was found. Rather, it was realized that the CO_2 acceptor was a C5 compound, ribulose 1,5-bisphosphate (RuBP), and that the product of carboxylation was two molecules of 3-PG. This crucial insight allowed the pathway of carbon flow to be determined.

While the CBB cycle involves a total of 13 individual enzymatic reactions, only two enzymes are unique to this pathway: ribulose-1,5-bisphosphate carboxylase/oxygenase (RuBisCO) and phosphorybulokinase (PRK). All other enzymes involved also perform functions in heterotrophic metabolism. PRK catalyzes the phosphorylation of ribulose-monophosphate to ribulose-1,5-bisphosphate (RUBP). RUBP in turn is the substrate for RuBisCO, which catalyzes the actual carbon fixation reaction.

RuBisCO

As a result, the RuBisCO enzyme alone represents the most important pathway by which inorganic carbon enters the biosphere. It has also been described as the most abundant protein on Earth. It is thought that as much as 95% of all carbon fixations by C3 organisms (that includes all phytoplankton) occur through RuBisCO.

RuBisCO is known to catalyze at least two reactions: the reductive carboxylation of ribulose 1,5-bisphosphate (RuBP) to form two molecules of 3-phosphoglycerate and the oxygenation of RuBP to form one molecule of 3-PG and one molecule of 2-phosphoglycolate.

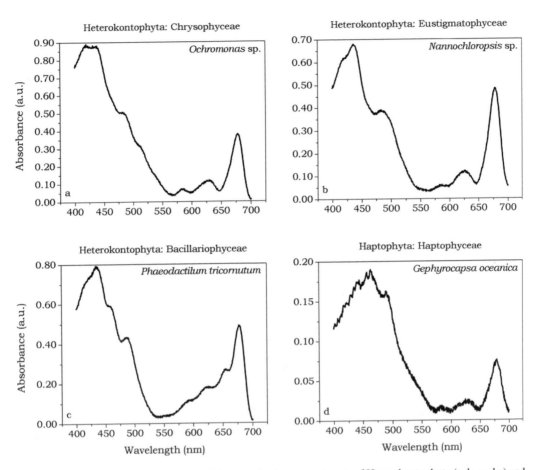

FIGURE 3.6 *In vivo* absorption spectra of photosynthetic compartments of Heterokontophyta (a, b, and c) and Haptophyta (d).

The oxygenation of RuBP is commonly referred to as photorespiration and has traditionally been seen as a wasteful process, in particular because the regeneration of RuBP in photorespiration leads to the evolution of CO_2 and requires free energy in the form of ATP. In addition RuBisCO suffers from several other inefficiencies. Both reactions (carboxylation and oxygenation) occur in the same active site and compete, making the enzyme extremely sensitive to local partial pressures of CO_2 and O_2. RuBisCO makes up 20–50% of the protein in chloroplasts. It acts very slowly, catalyzing three molecules per second. This is comparable to 1000 per second typical for enzymatic reactions. Large quantities are needed to compensate for its slow speed. It may be the most abundant protein on Earth. Lastly, RuBisCO rarely performs its function at a maximum rate (K_{max}), since the partial pressure of CO_2 in the vicinity of the enzyme is often smaller than even its Michaelis-Menten half-saturation constant (K_m). RuBisCO has been shown to occur in two distinct forms in nature termed forms I and II, respectively. Form I of the enzyme is an assemblage of eight 55 kDa large subunits (rbcL) and eight 15 kDa small subunits (rbcS). These subunits assemble into a 560 kDa hexadecameric protein-complex designated as L8S8. Most photosynthetic prokaryotes that depend on the CBB-cycle for carbon assimilation and all eukaryotic algae express a form I type RuBisCO. The exception to this rule is several marine dinoflagellates, which apparently contain a nuclear encoded form II of RuBisCO. Form II of RuBisCO is a dimer of large subunits (L2) and is otherwise found in many photosynthetic and chemoautotrophic bacteria. Phylogenetic analysis of large number of form I rbcL DNA sequences revealed the division of form I into four

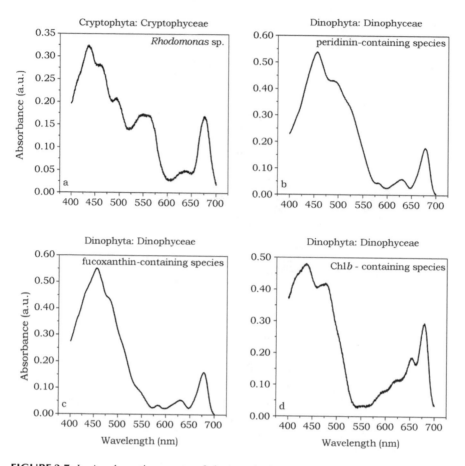

FIGURE 3.7 *In vivo* absorption spectra of photosynthetic compartments of Cryptophyta (a) and Dinophyta (b, c, and d).

major clades referred to as IA, IB, IC, and ID. Form IA is commonly found in nitrifying and sulfur oxidizing chemoautotrophic bacteria as well as some marine *Synechococcus* (marine type A) and all *Prochlorococcus* strains sequenced to date. All other cyanobacteria as well as all green algae possess a form IB type enzyme. Form IC of rbcL is expressed by some photosynthetic bacteria such as hydrogen oxidizers. Form ID encompasses a diverse group of eukaryotic lineages including essentially all chromophytic, eukaryotic algae such as Phaeophyceae, Rhodophyta, Bacillariophyceae, and Raphidophyceae.

The phylogeny of RuBisCO displays several interesting incongruencies with phylogenies derived from ribosomal DNA sequences. This has lead to the speculation that over evolutionary history numerous lateral gene transfers may have occurred, transferring RuBisCO among divergent lineages. For example the dinoflagellate *Gonyaulax polyhedra* contains a form II RuBisCO most similar to sequences found in proteobacteria. Within the form I clade as many as six lateral transfers have been suggested to explain the unusual phylogeny observed among the cyanobacteria, proteobacteria, and plastids. Some bacteria may have acquired a green-like cyanobacterial gene, while marine *Synechococcus* and *Prochlorococcus* almost certainly obtained their RuBisCO genes from a purple bacterium.

Three-dimensional structures of the RuBisCO enzyme are now known for a number of species, including *Synechococcus* and most recently the green alga *Chlamydomonas reinhardtii*. On the basis of these data and other studies it is now believed that the primary catalytic structure of

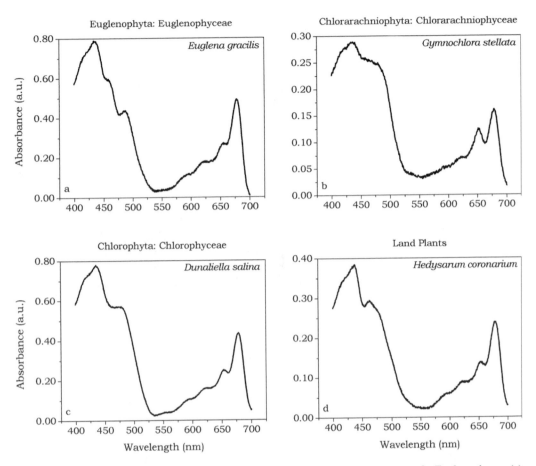

FIGURE 3.8 *In vivo* absorption spectra of photosynthetic compartments of Euglenophyta (a), Chlorarachnophyta (b), Chlorophyta (c), and Land Plants (d).

RuBisCO is a dimer of two large subunits (L2). In form I RuBisCO four L2 dimers are cemented to form L8S8 hexadecameric superstructure whereby the major contacts between the L2 dimers are mediated by the small subunits. A Mg^{2+} cofactor as well as the carbamylation of Lys201 is required for the activity of the enzyme. A loop in the beta barrel and two other elements of the large subunit, one in the N and one in the C terminus of the protein form the active site in *Synechococcus*. Small subunits apparently do not contribute to the formation of the active site.

Calvin Benson Bassham Cycle

The reactions of the Calvin cycle can be thought of as occurring in three phases (Figure 3.9):

- Carboxylation: fixation of CO_2 into a stable organic intermediate
- Reduction: reduction of this intermediate to the level of carbohydrate
- Regeneration: regeneration of the CO_2 acceptor

Carboxylation

Carboxylation involves the addition of one molecule of CO_2 to a 5-carbon "acceptor" molecule, ribulose biphosphate (RuBP). This reaction is catalyzed by the enzyme RuBisCO. Plants invest

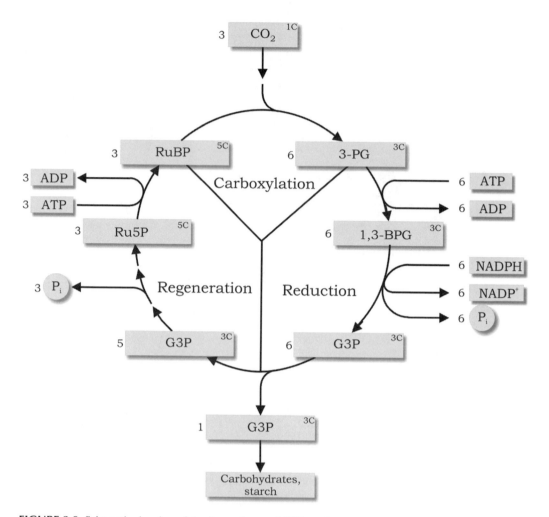

FIGURE 3.9 Schematic drawing of the three phases of CBB cycle.

a huge amount of their available nitrogen into making this one protein. As a result, RuBisCO is the most abundant protein in the biosphere. The resulting 6-carbon product splits into two identical 3-carbon products. These products are 3-phosphoglycerate, or simply 3-PG. At this point in the cycle, CO_2 has been "fixed" into an organic product but no energy has been added to the molecule.

Reduction

The second step in the Calvin cycle is the reduction of 3-PG to the level of carbohydrate. This reaction occurs in two steps: (1) phosphorylation of 3-PG by ATP to form a 1,3-biphosphoglycerate (1,3-BPG) and (2) reduction of 1,3-BPG by NADPH to form glyceraldehyde-3-phosphate (G3P), a simple 3-carbon carbohydrate, and its isomers collectively called triose phosphates. This reaction requires both ATP and NADPH, the high energy chemical intermediates formed in the light reactions.

The $NADP^+$ and ADP formed in this process return to the thylakoids to regenerate NADPH and ATP in the light reactions.

Regeneration

The final stage in the Calvin cycle is the regeneration of the CO_2 acceptor RuBP. This involves a series of reactions that convert triose phosphate first to the 5-carbon intermediate Ru5P (ribulose 5-phosphate), then phosphorylate Ru5P to regenerate RuBP (ribulose-bisphosphate). This final step requires ATP formed in the light reactions.

Overall, for every three turns of the cycle one molecule of product (triose phosphate) is formed (3CO_2:1G3P). The remaining 15 carbon atoms (5 G3P) re-enter the cycle to produce three molecules of RuBP.

The triose phosphate formed in the Calvin cycle can remain in the chloroplast where it is converted to starch. This is why chloroplasts form starch grains. Alternatively, triose phosphate can be exported from the chloroplast where it is converted to carbohydrates in the cytoplasm. Both reactions involve the release of phosphate. In the case of carbohydrates, the phosphate must be returned to the chloroplast to support continued photophosphorylation (ATP formation).

The net energy balance of six rounds of the Calvin cycle to produce one mole of hexose is thus:

$$6CO_2 + 18ATP + 12NADPH + 12H_2O \longrightarrow C_6H_{12}O_6 + 18ADP$$
$$+ 18P_i + 12NADP^+ + 6H^+ \qquad (3.4)$$

Photorespiration

Photosynthetic organisms must cope with a competing reaction that inhibits photosynthesis known as photorespiration. Unlike photosynthesis, this process involves the uptake of oxygen and the release of carbon dioxide.

Recall that mitochondrial respiration involves the uptake of O_2 and the evolution of CO_2 and is associated with the burning of cellular fuel to obtain energy in the form of ATP. In contrast, photorespiration starts in the chloroplast and wastes energy.

Photorespiration can be defined as the light-dependent uptake of O_2 in the chloroplast. It is caused by a fundamental "inefficiency" of RuBisCO.

During photosynthesis RuBisCO catalyzes the carboxylation of RuBP to give two molecules of PGA. However, it can also catalyze the oxygenation of RuBP to give one molecule of PGA and one molecule of a 2-carbon compound called phosphoglycolate. This reaction occurs because O_2 can compete with CO_2 at the active site of RuBisCO. As oxygenation of RuBP competes with carboxylation, it lowers the efficiency of photosynthesis. A significant portion (25%) of the carbon in phosphoglycolate is lost as CO_2. Algae must use energy to "recover" the remaining 75% of this carbon, which further limits the efficiency of photosynthesis.

If photorespiration lowers the yield of photosynthesis, why has such a process been maintained throughout the course of evolution? The answer to this intriguing question has to do with the origin of RuBisCO and the CBB cycle. RuBisCO is an ancient enzyme, having evolved over 2.5 billion years ago in cyanobacteria. During this period in Earth's history, the atmosphere contained high levels of CO_2 and very little oxygen. Thus, photorespiration did not present a problem for early photosynthetic organisms. By the time oxygen levels accumulated to significant levels in the atmosphere (ironically, by the process of photosynthesis!), the catalytic mechanism of RuBisCO was apparently "fixed." In other words, because both O_2 and CO_2 compete for the same active site of the enzyme, algae could not decrease the efficiency of oxygenation without also decreasing the efficiency of carboxylation. To compensate, algae evolved an elaborate pathway, known as the photorespiratory pathway, to recover at least some of the carbon that would otherwise be lost. This pathway involved biochemical reactions in the chloroplast, mitochondria, and peroxisome. The importance of photorespiration is easily demonstrated by the fact that nearly all plants grow better under high CO_2 versus low CO_2. Conditions that favor carboxylation (photosynthesis) over oxygenation (photorespiration) include high CO_2, moderate light intensities, and moderate temperatures. Conditions that favor oxygenation over carboxylation include low CO_2 levels, high temperatures, and high light intensities.

ENERGY RELATIONSHIPS IN PHOTOSYNTHESIS: THE BALANCE SHEET

The consumption of a mole of glucose releases 686 kcal of energy. This value represents the difference between the energy needed to break the bonds of the reactants (glucose and oxygen) and the energy liberated when the bonds of the products (H_2O and CO_2) form. Conversely, the photosynthesis of a mole of glucose requires the input of 686 kcal of energy. The overall equation for each process is the same; only the direction of the arrow differs:

$$C_6H_{12}O_6 + 6O_2 + < - > 6H_2O + 6CO_2 \qquad (3.5)$$

The average bond energies between common atoms are the following: C—H has 98 kcal/mol; O—H has 110 kcal/mol; C—C has 80 kcal/mol; C—O has 78 kcal/mol; H—H has 103 kcal/mol; C—N has 65 kcal/mol; O=O has 116 kcal/mol; C=C has 145 kcal/mol; C=O (as found in CO_2) has 187 kcal/mol.

The 24 covalent bonds of glucose require a total of 2182 kcal to be broken. The six double bonds of oxygen require another 696 kcal. Thus a grand total of 2878 kcal is needed to break all the bonds of the reactants in cellular respiration.

As for the products, the formation of six molecules of CO_2 involves the formation of 12 double polar covalent bonds each with a bond energy of 187 kcal; total = 2244. The formation of six molecules of H_2O involves the formation of 12 O—H bonds each with an energy of 110 kcal; total = 1320. Thus a grand total of 3564 kcal is released as all the bonds of the products form.

Subtracting this from the 2878 kcal needed to break the bonds of the reactants, we arrive at −686 kcal, the free energy change of the oxidation of glucose. This value holds true whether we oxidize glucose quickly by burning it or in the orderly process of cellular respiration in mitochondria. The minus sign indicates that free energy has been removed from the system. The details of the energy budget are just the same. The only difference is that now it takes 3564 kcal to break the bonds of the reactants and only 2878 kcal are released in forming glucose and oxygen. So we express this change in free energy (+686 kcal) with a plus sign to indicate that energy has been added to the system. The energy came from the Sun and now is stored in the form of bond energy that can power the needs of all life.

Photosynthetic reduction of CO_2 can be summarized by the equations:

$$2H_2O \rightarrow O_2 + 4H^+ + 4e^- \qquad (3.6)$$

$$CO_2 + 4H^+ + 4e^- \rightarrow (CH_2O) + H_2O \qquad (3.7)$$

Four electrons are required to be transferred from water, through a redox span of 1.24 eV, to reduce one molecule of CO_2. The energy required for the reduction of 1 mol of CO_2 is therefore 4 mol × 1.25 eV × 1.60 × 10^{-19} J eV^{-1} × 6.02 × 10^{23} mol^{-1} = 47.77 × 10^4 J (115.10 kcal). Theoretically, the energy requirement could be satisfied by the capture of 4 mol of photons of PAR light, say a red photon of 700 nm, which have an energy content of 4 mol × 2.84 × 10^{-19} J × 6.02 10^{23} mol^{-1} = 68.4 × 10^4 J (163.40 kcal). However, due to the thermodynamic losses during energy conversion, the fraction of absorbed photon energy converted into chemical energy seldom exceeds 0.35. Thus, eight moles of photons are required for the reduction of 1 mol of CO_2 (8 mol × 2.84 × 10^{-19} J × 6.02 × 10^{23} mol^{-1} = 136.82 × 10^4 J × 0.35 = 48.20 × 10^4 J (115.20 Kcal). A mole of glucose (formed by the addition of six CO_2 molecules) requires 6 × 115.20 kcal equal to 691.20 kcal. This value is calculated by taking in account the balance of the energy of bonds previously described.

SUGGESTED READING

Allen, J. F., Cyclic, pseudocyclic and noncyclic photophosphorylation: new links in the chain, *Trends in Plant Science*, 8, 15–19, 2003.

Allen, J. F. and Forsberg, J., Molecular recognition in thylakoid structure and function, *Trends in Plant Science*, 6, 317–326, 2001.

Barber, J., Photosystem II: a multisubunit membrane protein that oxidases water, *Current Opinion in Structural Biology*, 12, 523–530, 2002.

Bassi, R. and Caffarri, S., Lhc proteins and the regulation of photosynthetic light harvesting function by xanthophylls, *Photosynthesis Research*, 64, 243–256, 2000.

Bibby, T. S., Nield, J., Partensky, F., and Barber, J., Antenna ring around photosystem I, *Nature*, 413, 590, 2001.

Bibby, T. S., Nield, J., Chen, M., Larkum, A. W. D., and Barber, J., Structure of a photosystem II supercomplex isolated from *Prochloron didemni* retaining its chlorophyll a/b light-harvesting system, *Proceedings of the National Academy of Science*, 100, 9050–9054, 2003.

Brettel, K. and Leibl, W., Electron transfer in photosystem I, *Biochimica Biophysica Acta*, 1507, 100–114, 2001.

Line, M. A., The enigma of the origin of life and its timing. *Microbiology* 148; 21–27 2002. Catling, D. C., Zahnle, K. J., and McKay, C., Biogenic methane, hydrogen escape, and the irreversible oxidation of early earth, *Science*, 293, 839–843, 2001.

Chitnis, P. R., Photosystem I: function and physiology, *Annual Review of Plant Molecular Biology*, 52, 593–626, 2001.

De Martino, A., Douady, D., Quinet-Szely, M., Rousseau, B., Crepineau, F., Apt, K., and Caron, L., The light harvesting antenna of brown algae, *European Journal of Biochemistry*, 267, 5540–5549, 2000.

Falkowski, P. G. and Raven, J. A., *Aquatic Photosynthesis*, Blackwell Science, Oxford, 1997.

Ferreira, K. N., Iverson T. M., Maghlaoui, K., Barber, J., and Iwata, S., Architecture of the photosynthetic oxygen-evolving center, *Science*, 19(303), 1831–1838, 2004.

Fromme, P., Jordan, P., and Krauss, N. Structure of PSI, *Biochemical Biochimica Acta*, 1507, 5–31, 2001.

Germano, M., Yakushevska, A. E., Keegstra, W., van Gorkom, H. J., Dekker, J. P., and Boekema, E. J., Supramolecular organization of photosystem I and light-harvesting complex I in *Chlamydomonas reinhardtii*, *FEBS Letters*, 525, 121–125, 2002.

Govindjee, Chlorophyll *a* fluorescence: a bit of basics and history, in G. Papageorgiou and Govindjee, Eds. *Chlorophyll a Fluorescence: A Probe of Photosynthesis*. pp. 2–42 Kluwer Academic, Dordrecht, The Netherlands, 2004.

Govindjee and Krogmann, D. W., Discoveries in oxygenic photosynthesis (1727–2003): A perspective. *Photosynthesis Research*, 80, 15–57, 2004.

Govindjee, Allen, J. F., and Beatty, J. T., Celebrating the millennium: historical highlights of photosynthesis research, Part 3, *Photosynthesis Research*, 80, 1–13, 2004.

Hall, D. O. and Rao, K. K., *Photosynthesis*, Cambridge, U.K., Cambridge University Press, 1999.

Hankamer, B., Morris, E., Nield, J., Carne, A., and Barber, J., Subunit positioning and transmembrane helix organization in the core dimmer of photosystem II, *FEBS Letters*, 504, 142–151, 2001.

Heathcote, P., Fyfe, P. K., and Jones, M. R., Reaction centers: the structure and evolution of biological solar power, *Trends in Biochemical Sciences*, 27, 79–84, 2002.

Jordan, P., Fromme, P., Witt, H. T., Klukas, O., Saenger, W., and Krauß, N. Three-dimensional structure of cyanobacterial photosystem I at 2.5 Å resolution, *Nature*, 411, 909, 917, 2001.

Kamiya, N. and Shen, J.-R., Crystal structure of oxygen-evolving photosystem II from *Thermosynechococcus vulcanus* at 3.7 Å resolution, *Proceedings of the National Academy of Science*, 100, 98–103, 2003.

Kargul, J., Nield, J., and Barber, J., Three dimensional reconstruction of a light-harvesting complex I — photosystem I supercomplex from *Chlamydomonas reinhardtii, Journal of Biological Chemistry*, 278, 16135–16141, 2003.

Kramer, D. M., Avenson, T. J., and Edwards, G. E., Dynamic flexibility in the light reaction of photosynthesis governed by both electron and proton transfer reaction, *Trends in Plant Science*, 9, 349–357, 2004.

Kurisu, G., Zhang, H., Smith, J. L., and Cramer, W. A., Structure of the Cytochrome b6f complex of oxygenic photosynthesis: tuning the cavity, *Science*, 302, 1009–1014, 2003.

Larkum, A. W. D., Raven, J., and Douglas, S., Eds., *Photosynthesis of Algae*, Vol. 14, Govindjee, Series Ed. Kluwer Academic Publishers, Dordrecht, 2003.

Markovic, D., Energy storage in the photosynthetic electron-transport chain. An analogy with Michaelis-Menten kinetics, *Journal of Serbian Chemical Society*, 68, 615–618, 2003.

Nelson, N. and Ben-Shem, A., The complex architecture of oxygenic photosynthesis, *Nature Reviews of Molecular Cell Biology*, 5, 1–12, 2004.

Nugent, J. H. A. and Evans, M. C. W., Structure of biological solar energy converters — further revelations, *Trends in Plant Sciences*, 9, 368–370, 2004.

Pfannschmidt, T., Chloroplast redox signal: how photosynthesis controls its own genes, *Trends in Plant Sciences*, 8, 33–41, 2003.

Raven, J. A. Putting C in Phycology, *European Journal of Phycology*, 32, 319–333, 1997.

Reed, B., ATP synthase: powering the movement of life, *Harvard Science Review*, Spring, 8–10, 2002.

Saenger, W., Jordan, P., and Krauß, N., The assembly of protein subunits and cofactors in photosystem I., *Current Opinion in Structural Biology*, 12, 244–254, 2002.

Samsonoff, W. A. and MacColl, R., Biliproteins and phycobilisomes from cyanobacteria and red algae at the extremes of habitat, *Archives of Microbiology*, 176, 400–405, 2001.

Schmid, V. H. R., Potthast, S., Wiener, M., Bergauer, V., Paulsen, H., and Storf, S., Pigment binding of photosystem I in light-harvesting proteins, *Journal of Biological Chemistry*, 277, 37307–37314, 2002.

Stomp, M., Huisman, J., de Jongh, F., Veraart, A. J., Gerla, D., Rijkeboer, Ibelings, B. W., Wollenzlen, U. I. A., and Stal, L. J., Adaptive divergence in pigment composition promotes phytoplankton biodiversity, *Nature*, 432, 104–107, 2004.

Strzepek, R. F. and Harrison, P. J., Photosynthetic architecture differs in coastal and oceanic diatoms. *Nature*, 431, 689–692, 2004.

Tabita, F. R., Microbial ribulose 1,5–bisphosphate carboxylase/oxygenase: a different perspective, *Photosynthesis Research*, 60, 1–28, 1999.

Tice, M. M. and Lowe, D. R., Photosynthetic microbial mats in the 3,416-Myr-old ocean, *Nature*, 431, 549–552, 2004.

Ting, C. S., Rocap, G., King, J., and Chisholm, S. W., Cyanobacterial photosynthesis in the ocean: the origins and significance of divergent light-harvesting strategies. *Trends in Microbiology*, 10, 134–142, 2002.

Towe, K. M., Catling, D., Zahnle, K., and McKay, C., The problematic rise of archean oxygen, *Science*, 295, 1419a, 2002.

Walker, J. E. (Ed.), The mechanism of F_1F_0-ATPase, *Biochimica Biophysica Acta*, 1458, 221–510, 2000.

Wynn-Williams, D. D., Edwards, H. G. M., Newton, E. M., and Holder, J. M., Pigmentation as a survival strategy for ancient and modern photosynthetic microbes under high ultraviolet stress on planetary surfaces. *International Journal of Astrobiology*, 1, 39–49, 2002.

Zouni, A., Witt, H.-T., Kern, J., Fromme, P., Krauß, N., Saenger, W., and Orth, P., Crystal structure of photosystem II from Synechococcus elongates at 3.8 Å resolution, *Nature*, 409, 739–743, 2001.

http://www.bio.ic.ac.uk/research/barber/people/jbarber.html
http://www.bio.ic.ac.uk/research/nield/welcome.html
http://www.atpase.de/
http://www.marine.usf.edu/microbiology/regulation-rubisco.shtml
http://www.life.uiuc.edu/govindjee/

4 Biogeochemical Role of Alg

ROLES OF ALGAE IN BIOGEOCHEMISTRY

Biogeochemical cycling can be defined as the movement and exchange of both matter and energy between the four different components of the Earth, namely the atmosphere (the air envelope that surrounds the Earth), the hydrosphere (includes all the Earth's water that is found in streams, lakes, seas, soil, groundwater, and air), the lithosphere (the solid inorganic portion of the Earth, including the soil, sediments, and rock that form the crust and upper mantle, and extending about 80 km deep) and the biosphere (all the living organisms, plants and animals). The four spheres are not mutually exclusive, but overlap and intersect in a quite dynamic way. Soils contain air and exchange gases with the atmosphere, thus causing the geosphere and atmosphere to overlap; but they also contain water, so the geosphere and hydrosphere overlap. Dust from the geosphere and water from the hydrosphere occur in the atmosphere. Organisms are present in water bodies, soils, aquifers, and the atmosphere, so the biosphere overlaps with the other three spheres. Chemical elements are cyclically transferred within and among the four spheres, with the total mass of the elements in all of the spheres being conserved, though chemical transformations can change their form. The biogeochemical cycle of any element describes pathways that are commensurate with the move-ment of the biologically available form of that element throughout the biosphere (where the term biological availability is used to infer the participation of a substance in a "biological" reaction as opposed to its simple presence in biota). The most efficient cycles are often equated with a high atmospheric abundance of the element. These cycles ensure a rapid turnover of the element and have the flexibility to process the element in a number of different forms or phases (i.e., solid, liquid, gaseous). Except in a few rare but interesting situations (e.g., geothermal/tectonic systems), all biogeochemical cycles are driven directly or indirectly by the radiant energy of the sun. Energy is absorbed, converted, temporarily stored, and eventually dissipated, essentially in a one-way process (which is fundamental to all ecosystem function). In contrast to energy flow, materials undergo cyclic conversions. Through geologic time, biogeochemical cycling processes have fundamentally altered the conditions on Earth in a unidirectional manner, most crucially by decomposition of abiotically-formed organic matter on the primitive Earth by early heterotrophic forms of life, or changing the originally reducing atmosphere to an oxidized one via the evolution of oxygenic phototrophs. Contemporary biogeochemical cycles, however, tend to be cycling rather than unidirectional, leading to dynamic equilibria between various forms of cycled materials.

Like all organisms, algae provide biogeochemical as well as ecological services; that is, they function to link metabolic sequences and properties to form a continuous, self-perpetuating network of chemical element fluxes. They have played key roles in shaping Earth's biogeochem-istry and contemporary human economy, and these roles are becoming ever more significant as human impacts on ecosystems result in massive alteration of biogeochemical cycling of chemical elements. Just think about the Earth's initial atmosphere, 80% N_2, 10% CO/CO_2, 10% H_2 (by volume): no free O_2 appeared until the development of oxygenic photosynthesis by cyanobacteria, transforming the atmosphere composition in the actual 78% N_2, 21% O_2, 0.036% CO_2 and other minor gases (by volume); or the petroleum and natural gas we consume as fuels, plastics, dyes, etc. in our everyday life; these fossilized hydrocarbons are mostly formed by the deposition of organic matter consisting of the remains of several freshwater marine microalgae from the classes Eustigmatophyceae, Dinophyceae, and Chlorophyceae. These remains contain bacterially and chemically resistant, high aliphatic biopolymers (algaenans) and long-chain hydrocarbons that are selectively preserved upon sedimentation and diagenesis and make significant contribution to kerogens, a source of petroleum under appropriate geochemical condition. Moreover, we are still using the remains of calcareous microorganisms, deposited over millions of years in ancient ocean basins, for building material. Diatomaceous oozes are mined as additives for reflective paints, polishing materials, abrasives, and for insulation. The fossil organic carbon, skeletal remains, and oxygen are the cumulative remains of algae export production that has occurred uninterrupted for over 3 billions years in the upper ocean.

In total, 99.9% of the biomass of algae is accounted for by six major elements such as carbon (C), oxygen (O), hydrogen (H), nitrogen (N), sulfur (S) and phosphorus (P), plus calcium (Ca), potassium (K), sodium (Na), chlorine (Cl), magnesium (Mg), iron (Fe), and silicon (Si). The remaining elements occur chiefly as trace elements, because they are needed only in catalytic quantities.

All elements that become incorporated in organic material are eventually recycled, but on different time scales. The process of transforming organic materials back to inorganic forms of elements is generally referred to as mineralization. It takes place throughout the water column as well as on the bottom of water bodies (lakes, streams, and seas), where much of the detrital material from overlying waters eventually accumulates. Recycling of minerals may take place relatively rapidly (within a season) in the euphotic zone (i.e., the portion of water column supporting net primary production) or much more slowly (over geological time) in the case of refractory materials which sink and accumulate on the seabed. In the water column, where there is usually plenty of oxygen, decomposition of organic material takes place via oxidative degradation through the action of heterotrophic bacteria. Carbon dioxide and nutrients are returned for reutilization by the phytoplankton. Ecologically, the most important aspect of recycling in the water bodies is the rate at which growth-limiting nutrients are recycled. Among the nutrients that are in short supply, nitrate (NO_3^-), iron (bioavailable Fe), phosphate (PO_4^{3-}), and dissolved silicon [$Si(OH)_4$] are most often found in a concentration well below the half-saturation levels required for maximum phytoplankton growth. In particular, algae are important for the biogeochemical cycling of the chemical elements they uptake, assimilate, and produce, such as carbon, oxygen, nitrogen, phosphorus, silicon, and sulfur. We will now briefly consider some aspects of these elements in relation to biosynthesis and photosynthesis.

LIMITING NUTRIENTS

Among the elements required for algal growth, there are some that can become limiting. The original notion of limitation was introduced by von Liebig more than a century ago to establish a correlation between the yield of a crop and the elemental composition of the substrate required for the synthesis of that crop. Von Liebig stated that if one crop nutrient is missing or deficient, plant growth will be poor, even if the other elements are abundant. That nutrient will be defined "limiting nutrient." This concept is known as Liebig's "Law of the Minimum." Simply stated, Liebig's law means that growth is not controlled by the total of nutrients available but by the nutrient available in the smallest quantity with respect to the requirements of the plant. Liebig likens the potential of a crop to a barrel with staves of unequal length. The capacity of this barrel is limited by the length of the shortest stave and can only be increased by lengthening that stave. When that stave is lengthened, another one becomes the limiting factor.

The concentration of a nutrient will give some indication whether the nutrient is limiting, but the nutrient's supply rate or turnover time is more important in determining the magnitude or degree of limitation. For example, if the concentration of a nutrient is limiting, but the supply rate is slightly less than the uptake rate by the algae, then the algae will only be slightly nutrient-limited. Not all the algae are limited by the same nutrient, but it occurs at the species level, for example, all the diatoms are limited by silicate. Moreover, there is a considerable variation in the degree, kind, and seasonality of nutrient limitation, which is related to variations in riverine input, but also to conditions and weather in the outflow area.

In general, growth rate of a population of organisms would be proportional to the uptake rate of that one limiting factor. Nutrient-limited growth is usually modeled with a Monod (or Michaelis-Menten) equation:

$$\mu = \frac{\mu_{\text{max}}[\text{LN}]}{[\text{LN}] + K_{\text{m}}} \tag{4.1}$$

where μ is the specific growth rate of the population as a function of [LN]; [LN] is the concentration of limiting nutrient; μ_{max} is the maximum population growth rate (at "optimal" conditions) and K_{m} is the Monod coefficient, also called the half-saturation coefficient because it corresponds to the concentration at which μ is one-half of its maximum. When the concentration of limiting nutrient [LN] equals K_{m}, the population growth rate is $\mu_{\text{max}}/2$.

As [LN] increases, μ increases and so the algal population (number of cells) increases. Beyond a certain [LN], μ tends asymptotically to its maximum (μ_{max}), and the population tends to its maximum yield. If this concentration is not maintained, rapidly primary productivity returns to a level comparable to that prior to the nutrient enrichment. This productivity variation is the seasonal blooming. Normal becomes abnormal when there is a continuous over-stimulation of the system by excess supply of one or more limiting nutrients, which leads to intense and prolonged algal blooms throughout the year. The continuous nutrient supply sustains a constant maximum algal growth rate (μ_{max}). Therefore, instead of peaks of normal blooms, followed by periods when phytoplankton is less noticeable, we have a continuous primary production. When this occurs, we refer to it as eutrophication. In this process, the enhanced primary productivity triggers various physical, chemical, and biological changes in autotroph and heterotroph communities, as well as changes in processes in and on the bottom sediments and changes in the level of oxygen supply to surface water and oxygen consumption in deep waters. Eutrophication is considered to be a natural aging process for lakes and some estuaries, and it is one of the ways in which a water body (lake, rivers, and seas) transforms from a state where nutrients are scarce (oligotrophic), through a slightly richer phase (mesotrophic) to an enriched state (eutrophic).

Eutrophication can result in a series of undesirable effects. Excessive growth of planktonic algae increases the amount of organic matter settling to the bottom. This may be enhanced by changes in the species composition and functioning of the pelagic food web by stimulating the growth of small flagellates rather than larger diatoms, which leads to lower grazing by copepods and increased sedimentation. The increase in oxygen consumption in areas with stratified water masses can lead to oxygen depletion and changes in community structure or death of the benthic fauna. Bottom dwelling fish may either die or escape. Eutrophication can also promote the risk of harmful algal blooms that may cause discoloration of the water, foam formation, death of benthic fauna and wild or caged fish, or shellfish poisoning of humans. Increased growth and dominance of fast growing filamentous macroalgae in shallow sheltered areas are yet another effect of nutrient overload, which will change the coastal ecosystem, increase the risk of local oxygen depletion, and reduce biodiversity and nurseries for fish.

Human activities can greatly accelerate eutrophication by increasing the rate at which nutrients and organic substances enter aquatic ecosystems from their surrounding watersheds, for example introducing in the water bodies detergents and fertilizers very rich in phosphorus. The resultant aging, which occurs through anthropogenic activity, is termed cultural eutrophication.

Globally, nitrogen and phosphorus are the two elements that immediately limit, in a Liebig sense, the growth of photosynthetic organisms. Silicon could also become a more generally limiting nutrient, particularly for diatom growth. These nutrients are present in algal cells in a species-specific structural ratio, the so-called *Redfield ratio*, which determines the nutrient requirement of the species, and whose value depends on the conditions under which species grow and compete. Consequently, the species composition of an environment will be determined not only by nutrient availability but also by their proper relation, because changes in nutrient ratio cause shifts in phytoplankton communities and subsequent trophic linkages. Nitrogen generally limits overall productivity in the marine system. Nitrogen limitation occurs most often at higher salinities and during low flow periods. However, because marine system is in stoichiometric balance, any nutrient can become limiting. Phosphorus limitation occurs most often in freshwater system, in environments of intermediate salinities, and along the coasts during periods of high fresh water input. The occurrence of silicon limitation appears to be more spatially and temporally variable than phosphorus or nitrogen limitation, and is more prevalent in spring than summer.

In the case of phosphorus, the limitation of algal growth can be at least twofold. First, there is a limitation of nucleic acid synthesis. This limitation can be at the level of genome replication or at the level of RNA synthesis (a form of transcriptional control). The limitation can affect photosynthetic energy conversion by reducing the rate of synthesis of proteins in the photosynthetic apparatus, which is effectively a negative feedback on photosynthesis. This inhibition of protein synthesis may thus have effects on cell metabolism and oxidative stress similar to those for inhibition of protein synthesis under N limitation, except that the effect is indirect and less immediate. Secondly, a more immediate response to phosphorus limitation is on the rate of synthesis and regeneration of substrates in the Calvin-Benson cycle, thereby reducing the rate of light utilization for carbon fixation. Cells can undergo also a decrease in membrane phospholipids; moreover, the inability to produce nucleic acids under P limitation limits cell division, leading to an increased cell volume.

On a biochemical level, nitrogen limitation directly influences the supply of amino acids, which in turn limits the translation of mRNA and hence reduces the rate of protein synthesis. Under nitrogen-limited conditions also the efficiency of PSII decreases, primarily as a consequence of thermal dissipation of absorbed excitation energy in the pigment bed. This appears to be due mainly to a decrease in the number of PSII reaction centers relative to the antennae. The functional absorption cross-section of PSII increases under nitrogen-limiting conditions, while the probability of energy transfer between PSII reaction centers decreases. From a structural point of view, the reaction centers behave as if they were energetically isolated with a significant portion of the light-harvesting antenna disconnected from the photochemical processes. As nitrogen limitation leads to a reduction of growth and photosynthetic rates, it also leads to a reduction in respiratory rates. The relationship between the specific growth rate and specific respiration rate is linear with a positive intercept at zero growth, which is termed *maintenance respiration*. The molecular basis of the alterations is unclear; however, the demands for carbon skeletons and ATP, two of the major products of the respiratory pathways, are markedly reduced if protein synthesis is depressed.

The requirement for silicon for the construction of diatom frustule makes this group uniquely subject to silicate limitation. As silicic acid uptake, silica frustule formation, and the cell division cycle are all tightly linked, under silica limitation, the diatom cell cycle predominantly stops at the G_2 phase, before the completion of cell division. Thus, an inhibition of cell division linked to an inability to synthesize new cell wall material under silicon limitation can lead to an increase in the volume per cell. This increase could also be partly explained by the formation of auxospores with a larger cell diameter.

ALGAE AND THE PHOSPHORUS CYCLE

The phosphorus cycle is the simplest of the biogeochemical cycles. Phosphorus is the eleventh-most abundant mineral in the Earth's crust and does not exist in a gaseous state. Natural inorganic phosphorus deposits occur primarily as phosphates, that is, a phosphorous atom linked to four oxygen atoms, in the mineral apatite. The heavy molecule of phosphate never makes its way into the atmosphere; it is always a part of an organism, dissolved in water, or in the form of rock. Cycling processes of phosphorus are the same in both terrestrial and aquatic systems. When a rock with phosphate is exposed to water (especially water with a little acid in it), the rock is weathered out and goes into solution. Autotrophs (algae and plants) assimilate this dissolved phosphorus up and alter it to organic phosphorus using it in a variety of ways. It is an important constituent of lipid portion of cell membranes, many coenzymes, DNA, RNA, and, of course ATP. Heterotrophs obtain their phosphorus from the autotrophs they eat. When heterotrophs and autotrophs die (or when heterotrophs defecate), the phosphate may be returned to the soil or water by the decomposers. There, it can be taken up by other autotrophs and used again. This cycle will occur over and over until at last the phosphorus is lost at the bottom of the deepest parts of the ocean, where it becomes part of the sedimentary rocks forming there. If the rock is brought to the surface and weathered, this phosphorus will be released. During the natural process of weathering, the rocks gradually release the phosphorus as phosphate ions, which are soluble in water and the mineralized phosphate compounds breakdown. Phosphates PO_4^{3-} are formed from this element. Phosphates exist in three forms: orthophosphate, metaphosphate (or polyphosphate), and organically bound phosphate, each compound containing phosphorus in a different chemical arrangement. These forms of phosphate occur in living and decaying plant and animal remains, as free ions or weakly chemically bounded in aqueous systems, chemically bounded to sediments and soils, or as mineralized compounds in soil, rocks, and sediments.

Orthophosphate forms are produced by natural processes, but major man-influenced sources include: partially treated and untreated sewage, runoff from agricultural sites, and application of some lawn fertilizers. Orthophosphate is readily available to the biological community and typically found in very low concentrations in unpolluted waters. Polyforms are used for treating boiler waters and in detergents. In water, they are transformed into orthophosphate and become available for autotrophs uptake. The organic phosphate is the phosphate that is bound or tied up in autotrophs, waste solids, or other organic materials. After decomposition, this phosphate can be converted to orthophosphate.

Algae and plants are the key elements to passing on phosphates to other living organisms, but their importance in phosphorus cycle is connected mainly to the impact of this element on their growth. As already remarked, both phosphorus and nitrogen are among the nutrients that can become limiting, hence an overloading of these two elements leads to dramatic changes in the structure and functioning of an ecosystem.

Phosphorus, in the form of orthophosphate, is generally considered the main limiting nutrient in freshwater aquatic systems; that is, if all the phosphorus is used, autotroph growth will cease, no matter how much nitrogen is available. In phosphorus limited systems, excess phosphorus will trigger eutrophic condition. In these situations the natural cycle of the nutrient becomes overwhelmed by excessive inputs, which appear to cause an imbalance in the "production versus consumption" of living material (biomass) in an ecosystem. The system then reacts by producing more phytoplankton/vegetation than can be consumed by the ecosystem. This overproduction triggers the series of events determining the aging process of the water body.

Under aerobic conditions, as water plants and algae begin to grow more rapidly than normal, there is also an excess die off of the plants and algae as sunlight is blocked at lower levels. Bacteria try to decompose the organic waste, consuming the oxygen and releasing more phosphate, which is known as "recycling or internal cycling." Some of the phosphates may be precipitated as iron phosphate and stored in the sediment where it can then be released if anoxic conditions develop. In deeper environments, the phosphate may be stored in the sediments and then recycled through the natural process of lithotrophication, uplift, and erosion of rock formations. In anaerobic conditions, as conditions worsen as more phosphates and nitrates may be added to the water, all of the oxygen may be used up by bacteria in trying to decompose all of the waste. Different bacteria continue to carry on decomposition reactions; however, the products are drastically different. The carbon is converted to methane gas instead of CO_2; sulfur is converted to hydrogen sulfide gas. Some of the sulfide may be precipitated as iron sulfide. Under anaerobic conditions the iron phosphate precipitates in the sediments may be released from the sediments making the phosphate bioavailable. This is a key component of the growth and decay cycle. The water body may gradually fill with decaying and partially decomposed plant materials to make a swamp, which is the natural aging process. The problem is that this process can been significantly accelerated by man's activities.

Phosphates were once commonly used in laundry detergents, which contributed to excessive concentrations in rivers, lakes, and streams. Most detergents no longer contain phosphorus. Currently, the predominant outside sources of phosphorus are agricultural and lawn fertilizers and improperly disposed animal wastes.

ALGAE AND THE NITROGEN CYCLE

The growth of all organisms depends on the availability of mineral nutrients, and none is more important than nitrogen, which is required in large amounts as an essential component of peptides, proteins, enzymes, chlorophylls, energy-transfer molecules (ATP, ADP), genetic materials (RNA, DNA), and other cellular constituents.

Nitrogen is present in all the four different spheres of the Earth: the lithosphere contains about 98% of the global N (1.7×10^{17} tons), distributed among its different compartments (soils and sediments of the crust, mantle, and core). The core and the mantle have been estimated to contain a total of over 1.6×10^{17} tons of N. However, this N is not readily available to be cycled in the near-surface Earth environment. Some periodically enters the atmosphere and hydrosphere through volcanic eruptions, primarily as ammonia (NH_3) and nitrogen (N_2) gas. Most of the remainder ($\sim 2\%$, 4×10^{15} tons) is found in the atmosphere, where nitrogen gas (N_2) comprises more than 78% of the volume. The hydrosphere and the biosphere together contain relatively little N compared with the other spheres ($\sim 0.015\%$, 3×10^{12} tons).

Nitrogen has many chemical forms, both organic and inorganic, in the atmosphere, biosphere, hydrosphere, and lithosphere. It occurs in the gas, liquid (dissolved in water), and solid phases. N can be associated with carbon (organic species) and with elements other than carbon (inorganic species). Important inorganic species include nitrate (NO_3^-), nitrite (NO_2^-), nitric acid (HNO_3), ammonium (NH_4^+), ammonia (NH_3), the gas N_2, nitrous oxide (N_2O), nitric oxide (NO), and nitrogen dioxide (NO_2). Most organic N species in the four spheres are biomolecules, such as proteins, peptides, enzymes, and genetic materials. The presence of these many chemical forms make the N cycle more complex with respect to the cycle of other nutrients. The key processes linking the major pathways of the nitrogen cycle are the following:

- N-fixation, that is, reduction of atmospheric N_2 into ammonia NH_3
- Assimilation, that is, conversion of NO_3^- and NH_4^+ to organic nitrogen
- Mineralization or ammonification, that is, conversion of organic nitrogen to NH_4^+
- Nitrification, that is, conversion of NH_4^+ to NO_2^- and successively NO_3^-
- Denitrification, that is, conversion of NO_3^- to gaseous forms of nitrogen (NO, N_2O, N_2)

Though complex microbial relationships regulate these processes, we can assume that fixation, mineralization, nitrification, and denitrification are carried out almost exclusively by bacteria, whereas algae play a main active role only in nitrogen fixation and assimilation. Greatly simplifying the overall nitrogen cycle, and from an algal point of view, atmospheric molecular nitrogen is converted by prokaryotic algae (cyanobacteria) to compounds such as ammonia (fixation), which are in part directly converted into amino acids, proteins, and other nitrogen-containing cell constituents of the fixators, and in part excreted into the open environment. Eukaryotic algae, unable to perform fixation, incorporate fixed nitrogen, either ammonium or nitrate, into organic N compounds by assimilation. When organic matter is degraded, organic compounds are broken down into inorganic compounds such as NH_3 or NH_4^+ and CO_2 through the mineralization process. The resultant ammonium can be nitrified by aerobic chemoautotrophic bacteria that use it as electron donor in the respiration process. The cycle is completed by denitrification carried out usually by facultative anaerobic bacteria that reduce nitrate used as electron acceptor in respiration to nitrogen gas. We must stress that a realistic depiction of the N cycle would have an almost infinite number of inter-mediary steps along the circumference of a circle, connected by a spider web of internal, crisscross-ing complex connections.

It is one of nature's great ironies that though all life forms require nitrogen compounds the most abundant portion of it (98%) is buried in the rocks, therefore deep and unavailable, and the rest of nitrogen, the N_2 gas (2%), can be utilized only by very few organisms. This gas cannot be used by most organisms because the triple bond between the two nitrogen atoms makes the molecule almost inert. In order for N_2 to be used for growth this gas must be "fixed" in the forms directly accessible to most organisms, that is, ammonia and nitrate ions.

A relatively small amount of fixed nitrogen is produced by atmospheric fixation (5–8%) by means of the high temperature and pressure associated with lightning. The enormous energy of this phenomenon breaks nitrogen molecules and enables their atoms to combine with oxygen in the air forming nitrogen oxides. These molecules dissolve in rain, forming nitrates that are carried to the Earth. Another relatively small amount of fixed nitrogen is produced industrially (industrial fixation) by the Haber-Bosch process, in which atmospheric nitrogen and hydrogen (usually derived from natural gas or petroleum) can be combined to form ammonia (NH_3) using an iron-based catalyst, very high pressures and a temperature of about 600°C. Ammonia can be used directly as fertilizer, but most of it is further processed to urea ($(NH_2)_2CO$) and ammonium nitrate (NH_4NO_3).

The major conversion of N_2 into ammonium, and then into proteins, is a biotic process achieved by microorganisms, which represents one of the most metabolically expensive processes in biology. Biological nitrogen fixation can be represented by the following equation, in which two moles of ammonia (but in solution ammonia exists only as ammonium ion, that is, $NH_3 + H_2O \leftrightarrow NH_4^+ + OH^-$) are produced from one mole of nitrogen gas, at the expense of 16 moles of ATP and a supply of electrons and protons (hydrogen ions):

$$N_2 + 8H^+ + 8e^- + 16ATP \longrightarrow 2NH_3 + H_2 + 16ADP + 16P_i \qquad (4.2)$$

All known nitrogen-fixing organisms (diazotrophs) are prokaryotes, and the ability to fix nitrogen is widely, though paraphyletically, distributed across both the bacterial and archaeal domains. In cyanobacteria nitrogen-fixation is an inducible process, triggered by low environmental levels of fixed nitrogen.

The capacity of nitrogen fixation in diazotrophs relies solely upon an ATP-hydrolyzing, redox active enzyme complex termed nitrogenase. In many of these organisms nitrogenase comprises about 10% of total cellular proteins and consists of two highly conserved components, an iron protein (Fe-protein) and a molybdenum-iron (MoFe-protein). The Fe-protein is a γ_2 homodimer composed of a single Fe_4S_4 cluster bound between identical 32–40 kDa subunits. The Fe_4S_4 cluster is redox-active and is similar to those found in small molecular weight electron carrier proteins such as ferredoxins. It is the only known active agent capable of obtaining more than two oxidative states and transfers electrons to the MoFe-protein. The MoFe-protein is a $\alpha_2\beta_2$ heterotetramer; the ensemble is approximately 250 kDa. The α subunit contains the active site for dinitrogen reduction, typically a $MoFe_7S_9$ metal cluster (termed FeMo-cofactor), although some organisms contain nitrogenases wherein Mo is replaced by either Fe or V. These so-called alternative nitrogenases are found only in a limited subset of diazotrophs and, in all cases studied so far, are present secondary to the MoFe-nitrogenase. The MoFe-nitrogenase has been found to be more specific for and more efficient in binding N_2 and reducing it to ammonia than either of the alternative nitrogenases. The catalytic efficiency of these alternative nitrogenases is lower than that of the MoFe-nitrogenase. In addition to variations in metal cofactors, the nitro-genase complex is non-specific and reduces triple and double bond molecules other than N_2. These include hydrogen azide, nitrous oxide, acetylene, and hydrogen cyanide. The non-specificity of this enzyme and the alternative nitrogenases containing other metal co-factors implicate the role of varying environmental pressures on the evolutionary history of nitrogenase that could have selected for different functions of the ancestral enzyme. The nitrogenase enzyme system is extremely O_2 sensitive, because oxygen not only affects the protein structure but also inhibits the synthesis of nitrogenase in many diazothrops. The repression is both transient (lasting only a few hours) and permanent. The reactions occur while N_2 is bound to the nitrogenase enzyme complex. The Fe-protein is first reduced by electrons donated by ferredoxin. Then the reduced Fe-protein binds ATP and reduces the MoFe-protein, which donates electrons to N_2, producing diimide (HN=NH). In two further cycles of this process (each requiring electrons donated by ferredoxin) HN=NH is reduced to imide (H_2N—NH_2), and this in turn is reduced to $2NH_3$. Depending on the type of microorganism, the reduced ferredoxin that supplies electrons for this process is generated by fermentation or photosynthesis and respiration.

As already stated, nitrogenase is highly sensitive to molecular oxygen (*in vitro* it is irreversibly inhibited by exposure to O_2). Therefore, during the course of planetary evolution, cyanobacteria have co-evolved with the changing oxidation state of the ocean and atmosphere to accommodate the machinery of oxygenic photosynthesis and oxygen-sensitive N_2 fixation within the same cell or colony of cells. As nitrogen fixation occurs in a varied metabolic context in both anaerobic and aerobic environments, strategies have followed a very complex pattern of biochemical and physiological mechanisms for segregation that can be simplified in some spatial and/or temporal separation of the two pathways.

Nitrogenase is an ancient enzyme that almost certainly arose in the Archean ocean before the oxidation of the atmosphere by oxygenic photoautotrophs. An attractive hypothesis of the development of biological nitrogen fixation is that it arouse in response to changes in atmospheric composition that resulted in the reduction in the production of abiotically fixed nitrogen. On the early Earth, concentrations of CO_2 in the atmosphere were high, because of the oxidation of CO produced by impacts of extraterrestrial bodies and only slow removal of CO_2 by weathering (the continents were smaller at this time, meaning that a smaller area of minerals was exposed for weathering). With these CO_2 conditions, the initial production rate of NO was estimated to be about 3×10^{11} g yr^{-1}. Atmospheric CO_2 levels declined with time, however, as the impact rate dropped and the continents grew. A rise in atmospheric CH_4 produced by methanogenic, methane-generating, bacteria may have warmed the Archaean Earth and speeded the removal of CO_2 by silicate weathering. As this happened, the production rate of NO by lightning dropped to below 3×10^9 g yr^{-1} because of the reduced availability of oxygen atoms from the splitting of CO_2 and H_2O. The resulting crisis in the availability of fixed nitrogen for organisms triggered the evolution of biological nitrogen fixation about 2.2 billion years ago. Under the prevailing anaerobic conditions of that period in Earth's history anaerobic heterotrophs, such as *Clostridium*, developed. With the evolution of cyanobacteria and the subsequent generation of molecular oxygen, oxygen-protective mechanisms would be essential. A semitemporal separation of nitrogen fixation and oxygenic photosynthesis combined with spatial heterogeneity was the first oxygen-protective mechanism developed by marine cyanobacteria such as *Trichodesmium* sp. and *Katagnymene* sp. A full temporal separation, in which nitrogen is only fixed at night, then developed in unicellular cyanobateria diazotrophs and in some non-heterocystous filamentous diazotrophs (e.g., *Oscillatoria limosa* and *Plectonema boryanum*). Finally, in yet other filamentous organisms, complete segregation of N_2 fixation and photosynthesis was achieved with the cellular differentiation and evolution of heterocystous cyanobacteria (e.g., *Nostoc* and *Anabaena*).

The non-heterocystous filamentous cyanobacteria *Trichodesmium* sp. and *Katagnymene* sp., unlike all other non-heterocystous species fix nitrogen only during the day. Nitrogenase is compartimentalized in 15–20% of the cells in *Trichodesmium* sp., and 7% of the cells in *Katagnymene* sp. often arranged consecutively along the trichome, but active photosynthetic components (PSI, PSII, RuBisCo, and carboxysomes) are found in all cells, even those harboring nitrogenase. A combined spatial and temporal segregation of nitrogen fixation from photosynthesis, and a sequential progression of photosynthesis, respiration, and nitrogen fixation over a diel cycle are the strategies used by these cyanobacteria. These pathways are entrained in a circadian pattern that is ultimately controlled by the requirement for an anaerobic environment around nitrogenase. Light initiates photosynthesis, providing energy and reductants for carbohydrate synthesis and storage, stimulating electron cycling through PSI, and poising the plastoquinone (PQ) pool at reduced levels. High respiration rates early in the photoperiod supply carbon skeletons for amino acid synthesis (the primary sink for fixed nitrogen) but simultaneously reduce the PQ pool further. Linear electron flow to PSI is never abolished. The reduced PQ pool leads to a downregulation of PSII, which opens a window for N_2 fixation during the photoperiod, when oxygen consumption exceeds oxygen production. As the carbohydrate pool is consumed, respiratory electron flow through the PQ pool diminishes, intracellular oxygen concentrations rise, the PQ pool becomes increasingly oxidized, and net oxygenic production exceeds consumption. Nitrogenase activity is lost until the following day.

A full temporal separation between oxygenic photosynthesis and nitrogen fixation occurs in *P. boryanum* and *O. limosa*. Transcription of nitrogenase and photosynthetic genes are temporally separated within the photoperiod, that is, nitrogenase is expressed primarily during the night. Nitrogenase is contained in all cells in equal amounts. The onset of nitrogen fixation is preceded by a depression in photosynthesis that establishes a sufficiently low level of dissolved oxygen in the environment. *Plectonema* sp. has a versatile physiology that allows it to reversibly modulate uncoupling of the activity of the two photosystems in response to intracellular nitrogen status. *Oscillatoria* sp. initiates nitrogen fixation in the dark and performs it primarily in the absence of light.

In non-heterocystous cyanobacteria such as the filamentous *Symploca* sp. and *Lyngbya maiuscola*, and the unicellular *Gloeothece* sp. and *Cyanothece* sp., the temporal separation does not need a microaerobic environment. *Phormidium* sp. and *Pseudoanabaena* sp. are other examples of cyanobacteria fixing only under microanaerobic conditions. Under contemporary oxygen levels, all of these organisms are relegated to narrow environmental niches.

In heterocystous cyanobacteria, such as *Anabaena* sp. and *Nostoc* sp., a highly refined specialization spatially separates oxygenic photosynthesis from N_2 fixation. Here, nitrogenase is confined to a microaerobic cell, the heterocyst, characterized by a thick membrane that slows the diffusion of O_2, high PSI activity, loss of division capacity, absence of PSII (that splits the water forming O_2). This cell differentiates completely and irreversibly 12–20 h after combined nitrogen sources are removed from the medium. The development of these cells, formed at intervals between vegetative cells, is a primitive form of cell differentiation. In this process, all PSII activities are gradually lost, and the proteins involved in oxygenic photosynthesis are degraded, whereas PSI activity is maintained. Simultaneously, the production of active nitrogenase is triggered. Nitrogen fixation is localized specifically in heterocysts, and light is used for cyclic electron flow around PSI to maintain a supply of ATP for the process. The primary organic nitrogen product (glutamate) is exported to adjacent vegetative and photooxygenic cells, while carbon skeletons, formed by the respiratory and photosynthetic processes in the latter cells, are translocated to the heterocysts. In some heterocystous cyanobacteria such as *Anabaena variabilis*, under anaerobic conditions, a different Mo-dependent nitrogenase can be synthesized inside vegetative cells. This nitrogenase expresses shortly after nitrogen depletion, but prior to heterocysts formation, and can support the fixed N needs of the filaments independent of the nitrogenase in the heterocysts.

The biotic nitrogen fixation is estimated to produce about 1.7×10^8 tons of ammonia per year, whereas atmospheric and industrial nitrogen fixation produce about 1.7×10^7 tons of ammonia per year.

For all eukaryotic algae, the only forms of inorganic nitrogen that are directly assimilable are nitrate (NO_3^-), nitrite (NO_2^-), and ammonium (NH_4^+). The more highly oxidized form, nitrate, is the most thermodynamically stable form in oxidized aquatic environments, and hence is the predominant form of fixed nitrogen in aquatic ecosystems, though not necessarily the most readily available form. Following translocation across the plasmalemma (which is an energy-dependent process), the assimilation of NO_3^- requires chemical reduction to NH_4^+. This process is mediated by two enzymes, namely, nitrate reductase and nitrite reductase. Nitrate reductase is located in the cytosol and uses NADPH to catalyze the two-electron transfer:

$$NO_3^- + 2e^- + 2H^+ \longrightarrow NO_2^- + H_2O \tag{4.3}$$

In cyanobacteria nitrate reductase is coupled to the oxidation of ferredoxin rather than a pyridine nucleotide as in eukaryotic algae. The nitrite formed by nitrate reductase is reduced in a six-electron transfer reaction:

$$NO_2^- + 6e^- + 8H^+ \longrightarrow NH_4^+ + 2H_2O \tag{4.4}$$

Nitrite reductase utilizes ferredoxin in both cyanobacteria and eukaryotic algae; in the latter, the enzyme is localized in the chloroplasts. In both cyanobacteria and eukaryotic algae photosynthetic electron flow is an important source of reduced ferredoxin for nitrite reduction. The overall stoichiometry for the reduction of nitrate to ammonium can be written as:

$$NO_3^- + 8e^- + 10H^+ \longrightarrow NH_4^+ + 3H_2O \tag{4.5}$$

The incorporation of ammonium into ammino acids is primarily brought about by the sequential action of glutamine synthetase (GS) and glutamine 2-oxoglutarate aminotransferase (GOGAT). Ammonium assimilation by GS requires glutamate as substrate and ATP, and catalyzes the irreversible reaction:

$$\text{Glutamate} + NH_4^+ + \text{ATP} \longrightarrow \text{Glutamine} + \text{ADP} + P_i \tag{4.6}$$

The amino nitrogen of glutamine is subsequently transferred to 2-oxoglutarate, and reduced, forming two moles of glutamate:

$$\text{2-Oxoglutarate} + \text{Glutamine} + \text{NADPH} \longrightarrow 2[\text{Glutamate}] + \text{NADP}^+$$

Both GS and GOGAT are found in chloroplasts, although isoenzymes (multiple forms of an enzyme with the same substrate specificity, but genetic differences in their primary structures) of both enzymes may also be localized in the cytosol. Whatever the location of the enzymes, however, glutamate must be exported from the chloroplast to the cytosol where transamination reaction (the reversible transfer of an amino group of a specific amino acid to a specific keto acid, forming a new keto acid and a new amino acid) can proceed, thereby facilitating the syntheses of other amino acids.

ALGAE AND THE SILICON CYCLE

The biogeochemical cycle of silicon (Si) might be interpreted as those processes that link sources and sinks of silicic acid [$Si(OH)_4$]. Silicic acid is the only precursor in the processing and deposition of silicon in biota. The biogeochemical cycle of silicon does not facilitate a high biospheric abundance of the element, in fact silicon cycle differs from the cycles of carbon, nitrogen, and sulfur and it is similar to phosphorus in that there is no atmospheric reservoir. The silicon cycle, like those for phosphorus and the divalent metals calcium and magnesium, has a significant abiotic drain. It actually consists of two parts: the terrestrial or freshwater cycle and the marine cycle, the former feeding the latter. However, its replenishment can only occur via the marine sedimentary cycle. This is dependent on geotectonic processes, such as mountain building and subduction, and, as such, will incur delays of tens to hundreds of millions years before marine silicon is returned to the terrestrial environment.

The substantial losses of biospheric silicic acid to abiotic sinks may be compensated for in nature by its overall abundance in the Earth's crust. It is the second most abundant element in the lithosphere (28%), the iron being the first one (35%). It is found in the Earth's crust in silicate minerals; the most prevalent of which are quartz, the alkali feldspars, and plagioclase. The latter two minerals are aluminosilicates and contribute significantly to the aluminium content of the crust. All of these minerals are broken down by the process of weathering. Important feedbacks exist between autotrophs (algae and plants), weathering, and CO_2.

The dominant form of weathering is the carbonation reaction involving carbonic acid (H_2CO_3), which results in enhanced removal of CO_2 from the atmosphere, because the net effect of silicate mineral weathering is to convert soil carbon, derived ultimately from photosynthesis, into dissolved HCO_3^-. On a geological timescale, this transfer is an important control on the CO_2 content of the atmosphere and hence the global climate. Weathering is a complex function of rainfall, runoff, lithology, temperature, topography, vegetation, and magnitudes. Algae, plants, and their associate microbiota directly affect silicate mineral weathering in several ways: by the generation of organic substances, known as chelates, that have the ability to decompose minerals and rocks by the removal of metallic cations; by modifying pH through the production of CO_2 or organic acids such as acetic, citric, phenolic, etc., and by altering the physical properties of the soil, particularly the exposed surface areas of minerals and the residence time of water. The significance of this natural process for biota can be found in the detailed geochemistry of the weathering reactions and, in particular, in the rates at which these reactions occur. The rate of mineral weathering is dependent on a number of factors including the temperature, pH, ionic composition of the solvent (or leachate), and hydrogeological parameters such as water flow.

Silicification occurs in three clades of photosynthetic heterokonts: Chrysophyceae (Parmales), Bacillariophyceae, and Dictyochophyceae, with diatoms being the world's largest contributors to biosilicification. Because amorphous silica is an essential component of the diatom cell wall, silicon availability is a key factor in the regulation of diatom growth in nature; in turn, the use of silicon by diatoms dominates the biogeochemical cycling of silicon in the sea, with each atom of silicon weathered from land passing through a diatom on an average of 39 times before burial in the sea bed.

Several thousand million years ago little if any of the life on Earth was involved in the processing of silicic acid to amorphous silica ($SiO_2 * nH_2O$). The concentration of silicic acid in the aqueous environment was high, of the order of millimolar, and reflected equilibration according to the dominant mineral weathering reactions at that time. The prevalence of these environments rich in silicic acid is indicated in the fossil record by evidence of blue-green algae found encased in silica cherts. It is important to recognize that implicit in this observation is the acceptance that early biochemical evolution proceeded within environments that, relative to the conditions which prevail today, were extremely rich in silicic acid. Concomitant with the advent of dioxygen, and its subsequent gradual increase in atmospheric concentration from approximately 1% towards the level of 21% which is characteristic of today, an increasing number of organisms occurred within which silicic acid was processed to silica. The most important of these, in the terms of their diversity, ubiquity (both freshwater and marine species) and biomass, were the diatoms. The diatoms are characterized by a silica frustule that surrounds their cell wall. Silicic acid is freely diffusible across the cell walls and membranes and, in most cell types of most organisms, the intracellular concentration of silicic acid equilibrates with the extracellular environment according to a *Donnan equilibrium* (the equilibrium characterized by an unequal distribution of diffusible ions between two ionic solutions separated by a membrane, which is impermeable to at least one of the ionic species present). However, while the intracellular concentration of the silicic acid in the diatom has not been measured it is likely that it is under kinetic as opposed to thermodynamic control and that it is maintained at an extremely low level, probably less than $1~\mu mol~dm^{-3}$. This kinetic control of the intracellular silicic acid concentration may be achieved through the condensation and polymerization of silicic acid in a number of chemical (e.g., pH controlled) and physical (e.g., membrane-bound) compartments eventually resulting in amorphous silica. This biogenic silica is then deposited in a controlled manner to form the intricate and elaborate silica frustules. How all of these remarkable feats of chemistry are achieved within the diatom remains largely unknown. However, what is known, and is becoming more apparent, is the formative role played by diatoms and other silica-forming organisms, such as silicoflagellates, radiolarian, and sponges, in the biogeochemical cycle of silicon.

The reactions of condensation of silicic acid and subsequent polymerization to form biogenic silica eventually (i.e., upon the death of the organism) result in a net loss of silicic acid to the biosphere. The rate of the forward reaction (condensation and polymerization) is several orders of magnitude higher than that of the reverse (regeneration of silicic acid pool) with the result that concomitant with the rise of the diatoms and other silica-forming organisms was a significant reduction in the environmental silicic acid concentration. Silica frustules are formed in a matter of hours to days whereas the rate at which silicic acid is returned to the biosphere through the dissolution of the frustules of dead diatoms as they sink in the water column is of the order of $10^{-9}\mu mol~m^{-2}~sec^{-1}$. In addition, the dissolution of these sinking frustules can be greatly influenced by the chemistry of the water column. For example, the frustule is a highly adsorptive surface and is implicated in the removal of metal ions, for example, aluminium, from the water column. These adsorptive processes tend to stabilize the frustule surface towards dissolution and thereby reduce the amount of silicic acid returned to the biosphere during sedimentation. Once the silica frustules have settled to the bottom their silica enters the sedimentary cycle whereupon it is unlikely to reappear in the biosphere for tens of millions of years.

The biologically induced dramatic decline in the environmental silicic acid concentration had the effect of accelerating the rate of mineral weathering. This, in turn, consumed more carbon dioxide and precipitated a gradual reduction in the atmospheric concentration of this "greenhouse" gas. The impact of the emergence of diatoms and other silica-forming organisms, and latterly the spread of rooted vascular plants, on the biogeochemical cycle of silicon contributed significantly to the global cooling, which has resulted in the climate of today. The diatoms, in particular, are extremely successful organisms and will continue to deposit silica frustules of varying silica content at micromolar concentrations of environmental silicic acid. In this way they are a continuous accelerant of mineral weathering almost regardless of how low the environmental silicic acid concentration may fall. From the advent of the silica-forming organisms the process of biochemical evolution has continued in silicon-replete, though no longer of millimolar concentration, environments. Diatoms in sedimentary deposits of marine and continental, especially lacustrine, origin belong to different geologic ranges and physiographic environments. Marine diatoms range in age from Early Cretaceous to Holocene, and continental diatoms range in age from Eocene to Holocene; however, most commercial diatomites, both marine and lacustrine, were deposited during the Miocene. Marine deposits of commercial value generally accumulated along continental margins with submerged coastal basins and shelves where wind-driven boundary currents provided the nutrient-rich upwelling conditions capable of supporting a productive diatom habitat. Commercial freshwater diatomite deposits occur in volcanic terrains associated with events that formed sediment-starved drainage basins. Marine habitats generally are characterized by stable conditions of temperature, salinity, pH, nutrients, and water currents, in contrast to lacustrine habitats, which are characterized by wide variations in these conditions. Marine deposits generally are of higher quality and contain larger resources, owing to their greater areal extent and thickness, whereas most of the world's known diatomites are of lacustrine origin.

Unlike many other algae, whose division cycles are strongly coupled to the diel light cycle, diatoms are capable of dividing at any point of the diel cycle. This light independence extends to their nutrient requirements, with nitrate and silicic acid uptake and storage continuing during the night through the use of excess organic carbon synthesized during the day. Moreover, the silica frustule has recently been shown to play a role in CO_2 acquisition, which indicates that Si limitation can induce CO_2 limitation in diatoms. Specifically, the silica frustule facilitates the enzymatic conversion of bicarbonate to CO_2 at the cell surface by serving as a pH buffer thus enabling more efficient photosynthesis. This control by diatoms and the reduction in silicate input from rivers due to cooling and drying of the climate offers a feedback mechanism between climate variability, diatom productivity, and CO_2 exchange. These features of diatom physiology almost certainly contribute to the *in situ* observation that diatoms have greater maximum growth rates relative to comparable algae. Further, so long as silicic acid is abundant (and other nutrients non-limiting), diatoms are found to dominate algal communities. Diatoms are estimated to contribute up to 45% of total oceanic primary production, making them major players in the cycling of all biological elements. They globally uptake and process 240 Tmol Si yr^{-1}. Currently, risk to diatoms comes from both climate forced impacts and anthropogenic sources. Reduced input of silicate due to damming of rivers and changes in water use patterns, and increased input of inhibitory levels of ammonium to estuaries and adjacent coastal waters affect diatom success. The high ammonium concentrations prevalent in some estuaries, a result of anthropogenic inputs from sewage treatment plants and agricultural runoff inhibit the uptake of nitrate by diatoms which draw primarily on nitrate for high growth rates.

Aside from their role in the silicon cycle, the diatoms have also attracted attention because of their importance to the export of primary production to the ocean's interior. Aggregation and sinking is an important aspect in the life history of many diatom species, and high sinking velocities, whether as individuals, aggregates, or mats, allow diatoms to rapidly transport material out of the surface mixed layer. Additionally, mesozooplankton grazers that consume diatoms produce large, fast-sinking faecal pellets. These processes remove nutrients and carbon from the productive surface waters before they can be remineralized, making the diatoms crucial to "new" (or export) production. So long as silicic acid is available, diatoms act as a conduit for nutrients and carbon to deep waters, contrasting with the production of other algae, which "traps" nutrients in a regeneration loop at the surface.

Though diatoms are by far the most important organisms that take up silicic acid to form their encasing structures, silicoflagellates also deserve mentioning. These unicellular heterokont algae belong to a small group of siliceous marine phytoplankton. Silicoflagellates live in the upper part of the water column, and are adapted for life in tropical, temperate, and frigid waters. Silicoflagellates have a multi-stage life-cycle, not all stages of which are known. The best-known stage consists of a naked cell body with a single anterior flagellum and numerous plastids contained within an external lateral skeleton. This skeleton is composed of hollow beams of amorphous silica, forming a network of bars and spikes arranged to form an internal basket. The siliceous skeleton of silicoflagellates is very susceptible to dissolution, and therefore their preservation is often hindered by diagenetic processes; moreover, their abundance is relatively low compared with that of other siliceous microfossils, because they form a small component of marine sediments; both these reasons make their presence rare in the sedimentary record.

ALGAE AND THE SULFUR CYCLE

Sulfur is an essential element for autotrophs and heterotrophs. In its reduced oxidation state, the nutrient sulfur plays an important part in the structure and function of proteins. Three amino acids found in almost all proteins (cysteine, cystine, and methionine) contain carbon-bounded sulfur. Sulfur is also found in sulfolipids, some vitamins, sulfate esters, and a variety of other compounds.

In its fully oxidized state, sulfur exists as sulfate and is the major cause of acidity in both natural and polluted rainwater. This link to acidity makes sulfur important to geochemical, atmospheric, and biological processes such as the natural weathering of rocks, acid precipitation, and rates of denitrification. Sulfur cycle is also one of the main elemental cycles most heavily perturbed by human activity. Estimates suggest that emissions of sulfur to the atmosphere from human activity are at least equal or probably larger in magnitude than those from natural processes. Like nitrogen, sulfur can exist in many forms: as gases or sulfuric acid particles. The lifetime of most sulfur compounds in the air is relatively short (e.g., days). Superimposed on these fast cycles of sulfur are the extremely slow sedimentary-cycle processes or erosion, sedimentation, and uplift of rocks containing sulfur. Sulfur compounds from volcanoes are intermittently injected into the atmosphere, and a continual stream of these compounds is produced from industrial activities. These compounds mix with water vapor and form sulfuric acid smog. In addition to contributing to acid rain, the sulfuric acid droplets of smog form a haze layer that reflects solar radiation and can cause a cooling of the Earth's surface. While many questions remain concerning specifics, the sulfur cycle in general, and acid rain and smog issues in particular are becoming major physical, biological, and social problems.

The sulfur cycle can be thought of as beginning with the gas sulfur dioxide (SO_2) or the particles of sulfate (SO_4^{2-}) compounds in the air. These compounds either fall out or are rained out of the atmosphere. Algae and plants take up some forms of these compounds and incorporate them into their tissues. Then, as with nitrogen, these organic sulfur compounds are returned to the land or water after the algae and plants die or are consumed by heterotrophs. Bacteria are important here as well because they can transform the organic sulfur to hydrogen sulfide gas (H_2S). In the oceans, certain phytoplankton can produce a chemical that transforms organic sulfur to SO_2 that resides in the atmosphere. These gases can re-enter the atmosphere, water, and soil, and continue the cycle.

All living organisms require S as a minor nutrient, in roughly the same atom proportion as phosphorus. Sulfur is present in freshwater algae at a ratio of about 1 S atom to 100 C atoms (0.15–1.96% by dry weight), and the S content varies with species, environmental conditions, and season. Vascular plants, algae, and bacteria (except some anaerobes that require S^{2-}) have the ability to take up, reduce, and assimilate SO_4^{2-} into amino acids and convert SO_4^{2-} into ester sulfate compounds.

Reduced volatile sulfur compounds, which are released to the oxygen-rich atmosphere, are chemically oxidized during their atmospheric lifetime and end up finally as sulfur dioxide (oxidation state +4), sulfuric acid, particulate sulfate (oxidation state +6), and methane sulfonate (oxidation state +6). It is mainly these compounds that are removed from the atmosphere and brought back to the Earth by dry and wet deposition.

As the oxidation state of sulfur in sulfuric acid (oxidation state +6) is the most stable under oxic conditions, sulfate is the predominant form of sulfur in oxic waters and soils. Thus, the reduction of sulfate to a more reduced sulfur species is a necessary prerequisite for the formation of volatile sulfur compounds and their emission to the atmosphere. Biochemical processes which lead to this reduction can be considered as the driving force of the atmospheric sulfur cycle.

Two types of biochemical pathways of sulfate reduction are important in the global cycles: dissimilatory and assimilatory sulfate reduction. Dissimilatory reduction of sulfate is a strictly anaerobic process that takes place only in anoxic environments. Sulfate-reducing bacteria reduce sulfate and other sulfur oxides to support respiratory metabolism, using sulfate as a terminal electron acceptor instead of molecular oxygen. As the process is strictly anaerobic, dissimilatory sulfate reduction occurs largely in stratified, anoxic water basins and in sediments of wetlands, lakes, and coastal marine ecosystems. The process is particularly important in marine ecosystems, including salt marshes, because sulfate is easily available due to its high concentration in seawater (28 mM; 900 mg l^{-1} S).

In contrast to animals, which are dependent on organosulfur compounds in their food to supply their sulfur requirement, other biota (bacteria, cyanobacteria, fungi, eukaryotic algae, and vascular plants) can obtain sulfur from assimilatory sulfate reduction for synthesis of organosulfur compounds. Sulfate is assimilated from the environment, reduced inside the cell, and fixed into sulfur-containing amino acids and other organic compounds. The process is ubiquitous in both oxic and anoxic environments. Most of the reduced sulfur is fixed by the intracellular assimilation process and only a minor fraction of the reduced sulfur is released as volatile gaseous compounds, as long as the organisms are alive. However, after the death of organisms, microbial degradation liberates reduced sulfur compounds (mainly in the form of hydrogen sulfide [H_2S] and dimethyl sulfide [DMS], but also as organic sulfides) to the environment. During this stage, volatile sulfur compounds may escape to the atmosphere. However, as with sulfides formed from dissimilatory sulfate reduction, the sulfides released during decomposition are chemically unstable in an oxic environment and are reoxidized to sulfate by a variety of microorganisms.

Sulfate is taken up into the cell by an active transport mechanism, and inserted into an energetically activated molecule, APS (adenosine-5′-phosphosulfate), which can be further activated at the expense of one more ATP molecule to PAPS (3′-phosphoadenosine-5′-phosphosulphate). It is then transferred to a thiol carrier (a molecule with a -SH group) and reduced to the -2 oxidation state. In contrast to nitrate assimilation, where the various intermediates are present free in the cytoplasm, sulfur remains attached to a carrier during the reduction sequence. In a final step, the carrier-bound sulfide reacts with O-acetyl-serine to form cysteine. Cysteine serves as the starting compound for the biosynthesis of all other sulfur metabolites, especially the other sulfur-containing amino acids homocysteine and methionine. Cysteine and methionine are the major sulfur amino acids and represent a very large fraction of the sulfur content of biological materials.

One aspect of the sulfur metabolism of algae deserves special mention because of its atmospheric consequences: many types of marine algae including planktonic algae, such as prymnesiophytes, dinophytes, diatoms, chrysophytes, and prasinophytes, and macroalgae, such as chlorophytes and rhodophytes, produce large amount of dimethylsulfonium propionate (DMSP) from sulfur-containing amino acids (methionine). DMSP is the precursor of DMS, its enzymatic cleavage product, which is a gas with a strong smell, and in turn is a major source of atmospheric sulfur. Marine organisms generate about half the biogenic sulfur emitted to the atmosphere annually, and the majority of this sulfur is produced as DMS. Because reduced sulfur compounds such as DMS are rapidly oxidized to sulfur dioxides that function as cloud condensation nuclei (CCN), the production of DMS can potentially affect climate on a global scale. DMS has also a peculiar role: birds use DMS as a foraging cue, as algae being consumed by fish release DMS, as a consequence bringing the presence of the fish to the attention of the birds.

DMSP can function as osmoregulator, buoyancy controller, cryoprotectant, and antioxidant. Freshwater algae do not produce DMSP, because, being an osmolyte, it has no significance in freshwater systems. It has also been postulated that DMSP production is indirectly related to the nitrogen nutrition of algae, with DMSP being a store for excess sulfate taken up while assimilating the molybdenum necessary to synthesize nitrate reductase or nitrogenase (this could also be true for vascular plants with high nitrate-reductase activities or with symbiotic nitrogen-fixing bacteria). The cleavage of DMSP by means of the enzyme DMSP lyase produces gaseous DMS and acrylic acid. DMSP can also be released from phytoplankton cells during senescence, whereas zooplankton grazing on healthy cells is thought to facilitate the release of DMSP from ruptured algal cells during sloppy feeding. Once released from algal cells, DMSP undergoes microbially mediated conversion by cleavage into gaseous DMS and acrylate.

It had been known for many years that the global budget of sulfur could not be balanced without a substantial flux of this element from the oceans to the atmosphere and then to land. Once emitted from the sea, DMS is transformed in the atmosphere by free radicals (particularly hydroxyl and nitrate) to form a variety of products, most importantly sulfur dioxide and sulfate in the form of small particles. As already stated, these products are acidic and are responsible for the natural acidity of atmospheric particles; man's activities in burning fossil fuels add further sulfur acidity to this natural process. In addition, the sulfate particles (natural and man-made) can alter the amount of radiation reaching Earth's surface both directly by scattering of solar energy and indirectly by acting as the nuclei on which cloud droplets form (CCN), thereby affecting the energy reflected back to space by clouds (denser cloud albedo). The reduction of the amount of sunlight that reaches the Earth's surface leads to the consequent reduction of global temperature. This drop in temperature is suggested to cause a decrease in the primary production of DMSP (in other words DMS). Thus, DMS is considered to be counteractive to the behavior of greenhouse gases like CO_2 and CH_4. Removal of the algae-DMS-derived sulfur from the air by rain and deposition of particles is a significant source of this biologically important element for some terrestrial ecosystems.

ALGAE AND THE OXYGEN/CARBON CYCLES

The oxygen and carbon cycles are closely related, because they are directly associated with photosynthesis and respiration processes.

The natural oxygen cycle is determined by the aerobic respiration of glucose (taking place in all living organisms), which consumes oxygen in free form (O_2) using it as electron sink and produces carbon dioxide and water, and by photosynthesis, which consumes carbon dioxide and water to produce carbohydrates and molecular oxygen as a by-product, as we have seen in Chapter 3.

Today oxygen constitutes about 21% of the atmosphere, 85.8% of the ocean, and 46.7% by volume of the Earth's crust. It has not always been like that. The primordial atmosphere of the Earth is thought to have contained mainly CO_2, N_2, H_2O, and CO with traces of H_2, HCN, H_2S, and NH_3, but to have been devoid of O_2 (only small amounts were derived from the photolysis of water), thus being neutral to mildly reducing. Today, the atmosphere contains 78% N_2, 21% O_2, and 0.036% CO_2 by volume, and is strongly oxidizing. All of the molecular oxygen present in the Earth's atmosphere has been produced as the result of oxygenic photosynthesis, the source of the original O_2 being photosynthetic activity in the primordial oceans. The development of aquatic photosynthesis coincided with a long and reasonably steady drawdown of atmospheric CO_2, from concentration approximately 100-fold higher than in the present-day atmosphere to approximately half of the present levels. This drawdown was accompanied by a simultaneous evolution of oxygen from nil to approximately 21%, comparable to that of the present day. The current atmospheric oxygen concentration is maintained in equilibrium between the production by photosynthesis and the consumption by respiration, with annual fluctuations of $\pm 0.002\%$. Over geological time scales, the drawdown of CO_2 was not stoichiometrically proportional to the accumulation of O_2 because photosynthesis and respiration are but two of the many biological and chemical processes that affect the atmospheric concentrations of these two gases. The removal rate of CO_2 from the atmosphere by photosynthesis on land is about 60 gigatons C yr^{-1}, worldwide. The concentration of oxygen in the oceans (85.8%) is influenced horizontally and vertically by physical features such as the thermocline (i.e., a layer in a large body of water, such as a lake, that sharply separates regions differing in temperature), which isolates deep water from exchange with the atmosphere and can be a zone of significant decomposition causing an oxygen minimum. Oxygen is only sparingly soluble in water (oxygen solubility is inversely proportional to the temperature) and diffuses about 10^4 times more slowly in water than air. Deep water masses are produced at the sea surface in the polar zones where cooling gives rise to increased gas solubility and convection currents. These waters remain largely intact and move through the ocean basins with their oxygen concentration decreasing with time due to the decomposition of organic matter.

Carbon, the key element of all life on Earth, has a complex global cycle that involves both physical and biological processes, made up of carbon flows passing back and forth among four main natural reservoirs of stored carbon: the atmosphere, storing 735 gigatons (0.001%) of the world's carbon as carbon dioxide (CO_2), carbon monoxide (CO), methane (CH_4), longer chain volatile hydrocarbons, and halogen compounds (CFC and HCFC compounds); living organisms, storing 8000 gigatons (0.001%) of the world's carbon as compounds like fats, carbohydrates, and proteins; the hydrosphere, storing 39,000 gigatons (0.06%) of the world's carbon, as dissolved carbon dioxide; the lithosphere, storing 1000 gigatons (0.002%) of the world's carbon in the form of fossils (e.g., oil, natural gas, lignite, and coal), and 62,000,000 gigatons (99.9%) in sedimentary rocks (e.g., limestone and dolomite). Carbon is also present in the mineral soil, in the bottom sediments of water bodies, in peat, bogs and mires, in the litter and humus, which contain 3000 gigatons (0.005%) of the world's carbon.

Carbon dioxide enters the ocean from the atmosphere because it is highly soluble in water; in the sea, free dissolved CO_2 combines with water and ionizes to form bicarbonate and carbonate ions, according to the following equilibrium:

$$CO_2 + H_2O \longleftrightarrow H_2CO_3 \longleftrightarrow HCO_3^- + H^+ \longleftrightarrow CO_3^{2-} + H^+ \longleftrightarrow HCO_3^- \longleftrightarrow CO_2 + OH^- \quad (4.7)$$

These ions are bound forms of carbon dioxide, and they (especially bicarbonate) represent by far the greatest proportion of dissolved carbon dioxide in seawater. On average, there are about 45 ml of total CO_2 in 1 l of seawater, but because of the equilibrium of chemical reactions, nearly all of this occurs as bound bicarbonate and carbonate ions which thus act as a reservoir of free CO_2. The amount of dissolved CO_2 occurring as gas in 1 l of seawater is about 0.23 ml. When free CO_2 is removed by photosynthesis, the reaction shifts to the left and the bound ionic forms release more free CO_2; so even when there is a lot of photosynthesis, carbon dioxide is never a limiting factor to plant production. Conversely, when CO_2 is released by the respiration of algae, plants, bacteria, and animals, more bicarbonate and carbonate ions are produced.

According to the general chemical reactions presented earlier, the pH of seawater is largely regulated by the concentrations of bicarbonate and carbonate, and the pH is usually 8 ± 0.5. The seawater acts as a buffered solution, because when CO_2 is added to seawater due to mineralization processes and respiration, the number of hydrogen ions increases and the pH goes down (the solution becomes more acidic). If CO_2 is removed from water by photosynthesis, the reverse happens and the pH is elevated.

Some marine organisms combine calcium with carbonate ions in the process of calcification to manufacture calcareous skeletal material. The calcium carbonate ($CaCO_3$) may either be in the form of calcite or aragonite, the latter being a more soluble form. After death, this skeletal material sinks and is either dissolved, in which case CO_2 is again released into the water, or it becomes buried in sediments, in which case the bound CO_2 is removed from the carbon cycle. The amount of CO_2 taken up in the carbonate skeletons of marine organisms has been, over geological time, the largest mechanism for absorbing CO_2. At present, it is estimated that about 50×10^{15} tons of CO_2 occurs as limestone, 12×10^{15} tons in organic sediments, and 38×10^{12} tons as dissolved inorganic carbonate.

Calcification is not confined to a specific phylogenetically distinct group of organisms, but evolved (apparently independently) several times in marine organisms. Carbonate sediments blanket much of the Atlantic Basin, and are formed from the shells of both coccolithophorids and foraminifera. As the crystal structures of the carbonates in both groups is calcite (as opposed to the more diagenically susceptible aragonite), the preservation of these minerals and their co-precipitating trace elements provides an invaluable record of ocean history. Although on geological time scales, huge amounts of carbon are stored in the lithosphere as carbonates, on ecological time scales, carbonate formation depletes the oceans of Ca^{2+}, and in so doing, potentiates the efflux of CO_2 from the oceans to the atmosphere. This calcification process can be summarized by the following reaction:

$$Ca^{2+} + 2HCO_3^- \longleftrightarrow CaCO_3 + CO_2 + H_2O \quad (4.8)$$

Among the marine organisms responsible for calcification, coccolithophores play a major role, especially *Emiliania huxleyi*. When the blooms of this Haptophyta appear over large expanses of the ocean (white water phenomenon), myriad effects on the water and on the atmosphere above can be observed. Although each cell is invisibly small, there can be as many as a thousand billion billion (10^{21}) of them in a large bloom, and the population as a whole has an enormous impact. *E. huxley* blooms are processed through the food web, with viruses, bacteria, and zooplankton all contributing to the demise and decomposition of blooms. Some debris from the bloom survive to sink to the ocean floor, taking chemicals out of the water column. While they live and when they die, the phytoplankton cells leak chemicals into the water. A bloom can be thought of as a massive chemical factory, extracting dissolved carbon dioxide, nitrate, phosphate, etc. from the water, and at the same time injecting other chemicals such as oxygen, ammonia, DMS, and other dissolved organic compounds into the water. At the same time, the chemical factory pumps large volumes of organic matter and calcium carbonate into the deep ocean and to the ocean floor. Some of this calcium carbonate eventually ends up as chalk or limestone marine sedimentary rocks, perhaps to cycle through the Earth's crust and to reappear millions of years later as mountains, hills, and cliffs. Coccolithophorids are primarily found at low abundance in tropical and subtropical seas, and at higher concentrations at high latitudes in midsummer, following diatom blooms. Hence, export of inorganic carbon by diatoms in spring at high latitudes can be offset by an efflux of carbon to the atmosphere with the formation of coccolithophore blooms later in the year.

The contemporary ocean export of organic carbon to the interior is often associated with diatom blooms. This group has only risen to prominence over the past 40 million years.

Coccolithophorid abundance generally increases through the Mesozoic, and undergoes a culling at the Kretaceous/Tertiari (K/T) boundary, followed by numerous alterations in the Cenozoic. The changes in the coccolithophorid abundances appear to trace eustatic sea level variations, suggesting that transgressions lead to higher calcium carbonate fluxes. In contrast, diatom sedimentation increases with regressions and because of the K/T impact, diatoms have generally replaced coccolithophorids as ecologically important eukaryotic phytoplankton. On much finer time scales, during the Pleistocene, it would appear that interglacial periods favor coccolithophorids abundance, whereas glacial periods favor diatoms. The factors that lead to glacial-interglacial variations between these two functional groups are relevant to elucidating their distributions in the contemporary ecological setting of the ocean.

Coccolithophores influence regional and global temperature, because they can affect ocean albedo and ocean heat retention, and have a greenhouse effect. Coccoliths do not absorb photons, but they are still optically important because they act like tiny reflecting surfaces, diffusely reflecting the photons.

A typical coccolith bloom (containing 100 mg m^{-3} of calcite carbon) can increase the ocean albedo from 7.5 to 9.7%. If each bloom is assumed to persist for about a month, then an annual coverage of 1.4×10^6 km^2 will increase the global annual average planetary albedo by

$$(9.7 - 7.5) * \left(\frac{1}{12}\right) * \left(\frac{1.4}{510}\right) = 0.001\% \tag{4.9}$$

where 510×10^6 km^2 is the surface area of the Earth.

This is a lower bound on the total impact, because sub-bloom concentration coccolith light scattering will have an impact, over much larger areas (estimated maximum albedo impact $= 0.21\%$). A 0.001% albedo change corresponds to a 0.002 W m^{-2} reduction in incoming solar energy, whereas an albedo change of 0.21% causes a reduction of 0.35 W m^{-2}. These two numbers can be compared to the forcing due to anthropogenic addition of CO_2 since the 1700s, estimated to be about 2.5 W m^{-2}. Coccolith light scattering is therefore a factor of only secondary importance in the radiative budget of the Earth. However, the scattering caused by coccoliths causes more heat and light than usual to be pushed back into the atmosphere; it causes more of the remaining heat to be trapped near to the ocean surface, and only allows a much smaller fraction of the total heat to penetrate deeper in the water. Because it is the near-surface water that exchanges heat with the atmosphere, all three of the effects just described conspire to mean that coccolithophore blooms may tend to make the overall water column dramatically cooler over an extended period, even though this may initially be masked by a warming of the surface skin of the ocean (the top few meters).

All phytoplankton growth removes CO_2 into organic matter and reduces atmospheric CO_2 (by means of photosynthesis). However, coccolithophores are unique in that they also take up bicarbonate, with which to form the calcium carbonate of their coccoliths (calcification process). The coccolithophorid blooms are responsible for up to 80% of surface ocean calcification. In the equilibrium of calcification process, an increase in CO_2 concentration leads to calcium carbonate dissolution, whereas a decrease in CO_2 levels achieves the reverse. While photosynthetic carbon fixation decreases the partial pressure of CO_2 as dissolved inorganic carbon is being utilized, conditions favoring surface calcification by coccolithophorid blooms contribute to the increase of dissolved CO_2.

The relative abundance of the components of the carbonate system (CO_2, H_2CO_3, HCO_3^-, and CO_3^{2-}) depends on pH, dissolved inorganic carbon, and the total alkalinity, and the equilibrium between the components can shift very easily from being in one of these dissolved forms to being in another. How much of the total carbon in each form is determined mainly by the alkalinity and by the water temperature? When the seawater carbon system is perturbed by coccolithophore cells removing HCO_3^- to form coccoliths, this causes a rearrangement of how much carbon is in each dissolved form, and this rearrangement takes place more or less instantaneously. The removal of two molecules of HCO_3^- and the addition of one molecule of CO_2 change the alkalinity and this indirectly causes more of the dissolved carbon to be pushed into the CO_2 form. Although the total dissolved carbon is obviously reduced by the removal of dissolved carbon (bicarbonate ions) into solid calcium carbonate, yet the total effect, paradoxically, is to produce more dissolved CO_2 in the water. In this way, coccolithophore blooms tend to exacerbate global warming by causing increased atmospheric CO_2 (greenhouse effect), rather than to ameliorate it, as is the case when dissolved CO_2 goes into new organic biomass. However, recent work is showing that additional properties of coccoliths may make the situation yet more complicated. Coccolith calcite is rather dense (2.7 kg l^{-1} compared to seawater density of 1.024 kg l^{-1}), and the presence of coccoliths in zooplankton faecal pellets and marine snow (the two main forms in which biogenic matter sinks to the deep ocean) causes them to sink more rapidly. Slow-sinking organic matter may also adhere to the surfaces of coccoliths, hitching a fast ride out of the surface waters. If organic matter sinks faster then there is less time for it to be attacked by bacteria and so more of the locked-in carbon will be able to escape from the surface waters, depleting the surface CO_2. Probably this co-transport of organic matter with coccoliths offsets the atmospheric CO_2 increase that would otherwise be caused, and makes coccolithophore blooms act to oppose global warming, rather than to intensify it.

SUGGESTED READING

Ærtebjerg, G., Carstensen, J., Dahl, K., Hansen, J., Nygaard, K., Rygg, B., Sørensen, K., Severinsen, G., Casartelli, S., Schrimpf, W., Schiller, C., and Druon, J. N., Eutrophication in Europe's coastal waters. Topic report 7/2001, European Environment Agency, Copenhagen, 2001.

Anbar, A. D. and Knoll, A. H., Proteozoic ocean chemistry and evolution: a bioorganic bridge? *Science*, 297, 1137–1142, 2002.

Berman-Frank, I., Lundgren, P., Chen, Y., Kupper, H., Kolber, Z., Bergman, B., and Falkowski, P., Segregation of nitrogen fixation and oxygenic photosynthesis in the marine cyanobacterium *Thricodesmium, Science*, 294, 1534–1537, 2001.

Berman-Frank, I., Lundgren, P., and Falkowski, P., Nitrogen fixation and photosynthetic oxygen evolution in cyanobacteria, *Research in Microbiology*, 154, 157–164, 2003.

Bjerrum, C. J. and Canfield, D. E., Ocean productivity before about 1.9 Gyr ago limited by phosphorus adsorption onto iron oxides, *Nature*, 417, 159–162, 2002.

Bucciarelli, E. and Sunda, W. G., Influence of CO_2, nitrate, phosphate, and silicate limitation on intracellular dimethylsulfoniopropionate in batch cultures of the coastal diatom *Thalassiosira pseudonana*, *Limnology and Oceanography*, 48, 2256–2265, 2003.

Capone, D. G., Zehr, J. P., Paerl, H. W., Bergman, B., and Carpenter, E. J., *Tricodesmium*, a globally significant marine cyanobacterium, *Science*, 276, 1221–1229, 1997.

Conley, D. J., Biochemical nutrient cycles and nutrient management strategies, *Hydrobiologia*, 410, 87–96, 2000.

Conley, D. J., Terrestrial ecosystems and the global biogeochemical silica cycle, *Global Biogeochemical Cycles*, 16, 1121–1128, 2002.

Cornell, S., Randell, A., and Jickells, T., Atmospheric inputs of dissolved organic nitrogen to the oceans, *Nature*, 376, 243–246, 2002.

Cottingham, P., Hart, B., Adams, H., Doolan, J., Feehan, P., Grace, N., Grayson, R., Hamilton, D., Harper, M., Hibbert, B., Lawrance, I., Oliver, R., Robinson, D., Vollenbergh, P., and Whittington, J., Quantifying nutrient-algae relationships in freshwater systems: Outcomes of a workshop held at Monash University on the 8th August 2000, CRCFE Technical Report 8, CRCFE, Canberra, 2000.

Dismukes, G. C., Klimov, V. V., Baranov, S. V., Kozlov, Yu, N., DasGupta, J., and Tyryshkin, A. Special feature: The origin of atmospheric oxygen on Earth: the innovation of oxygenic photosynthesis, *Proceedings of the National Academy of Science*, 98, 2170–2175, 2001.

Edwards, A. M., Platt, T., and Sathyendranath, S., The high-nutrient, low-chlorophyll regime of the ocean: limits on biomass and nitrate before and after iron enrichment, *Ecological modeling*, 171, 103–125, 2004.

Exley, C., Silicon in life: a bioinorganic solution to bioorganic essentiality, *Journal of Inorganic Biochemistry*, 69, 139–144, 1998.

Falkowski, P. G., Rationalizing elemental ratio in unicellular algae, *Journal of Phycology*, 36, 3–6, 2000.

Fasham, M. J. R., *Ocean Biogeochemistry*, Springer, Berlin, 2003.

Gabric A., Gregg, W., Najjar, R., Erickson, D., and Matrai, P., Modeling the biogeochemical cycle of dimethylsulfide in the upper ocean: a review, *Chemosphere — Global Change Science*, 3, 377–392, 2001.

Gallon, J. R., N_2 fixation in phototrophs: adaptation to a specialized way of life, *Plant and Science*, 230, 39–48, 2001.

Galloway, J. N., Aber, J. D., Erisman, J. W., Seitzinger, S. P., Howarth, R. W., Cowling, E. B., and Cosby, B. J., The nitrogen cascade, *Bioscience*, 53, 341–356, 2003.

Galloway, J. N., Dentener, F. J., Capone, D. G., Boyer, E. W., Howarth, R. W., Seitzinger, S. P., Cleveland, C. C., Green, P. A., Holland, E. A., Karl, D. M., Porter, J. H., Townsend, A. R., and Vorosmarty, C. J., Nitrogen cycles: past, present and future, *Biogeochemistry*, 70, 153–226, 2004.

Holloway, J. A. M. and Dahlgren, R. A., Nitrogen in rock: occurrences and biogeochemical implications, *Global Biogeochemical Cycles*, 16, 1118–1134, 2002.

Horwarth, R. W., Stewart, J. W. B., and Ivanov, M. V., Sulphur cycling on the continent, *Scope*, 48, 1992.

Inokuchi, R., Kuma, K., Miyata, T., and Okada, M., Nitrogen-assimilating enzymes in land plants and algae: pylogenic and physiological perspectives, *Physiologia Plantarum*, 116, 1–11, 2002.

Kasamatsu, N., Hirano, T., Kudoh, S., Odate, T., and Fukuchi, M., Dimethylsulfopropionate production by psychrophilic diatom isolates. *Journal of Phycology*, 40, 874–878, 2004.

Klausmeier, C. A., Litchman, E., Daufresne, T., and Levin, S. A., Optimal nitrogen-to-phosporus stoichiometry of phytoplankton, *Nature*, 429, 171–174, 2004.

Kumazawa, S., Yumura, S., and Yoshisuji, H., Photoautrophic growth of a recently isolated N_2-fixing marine non-heterocystous filamentous cyanobacterium, *Symploca* sp., *Journal of Phycology*, 37, 482–487, 2001.

Lee, P. A. and de Mora, S. J., Intracellular dimethylsulfoxide in unicellular marine algae: speculation on its origin and possible biological role, *Journal of Phycology*, 35, 8–18, 1999.

Litchman, E., Steiner, D., and Bossard, P., Photosynthetic and growth response of three freshwater algae to phosphorus limitation and daylength, *Freshwater Biology*, 48, 2141–2148, 2003.

Lomans, B. P., Smolders, A. J. P., Intven, L. M., Pol, A., Op de Camp, H. J. M., and van der Drift, C., Formation of dimethyl sulfide and methanethiol in anoxic freshwater sediments, *Applied and Environmental Microbiology*, 63, 4741–4747, 1997.

Lomans, B. P., van der Drift, C., Pol, A., and Op den Camp, H. J. M., Microbial cycling of volatile organic sulfur compounds, *Cellular and Molecular Life Sciences*, 59, 575–588, 2002.

Lundgren, P., Soderback, E., Singer, A., Carpenter, E. J., and Bergman, B., *Katagnymene*: characterization of a novel marine diazotroph, *Journal of Phycology*, 37, 1052–1062, 2001.

Misra, H., Khairnar, S., and Mahajan, S. K., An alternate photosynthetic electron donor system for PSI supports light dependent nitrogen fixation in a non-heterocystous cyanobacterium, *Plectonema boryanum*. *Journal of Plant Physiology*, 160, 33–39, 2003.

NASA's Earth Science Enterprise Research Strategy for 2000–2010. Biology and Biogeochemistry NASA research on the biology and biogeochemistry of ecosystems and the global carbon cycle aims to understand and predict how terrestrial and marine ecosystems are changing. http://earth.nasa.gov/visions/researchstrat/

Navarro-Gonzales, R., McKay, C. P., and Mvondo, D. N., A possible nitrogen crisis for Archean life due to reduced nitrogen fixation by lightning. *Nature*, 412, 61–64, 2001.

Otsuki, T., A study of the biological CO_2 fixation and utilization system. *The Science of Total Environment*, 27, 121–25, 2001.

Raymond, J., Siefert, J. L., Staples, C. R., and Blankenship, R. E., The natural history of nitrogen fixation. *Molecular Biology and Evolution*, 21, 541–554, 2004.

Raven, J. A. and Falkowki, P. G., Oceanic sinks for atmospheric CO_2, *Plant and Cell Environment*, 22, 741–755, 1999.

Raven, J. A. and White, A. M., The evolution of silicification in diatoms: inesplicable sinking and sinking as escape? *New Phytologist*, 162, 45–61, 2004.

Sarmiento, J. L. and Gruber, N., *Ocean Biogeochemical Dynamics*, Princeton University Press, Princeton, 2004.

Sarmiento, J. L., Wofsy, S. C., Shea, E., Denning, A. S., Easterling, W., Field, C., Fung, I., Keeling, R., McCarthy, J., Pacala, S., Post, W. M., Schimel, D., Sundquist, E., Tans, P., Weiss, R., and Yoder, J., *A U.S. Carbon Cycle Science Plan*. University Corporation for Atmospheric Research, Boulder, Colorado, 1999.

Seibach, J., Ed., *Cellular Origin and Life in Extreme Habitats*, Vol. 3, Kluwer, Dordrecht, 2002.

Seitzinger, S. P., Sander, R. W., and Styles, R., Bioavalability of DON from natural and antropogenic sources to estuarine plankton, *Limnology and Oceanography*, 47, 353–366, 2002.

Sharpley, A. N., Daniel, T., Sims, T., Lemunyon, J., Stevens, R., and Parry, R., *Agricultural phosphorus and eutrophication*, United States Department of Agriculture, Agricultural Research Service ARS–149, 2003.

Sunda, W., Kleber, D. J., Keine, R., and Huntsman, S., An antioxidant function for DMSP and DMS in marine algae, *Nature*, 418, 317–320, 2002.

The Open University, *Ocean Chemistry and Deep-sea Sediments*, Butterworth-Heinemann, Oxford, 2001.

Toggweller, J. R., An ultimately limiting nutrient, *Nature*, 400, 511–512, 1999.

Tyrrel, T., The relative influence of nitrogen and phosphorus on oceanic primary production, *Nature*, 400, 525–531, 1999.

Van Alstyne, K. L. and Houser, L. T., Dimethylsulfide release during macroinvertebrate grazing and its role as an activated chemical defense, *Marine Ecology Progress Series*, 250, 175–181, 2003.

Wolfe, G. V., Steinke, M., and Kirst, G. O., Grazing-activated chemical defence in a unicellular marine algae, *Nature*, 387, 894–897, 1997.

Yool, A. and Tyrrel, T., Role of diatoms in regulating the ocean's silicon cycle, *Global Biogeochemical Cycles*, 17, 1103–1124, 2003

http://www.afsni.ac.uk/Research/eutrophication.htm
http://www.calspace.ucsd.edu/distance.htm
http://www.environment-agency.gov.uk/
http://www.essp.csumb.edu/esse/
http://www.freshwater.canberra.edu.au/
http://www.globalchange.umich.edu/
http://history.nasa.gov/CP-2156/contents.htm
http://www.icsu-scope.org/
http://www.icsu-scope.org/downloadpubs/scope13/
http://www.icsu-scope.org/downloadpubs/scope48/

http://www.icsu-scope.org/downloadpubs/scope54/
http://www.lifesciences.napier.ac.uk/
http://www.mp-docker.demon.co.uk/environmental_chemistry/topic_4b/index.html
http://www.physicalgeography.net/
http://www.soes.soton.ac.uk/staff/tt/eh/biogeochemistry.html
http://www.sws.uiuc.edu/nitro/nitrodesc.asp

5 Working with Light

WHAT IS LIGHT?

Light is an electromagnetic radiation, with wave and particle properties. The electromagnetic radiation has a spectrum or wavelength distribution from short wavelength (10^{-6} nm, gamma and x-rays) to long wavelength (10^{15} nm, long radio waves). About 99% of the Sun's radiation is in the wavelength region from 300 to 4000 nm and it is called the broadband or total solar radiation. Within this broadband, different forms of energy exist, which can be associated with specific phenomena such as harmful and potentially mutagen ultraviolet radiation (UV 100–400 nm), sight (visible light 400–700 nm), and heat (infrared radiation 700–4000 nm) (Figure 5.1). Therefore, what we see as visible light is only a tiny fraction of the electromagnetic spectrum; detecting the rest of the spectrum requires an arsenal of scientific instruments ranging from radio receivers to scintillation counters.

Ultraviolet light is arbitrarily broken down into three bands, according to its anecdotal effects. UV-A (315–400 nm) is the least harmful and the most commonly found type of UV light, because it has the least energy. UV-A light is often called black light, and is used for its relative harmlessness and its ability to cause fluorescent materials to emit visible light, thus appearing to glow in the dark. Most phototherapy and tanning booths use UV-A lamps. UV-B (280–315 nm) is typically the most destructive form of UV light, because it has enough energy to damage biological tissues, yet not quite enough to be completely absorbed by the atmosphere. UV-B is known to cause skin cancer. As most of the extraterrestrial UV-B light is blocked by the atmosphere, a small change in the ozone layer could dramatically increase the danger of skin cancer. Short wavelength UV-C (200–280 nm) is almost completely absorbed in air within a few hundred meters. When UV-C photons collide with oxygen molecules, the O—O bond is broken, and the released O atom reacts with O_2 molecule (and for energetic reasons with a collision partner M) and forms ozone (O_3). UV-C is almost never observed in nature, because it is absorbed very quickly. Germicidal UV-C lamps are often used to purify air and water, because of their ability to kill bacteria.

Infrared light contains the least amount of energy per photon of any other band. Because of this, an infrared photon often lacks the energy required to pass the detection threshold of a quantum detector. Infrared is usually measured using a thermal detector such as a thermopile, which measures temperature change due to absorbed energy. As heat is a form of infrared light, far infrared detectors are sensitive to environmental changes, such as someone moving in the field of view. Night vision equipment takes advantage of this effect, amplifying infrared to distinguish people and machinery that are concealed in the darkness. Little of the ultraviolet radiation (UV-A and UV-B) and infrared are utilized directly in photosynthesis.

Whether transmitted to a radio from the broadcast station, heat radiating from the oven, furnace or fireplace, x-rays of teeth, or the visible and ultraviolet light emanating from the Sun, the various forms of electromagnetic radiation all share fundamental wave-like properties. Every form of electromagnetic radiation, including visible light, oscillates in a periodic fashion with peaks and valleys, and displays a characteristic amplitude, wavelength, and frequency. The standard unit of measure for all electromagnetic radiation is the magnitude of the wavelength (λ) and is measured by the distance between one wave crest to the next. Wavelength is usually measured in nanometers (nm) for the visible light portion of the spectrum. Each nanometer represents one-thousandth of a micrometer. The corresponding frequency (ν) of the radiation wave, that is, the number of complete wavelengths that passes a given point per second, is proportional to the reciprocal of the

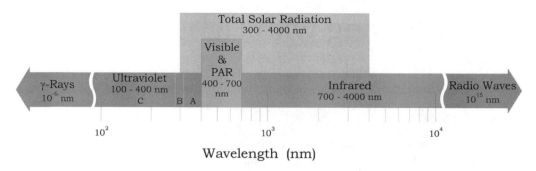

Wavelength (nm)

FIGURE 5.1 The electromagnetic spectrum from γ-rays (10^{-6}) to radio waves (10^{15}).

wavelength. Frequency is usually measured in cycles per second or Hertz (Hz). Thus, longer wavelengths correspond to lower frequency radiation and shorter wavelengths correspond to higher frequency radiation. A wave is characterized by a velocity (the speed of light) and phase. If two waves arrive at their crests and troughs at the same time, they are said to be in phase.

An electromagnetic wave, although it carries no mass, does carry energy. The amount of energy carried by a wave is related to the amplitude of the wave (how high is the crest). A high energy wave is characterized by high amplitude; a low energy wave is characterized by low amplitude. The energy transported by a wave is directly proportional to the square of the amplitude of the wave. The electromagnetic wave does not need any medium for its sustaining; unlike the sound, light can travel in the vacuum.

HOW LIGHT BEHAVES

During traveling light waves interact with matter. The consequences of this interaction are that the waves are scattered or absorbed. In the following, we describe the principal behaviors of light.

SCATTERING

Scattering is the process by which small particles suspended in a medium of a different density diffuse a portion of the incident radiation in all directions. In scattering, no energy transformation results, there is only a change in the spatial distribution of the radiation (Figure 5.2).

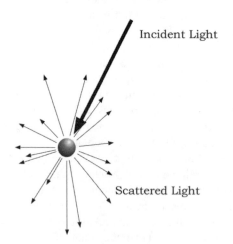

FIGURE 5.2 Light interaction with matter: the scattering process.

In the case of solar radiation, scattering is due to its interaction with gas molecules and suspended particles found in the atmosphere. Scattering reduces the amount of incoming radiation reaching the Earth's surface because significant proportion of solar radiation is redirected back to space. The amount of scattering that takes place is dependent on two factors: wavelength of the incoming radiation and size of the scattering particle or gas molecule. For small particles compared to the visible radiation, Rayleigh's scattering theory holds. It states that the intensity of scattered waves roughly in the same direction of the incoming radiation is inversely proportional to the fourth power of the wavelength. In the Earth's atmosphere, the presence of a large number of small particles compared to the visible radiation (with a size of about 0.5 μm) results such that the shorter wavelengths of the visible range are more intensely diffused. This factor causes our sky to look blue because this color corresponds to those wavelengths. When the scattering particles are very much larger than the wavelength, then the intensity of scattered waves roughly in the same direction of the incoming radiation become independent of wavelength and for this reason, the clouds, made of large raindrops, are white. If scattering does not occur in our atmosphere the daylight sky would be black.

ABSORPTION: LAMBERT–BEER LAW

Some molecules have the ability to absorb incoming light. Absorption is defined as a process in which light is retained by a molecule. In this way, the free energy of the photon absorbed by the molecule can be used to carry out work, emitted as fluorescence or dissipated as heat.

The Lambert–Beer law is the basis for measuring the amount of radiation absorbed by a molecule, a subcellular compartment, such as a chloroplast or a photoreceptive apparatus and a cell, such as a unicellular alga (Figure 5.3). A plot of the amount of radiation absorbed (absorbance, A_λ) as a function of wavelengths is called a spectrum. The Lambert–Beer law states that the variation of the intensity of the incident beam as it passes through a sample is proportional to the concentration of that sample and its thickness (path length). We have adopted this law to measure the absorption spectra in all algal photosynthetic compartments presented in Chapter 3.

The Lambert–Beer law states the logarithmic relationship between absorbance and the ratio between the incident (I_I) and transmitted light (I_T). In turn, absorbance is linearly related to the

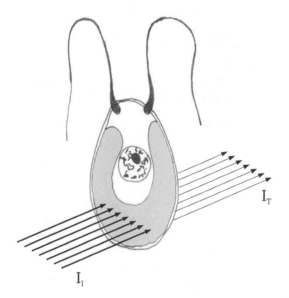

FIGURE 5.3 Light absorption by a unicellular alga: I_I, light incident on the cell and I_T, light transmitted by the cell.

pigment concentration C (mol l^{-1}), the path length l (cm) and the molar extinction coefficient ε_λ, which is substance-specific and a function of the wavelength.

$$A_\lambda = \log \frac{I_1}{I_T} = \varepsilon_\lambda C l \qquad (5.1)$$

Table 5.1 shows the comparison between transmitted light and absorbance values.

INTERFERENCE

Electromagnetic waves can superimpose. Scattered waves, which usually have the same frequency, are particularly susceptible to the phenomenon of interference, in which waves can add constructively or destructively. When two waves, vibrating in the same plane, meet and the crests of one wave coincide, with the crests of the other wave, that is, they are in phase, then constructive interference occurs. Therefore, the amplitude of the wave has been increased and this results in the light appearing brighter. If the two waves are out of phase, that is, if the crests of one wave encounter the troughs of the other, then destructive interference occurs. The two waves cancel out each other, resulting in a dark area (Figure 5.4). The interference of scattered waves gives rise to reflection, refraction, diffusion, and diffraction phenomena.

REFLECTION

Reflection results when light is scattered in the direction opposite to that of incident light. Light reflecting off a polished or mirrored flat surface obeys the law of reflection: the angle between the incident ray and the normal to the surface (θ_I) is equal to the angle between the reflected ray and the normal (θ_R). This kind of reflection is termed specular reflection. Most hard polished (shiny) surfaces are primarily specular in nature. Even transparent glass specularly reflects a portion of incoming light. Diffuse reflection is typical of particulate substances like powders. If you shine a light on baking flour, for example, you will not see a directionally shiny component. The powder will appear uniformly bright from every direction. Many reflections are a combination of both diffuse and specular components, and are termed spread (Figure 5.5), such as that performed by *Emiliana* blooms.

TABLE 5.1
Relationship between Transmitted Light Percentage and Absorbance Value

Transmittance	Absorbance
100	0.000
90	0.045
80	0.096
70	0.154
60	0.221
50	0.301
40	0.397
30	0.522
20	0.698
10	1.000
1	2.000
0.1	3.000

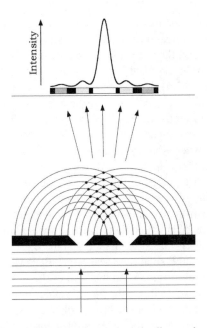

FIGURE 5.4 Interference of light passing through two narrow slits, each acting as a source of waves. The superimposition of waves produces a pattern of alternating bright and dark bands. When crest meets crest or trough meets trough, constructive interference occurs, which makes bright bands; when crest meets trough destructive interference occurs, which makes dark bands. The dots indicate the points of constructive interference. The light intensity distribution shows a maximum that corresponds to the highest number of dots.

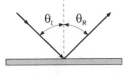

Specular

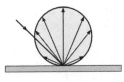

Diffuse

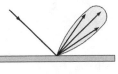

Spread

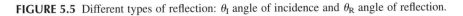

FIGURE 5.5 Different types of reflection: θ_I angle of incidence and θ_R angle of reflection.

Now we will turn attention to the topic of curved mirrors, and specifically curved mirrors that have the shape of spheres, the spherical mirrors. Spherical mirrors can be thought of as a portion of a sphere that was sliced away and then silvered on one of the sides to form a reflecting surface. Concave mirrors were silvered on the inside of the sphere and convex mirrors were silvered on the outside of the sphere (Figure 5.6). If a concave mirror were thought of as being a slice of a sphere, then there would be a line passing through the center of the sphere and attaching to the mirror in the exact center of the mirror. This line is known as the principal axis. The center of sphere from which the mirror was sliced is known as the center of curvature of the mirror. The point on the mirror's surface where the principal axis meets the mirror is known as the vertex. The vertex is the geometric center of the mirror. Midway between the vertex and the center of curvature is the focal point. The distance from the vertex to the center of curvature is known as the radius of curvature. The radius of curvature is the radius of the sphere from which the mirror was cut. Finally, the distance from the mirror to the focal point is known as the focal length. The focal point is the point in space at which light incident towards the mirror and traveling parallel to the principal axis will meet after reflection. In fact, if some light from the Sun was collected by a concave mirror, then it would converge at the focal point. Because the Sun is at such a large distance from the Earth, any light ray from the sun that strikes the mirror will essentially be traveling parallel to the principal axis. As such, this light should reflect through the focal point.

Unlike concave mirror, a convex mirror can be described as a spherical mirror with silver on the outside of the sphere. In convex mirrors, the focal point is located behind the convex mirror, and such a mirror is said to have a negative focal length value. A convex mirror is sometimes referred to as a diverging mirror due to its ability to take light from a point and diverge it. Any incident ray

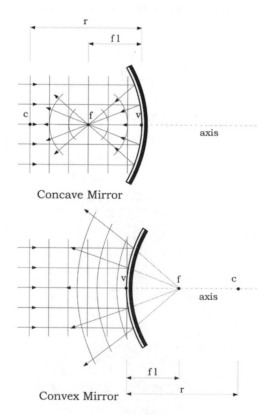

Concave Mirror

Convex Mirror

FIGURE 5.6 Curved mirrors: c, center of curvature of the mirror; v, vertex or geometric center of the mirror; f, focal point; r, radius of curvature; and fl, focal length.

traveling parallel to the principal axis on the way to a convex mirror will reflect in a manner that its extension will pass through the focal point. Any incident ray traveling towards a convex mirror such that its extension passes through the focal point will reflect and travel parallel to the principal axis.

REFRACTION: SNELL'S LAW

Refraction results when light is scattered in the same direction as that of incident light but passing between dissimilar materials, the rays bend and change velocity slightly. Refraction is dependent on two factors: the incident angle θ, that is, the angle between the incident light and the normal to the surface, and the refractive index, n of the material, defined as the ratio between the velocity of the wave in vacuum (c_v) and the velocity of the wave in the medium (c_s),

$$n = \frac{c_v}{c_s}.$$
(5.2)

The refraction results in the following relationship

$$_1n_2 = \frac{\sin(\theta_1)}{\sin(\theta_2)}$$
(5.3)

where $_1n_2$ is the refracting index in passing from Medium 1 to Medium 2 and θ_1 and θ_2 are the angles made between the direction of the propagated waves and the normal to the surface separating the two media.

For a typical air–water boundary, ($n_{air} = 1$, $n_{water} = 1.333$), a light ray entering the water at $45°$ from normal travels through the water at $32,11°$ (Figure 5.7).

The index of refraction decreases with increasing the wavelength. This angular dispersion causes blue light to refract more than red, causing rainbows and prisms to separate the spectrum (dispersion). Table 5.2 shows the refraction index of some common materials.

DISPERSION

Dispersion is a phenomenon that causes the separation of a light into components with different wavelenghts, due to their different velocities in a medium other than vacuum. As a consequence, the white light traveling through a triangular prism is separated into its color components, the spectrum of light. The red portion of the spectrum deviates less than the violet from the direction of propagation of the white light (Figure 5.8).

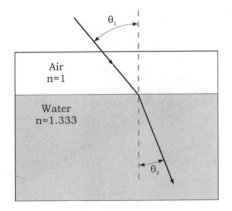

FIGURE 5.7 Refraction of a light ray passing from a medium with lower refraction index (air) to a medium with higher refraction index (water). θ_1, angle of incidence and θ_2, angle of refraction.

TABLE 5.2
Refraction Index of Some Common Materials

Material	Index
Vacuum	1.000
Air at STP	1.00029
Water at 0°C	1.333
Water at 20°C	1.332
Ice	1.309
Glycerin	1.473
Oil	1.466–1.535
Fluorite	1.434
Quartz	1.544
Glass, fused silica	1.459
Glass, Pyrex	1.474
Glass, Crown (common)	1.520
Glass, Flint 29% lead	1.569
Glass, Flint 55% lead	1.669
Glass, Flint 71% lead	1.805
Glass, Arsenic Trisulfide	2.040
Polypropylene	0.900
Polycarbonate	1.200
Plexiglas	1.488
Plastic	1.460–1.55
Nylon	1.530
Teflon	2.200
Salt	1.516

DIFFRACTION

Light waves change the progagation direction when they encounter an obstruction or edge, such as a narrow aperture or slit (Figure 5.9). Diffraction depends on both wavelength of incoming radiation (λ) and obstruction or edge dimensions (a). It is negligible when a/λ is sufficiently large, and becomes more and more important when the ratio tends to zero. This effect is almost absent in

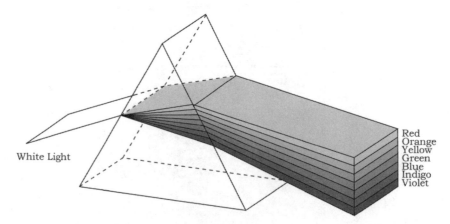

White Light

Red
Orange
Yellow
Green
Blue
Indigo
Violet

FIGURE 5.8 Dispersion of white light through a prism: the red portion of the spectrum deviates less than the violet from the direction of propagation of white light.

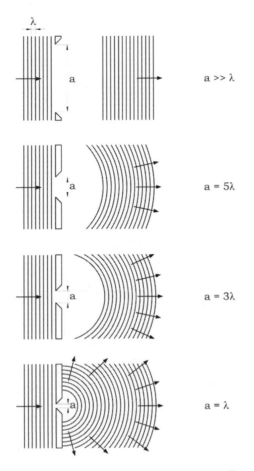

FIGURE 5.9 Diffraction of light from different width aperture; the effect increases with decreasing aperture width.

most optical systems, such as photographic and video cameras, with a large a/λ; but it is very important in all microscopes, where diffraction limits the resolution that microscope can ultimately achieve (a/λ tends to zero). The resolution is the smallest distance between two points to discriminate them as separate.

FIELD INSTRUMENTS: USE AND APPLICATION

Almost all light in the natural environment originates from the Sun. Its spectral distribution is similar to that of an efficient radiant surface known as a blackbody at a temperature of 5800 K, which ranges from 100 to 9000 nm, (Figure 5.10).

In passing through the atmosphere, a small portion of this light is absorbed, and some is scattered. Short wavelengths are strongly scattered, and ozone absorption effectively eliminates wavelengths less than 300 nm. At longer wavelengths, water vapor, carbon dioxide, and oxygen absorb light significantly at particular wavelengths, producing sharp dips in the spectrum. At still-longer wavelengths, beyond 4000 nm, all objects in the environment become significant sources of radiations, depending on their temperature, and surpass sunlight in intensity. These characteristics of the environment restrict the range of electromagnetic radiation. Solar radiant

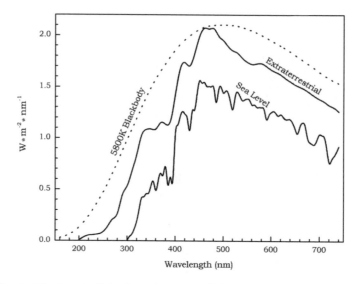

FIGURE 5.10 Spectral irradiance of the incoming sun radiation outside the atmosphere and at sea level compared with that of a perfect blackbody at 5800 K

energy that reaches the surface of the earth has a spectral range from about 300 nm (ultraviolet) to about 4000 nm (infrared). Photosynthetically active radiation (PAR) occurs between approximately 400 and 700 nm and is less than 50% of the total energy impinging on the Earth's surface.

Before describing the detectors used in the field application, a short lexicon of the terms and the conversion units on light measurements would be very useful because of the plethora of confusing terminology and units.

RADIOMETRY

Radiometry is the science of measuring light in any portion of the electromagnetic spectrum. In practice, the term is usually limited to the measurement of infrared, visible, and ultraviolet light using optical instruments, such as radiation thermocouples, bolometers, photodiodes, photosensitive dyes and emulsions, vacuum phototubes, charge-coupled devices, etc.

MEASUREMENT GEOMETRIES, SOLID ANGLES

One of the key concepts to understanding the relationships between measurement geometries is that of the solid angle (ω). This can be defined as the angle that, seen from the center of the sphere, includes a given area on the surface of that sphere. The value of a solid angle is numerically equal to the size of the area on the surface of the sphere (A) divided by the square of the radius (r) of that sphere:

$$\omega = \frac{A}{r^2} \tag{5.4}$$

The value of a solid angle is given in steradian. A sphere of radius r and surface of $4\pi r^2$ will contain 4π steradians. The steradian is a dimensionless quantity.

A surface can be described as a continuum of infinitesimal points, each occupying an infinitesimal area dA,

$$d\omega = \frac{dA}{r^2} \qquad (5.5)$$

where dω is the differential solid angle of the elemental cone containing a ray of light that is arriving at or leaving a infinitesimal surface dA. The symbol d stands for differential, the operator that reduces the applied variable to an infinitesimal quantity.

Most radiometric measurements do not require an accurate calculation of the spherical surface area. Flat area estimates can be substituted for spherical area when the solid angle is less than 0.03 steradians, resulting in an error of less than 1%. This roughly translates to a distance at least five times greater than the largest dimension of the detector. When the light source is the Sun, flat area estimates can be substituted for spherical area.

RADIANT ENERGY

Light is radiant energy. When light is absorbed by a physical object, its energy is converted into some other form. Visible light causes an electric current to flow in a light detector when its radiant energy is transferred to the electrons as kinetic energy. Radiant energy (denoted as Q) is measured in joules (J).

SPECTRAL RADIANT ENERGY

A broadband source such as the Sun emits electromagnetic radiation throughout most of the electromagnetic spectrum. However, most of its radiant energy is concentrated within the PAR. A single-wavelength laser, on the other hand, is a monochromatic source; all of its radiant energy is emitted at one specific wavelength. From this, we can define spectral radiant energy, which is the amount of radiant energy per unit wavelength interval at wavelength λ. It is defined as:

$$Q_\lambda = \frac{dQ}{d\lambda} \qquad (5.6)$$

Spectral radiant energy is measured in joules per nanometer (J nm^{-1}).

RADIANT FLUX (RADIANT POWER)

Energy per unit time is power, which we measure in joules per second (J sec^{-1}), or watts (W). Light "flows" through space and so radiant power is more commonly referred to as the flow rate of radiant energy with respect to time or radiant flux. It is defined as:

$$\Phi = \frac{dQ}{dt} \qquad (5.7)$$

where Q is radiant energy and t is time.

In terms of a light detector measuring PAR, the instantaneous magnitude of the electric current is directly proportional to the radiant flux. The total amount of current measured over a period of time is directly proportional to the radiant energy absorbed by the light detector during that time. For phycological purpose radiant flux is expressed also as micro moles of photons per second.

Spectral Radiant Flux (Spectral Radiant Power)

Spectral radiant flux at wavelength λ is radiant flux per unit wavelength interval. It is defined as:

$$\Phi_\lambda = \frac{d\Phi}{d\lambda} \qquad (5.8)$$

and is measured in watts per nanometer (W nm^{-1}).

Radiant Flux Density (Irradiance and Radiant Exitance)

Radiant flux density is the radiant flux per unit area at a point on a surface. Radiant flux density is measured in watts per square meter (W m^{-2}).

There are two possible conditions. The flux can be arriving at the surface in which case the radiant flux density is referred to as irradiance. Irradiance is defined as:

$$E = \frac{d\Phi}{dA} \qquad (5.9)$$

where Φ is the radiant flux arriving at the infinitesimal area dA.

As irradiance is the radiant flux per unit area, it can be expressed as mole of photons per area per unit time (μmol m^{-2} sec^{-1}). Modern instruments measure *in situ* irradiance directly in these units.

The flux can also be leaving the surface due to emission or reflection. The radiant flux density is then referred to as radiant exitance. As with irradiance, the flux can leave in any direction above the surface. The definition of radiant exitance is:

$$M = \frac{d\Phi}{dA} \qquad (5.10)$$

where Φ is the radiant flux leaving the infinitesimal area dA.

Typical Values for Irradiance	(in W m^{-2})
Sunlight	1000
Skylight	100
Overcast daylight	10
Moonlight	0.001
Starlight	0.0001

Spectral Radiant Flux Density

Spectral radiant flux density is radiant flux per unit wavelength interval at wavelength λ. When the radiant flux is arriving at the surface, it is called spectral irradiance and is defined as:

$$E_\lambda = \frac{dE}{d\lambda} \qquad (5.11)$$

When the radiant flux is leaving the surface, it is called spectral radiant exitance, and is defined as:

$$M_\lambda = \frac{dM}{d\lambda} \qquad (5.12)$$

Spectral radiant flux density is measured in watts per square meter per nanometer $(\text{W m}^{-2}\,\text{nm}^{-1})$.

RADIANCE

Imagine a ray of light arriving at or leaving a point on a surface in a given direction. Radiance is simply the amount of radiant flux contained in this ray (a cone of solid angle $d\omega$). If the ray intersects a surface at an angle θ with the normal to that surface, and the area of intersection with the surface has an infinitesimal cross-sectional area dA, the cross-sectional area of the ray is $dA\cos\theta$. The radiance of this ray is:

$$L = \frac{d^2\Phi}{dA * d\omega * \cos\theta} \tag{5.13}$$

Radiance is measured in watts per square meter per steradian $(\text{W m}^{-2}\,\text{sr}^{-1})$.

Unlike radiant flux density, the definition of radiance does not distinguish between flux arriving at or leaving a surface.

Another way of looking at radiance is to note that the radiant flux density at a point on a surface due to a single ray of light arriving (or leaving) at an angle θ to the normal to that surface is $d\Phi/dA * \cos\theta$. The radiance at that point for the same angle is then $d^2\Phi/(dA * d\omega * \cos\theta)$, or radiant flux density per unit solid angle.

The irradiance, E, at any distance from a uniform extended area source, is related to the radiance, L, of the source by the following the relationship, which depends only on the subtended central viewing angle, θ, of the radiance detector:

$$E = \pi * L * \sin^2\left(\frac{\theta}{2}\right) \tag{5.14}$$

Radiance is independent of distance for an extended area source, because the sampled area increases with distance, cancelling inverse square losses. The inverse square law defines the relationship between the irradiance from a point source and distance (d). It states that the intensity per unit area varies in inverse proportion to the square of the distance. In other words, if you measure $16\,\text{W m}^{-2}$ at 1 m, you will measure $4\,\text{W m}^{-2}$ at 2 m, and can calculate the irradiance at any other distance. An alternate form is often more convenient:

$$E_1 * d_1^2 = E_2 * d_2^2 \tag{5.15}$$

SPECTRAL RADIANCE

Spectral radiance is the radiance per unit wavelength interval at wavelength λ. It is defined as:

$$L_\lambda = \frac{d^3\Phi}{dA * d\omega * \cos\theta * d\lambda} \tag{5.16}$$

and is measured in watts per square meter per steradian per nanometer $(\text{W m}^{-2}\,\text{sr}^{-1}\,\text{nm}^{-1})$.

RADIANT INTENSITY

We can imagine a small point source of light that emits radiant flux in every direction. The amount of radiant flux emitted in a given direction can be represented by a ray of light contained in an elemental cone. This gives us the definition of radiant intensity:

$$I = \frac{d\Phi}{d\omega} \tag{5.17}$$

where $d\omega$ is the infinitesimal solid angle of the elemental cone containing the given direction. From the definition of an infinitesimal solid angle, we get:

$$I = r^2 * \left(\frac{d\Phi}{dA}\right) \quad \text{or} \quad I = r^2 * E \quad \text{or} \quad E = \frac{I}{r^2} \tag{5.18}$$

where the infinitesimal surface area dA is on the surface of a sphere centered on the source and at a distance r from the source and E is the irradiance of that surface. More generally, the radiant flux will intercept dA at an angle θ from the surface normal. This gives us the inverse square law for point sources:

$$E = \frac{I_\theta * \cos\theta}{d^2} \tag{5.19}$$

where I_θ is the intensity of the source in the θ direction and d is the distance from the source to the surface element dA.

Radiant intensity is measured in watts per steradian (W sr^{-1}). Combining the definitions of radiance [Equation (5.13)] and radiant intensity [Equation (5.17)] gives us an alternative definition of radiance:

$$L = \frac{d I_\theta}{dA * \cos\theta} \tag{5.20}$$

where dI_θ is the infinitesimal intensity of the point source in the θ direction with the surface normal.

Spectral Radiant Intensity

Spectral radiant intensity is radiant intensity per unit wavelength interval at wavelength λ. It is defined as:

$$I_\lambda = \frac{d I}{d\lambda} \tag{5.21}$$

and is measured in watts per steradian per nanometer (W sr^{-1} nm^{-1}).

PHOTOMETRY

Photometry is the science of measuring visible light in units that are weighted according to the sensitivity of the human eye. It is a quantitative science based on a statistical model of the human visual response to light under carefully controlled conditions. We cannot apply this model to the "perception" of light by algae, because we should substitute the sensitivity of the algal photoreception systems for that of the human eye as quantified by action spectroscopy in Chapter 2.

For the human perception, the Commission International d'Eclairage (CIE) photometric curves (photopic and scotopic) provide a weighting function that can be used to convert radiometric into photometric measurements. In scotopic curve, yellowish-green light receives the greatest weight because it stimulates the eye more than blue or red light of equal radiant power ($\lambda_{max} = 555$ nm) (Figure 5.11, Table 5.3); in photopic curve blue-green light receives the greatest

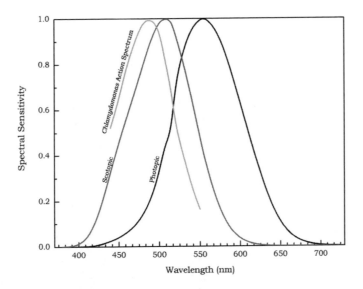

FIGURE 5.11 Curves of spectral sensitivity for the scotopic (dark adapted) and photopic (light adapted) human vision compared with the action spectrum of *Chlamydomonas*.

weight because it stimulates the eye more than other lights of equal radiant power ($\lambda_{max} = 507$ nm), (Figure 5.11, Table 5.3). For algae, action spectroscopy may be used for a similar purpose even though the spectra so far measured are contradictory, not very accurate, and very often are difficult to interpret (Figure 5.11).

Luminous Flux (Luminous Power)

Luminous flux is a radiant flux weighted to match the eye response of the "standard observer." Its unit of measurement is the lumen. The lumen (lm) is the photometric equivalent of the watt, and it is defined as $1/683$ W of radiant power at a frequency of 540×10^{12} Hertz, or better at a wavelength of 555 nm.

Luminous Intensity

The luminous intensity is the luminous flux emitted from a point source per unit solid angle into a given direction. The luminous intensity is measured in candela (the Latin word for "candle"). Together with the CIE photometric curve, the candela provides the weighting factor needed to convert between radiometric and photometric measurements. The candela (cd) is the luminous intensity, in a given direction, of a source that emits monochromatic radiation with a frequency of 540×10^{12} Hertz ($\lambda = 555$ nm) and has a radiant intensity of $1/683$ W sr^{-1} in that direction.

If a light source is isotropic, that is, its intensity does not vary with direction, the relationship between lumens and candelas is 1 cd = 4π lm. In other words, an isotropic source having a luminous intensity of 1 cd emits 4π lm into space, which just happens to be 4π sr. We can also state that 1 cd = 1 lm sr, analogous to the equivalent radiometric definition. If a source is not isotropic, the relationship between candelas and lumens is empirical. A fundamental method used to determine the total flux (lumens) is to measure the luminous intensity (candelas) in many directions using a goniophotometer, and then numerically integrate over the entire sphere. Because a steradian has a

TABLE 5.3
Luminous Efficacy of the Eye at the Different Wavelengths for Both the Light-Adapted (Photopic) Vision and the Dark-Adapted (Scotopic) Vision

λ (nm)	Photopic Luminous Sensitivity	Photopic lm W^{-1} Conversion[a]	Scotopic Luminous Sensitivity	Scotopic lm W^{-1} Conversion[a]
380	0.000039	0.027	0.000589	1.001
390	0.000120	0.082	0.002209	3.755
400	0.000396	0.270	0.009290	15.793
410	0.001210	0.826	0.034840	59.228
420	0.004000	2.732	0.096600	164.220
430	0.011600	7.923	0.199800	339.660
440	0.023000	15.709	0.328100	557.770
450	0.038000	25.954	0.455000	773.500
460	0.060000	40.980	0.567000	963.900
470	0.090980	62.139	0.676000	1149.200
480	0.139020	94.950	0.793000	1348.100
490	0.208020	142.078	0.904000	1536.800
500	0.323000	220.609	0.982000	1669.400
507	0.444310	303.464	**1.000000**	**1700.000**
510	0.503000	343.549	0.997000	1694.900
520	0.710000	484.930	0.935000	1589.500
530	0.862000	588.746	0.811000	1378.700
540	0.954000	651.582	0.650000	1105.000
550	0.994950	679.551	0.481000	817.700
555	**1.000000**	**683.000**	0.402000	683.000
560	0.995000	679.585	0.328800	558.960
570	0.952000	650.216	0.207600	352.920
580	0.870000	594.210	0.121200	206.040
590	0.757000	517.031	0.065500	111.350
600	0.631000	430.973	0.033150	56.355
610	0.503000	343.549	0.015930	27.081
620	0.381000	260.223	0.007370	12.529
630	0.265000	180.995	0.003335	5.670
640	0.175000	119.525	0.001497	2.545
650	0.107000	73.081	0.000677	1.151
660	0.061000	41.663	0.000313	0.532
670	0.032000	21.856	0.000148	0.252
680	0.017000	11.611	0.000072	0.122
690	0.008210	5.607	0.000035	0.060
700	0.004102	2.802	0.000018	0.030
710	0.002091	1.428	0.000009	0.016
720	0.001047	0.715	0.000005	0.008
730	0.000520	0.355	0.000003	0.004
740	0.000249	0.170	0.000001	0.002
750	0.000120	0.082	0.000001	0.001
760	0.000060	0.041	0.000001	0.001
770	0.000030	0.020	0.000001	0.001

[a]The conversion factors to use when reporting the photometric units (luminous flux) in radiometric units (radiant flux).

projected area of 1 m^2 at a distance of 1 m, a 1 cd light source will similarly produce 1 lm m^{-2}, (Figure 5.12).

Typical Values for Luminous Intensity	(in cd)
Flashtube	1,000,000
Automobile headlamp	100,000
100 W incandescent bulb	100
Light emitting diode (LED)	0.0001

LUMINOUS ENERGY

Luminous energy is a photometrically weighted radiant energy. It is measured in lumen per seconds (lm sec^{-1}).

LUMINOUS FLUX DENSITY (ILLUMINANCE AND LUMINOUS EXITANCE)

Luminous flux density is a photometrically weighted radiant flux density. Illuminance is the photometric equivalent of irradiance, whereas luminous exitance is the photometric equivalent of radiant exitance.

Luminous flux density is measured in lux (lx) or lumens per square meter (lm m^{-2}).

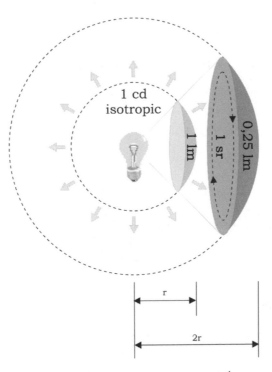

FIGURE 5.12 Luminous intensity: a 1 cd light source emits 1 lm sr^{-1} in all directions (isotropically). According to the inverse square law, the intensity varies in inverse proportion to the square of the distance, that is, at a distance of $2r$ from the source the intensity will be equal to $1/4$ lm sr^{-1}.

Typical Values for Illuminance	(in lux)
Sunlight	100,000
Skylight	10,000
Overcast daylight	1,000
Moonlight	0.1
Starlight	0.01

LUMINANCE

Luminance is a photometrically weighted radiance. In terms of visual perception, we perceive luminance. It is an approximate measure of how "bright" a surface appears when we view it from a given direction. Luminance used to be called "photometric brightness." Luminance is measured in lumens per square meter per steradian ($lm\ m^{-2}\ sr^{-1}$).

LAMBERTIAN SURFACES

A Lambertian surface is referred to as a perfectly diffusing surface, which adheres to Lambert's cosine law. This law states that the reflected or transmitted luminous intensity in any direction from an element of a Lambertian surface varies as the cosine of the angle between that direction and the normal of the surface. The intensity I_θ of each ray leaving the surface at an angle θ from the ray in a direction perpendicular to the surface (I_n) is given by:

$$I_\theta = I_n * \cos \theta \tag{5.22}$$

Therefore, even if the luminous intensity decreases with a factor $\cos(\theta)$ from the normal, the projected surface decreases with the same factor; as a consequence, the radiance (luminance) of a Lambertian surface is the same regardless of the viewing angle and is given by:

$$L = \frac{dI_n * \cos \theta}{dA * \cos \theta} = \frac{dI_n}{dA} \tag{5.23}$$

It is worthwhile to note that in a Lambertian surface the ratio between the radiant exitance and the radiance is π and not 2π:

$$\frac{M}{L} = \pi \tag{5.24}$$

This equation can be easily derived. Suppose we place an infinitesimal Lambertian emitter dA on the inside surface of an imaginary sphere S. The inverse square law [Equation (5.15)] provides the irradiance E at any point P on the inside surface of the sphere. However, $d = D * \cos \theta$, where D is the diameter of the sphere. Thus:

$$E = \frac{I_\theta}{D^2 * \cos \theta} \tag{5.25}$$

and from Lambert's cosine law [Equation (5.22)], we have:

$$E = \frac{I_n}{D^2} \tag{5.26}$$

which simply says that the irradiance (radiant flux density) of any point P on the inside surface of S is a constant.

This is interesting. From the definition of irradiance [Equation (5.9)], we know that $\Phi = E * A$ for constant flux density across a finite surface area A. As the area A of the surface of a sphere with radius r is given by:

$$A = 4\pi * r^2 = \pi * D^2 \tag{5.27}$$

we have:

$$\Phi = E * A = \pi * I_n \tag{5.28}$$

Given the definition of radiant exitance [Equation (5.10)] and radiance for a Lambertian surface [Eqation (5.23)], we have:

$$M = \frac{d\Phi}{dA} = \pi * L \tag{5.29}$$

This explains, clearly and without resorting to integral calculus, where the factor of π comes from.

UNITS CONVERSION

Radiant and Luminous Flux (Radiant and Luminous Power)

1 J (joule) = 1 W sec (watt * second)
1 W (watt) = 683.0 lm (photopic) at 555 nm = 1700.0 lm (scotopic) at 507 nm
1 lm = 1.464×10^{-3} W at 555 nm = $1/(4\pi)$ candela (cd) (only if isotropic)
1 lm sec^{-1} (lumen seconds^{-1}) = 1.464×10^{-3} J at 555 nm.

A monochromatic point source with a wavelength of 510 nm with a radiant intensity of 1/683 W sr^{-1} has a luminous intensity of 0.503 cd, as the photopic luminous efficiency at 510 nm is 0.503.

A 680 nm laser pointer with the power of 5 mW produces 0.005 W 0.017 683 lm W^{-1} = 0.058 lm, while a 630 nm laser pointer with a power of 5 mW produces 0.905 lm, a significantly greater power output.

Determining the luminous flux from a source radiating over a spectrum is more difficult. It is necessary to determine the spectral power distribution for the particular source. Once that is done, it is necessary to calculate the luminous flux at each wavelength, or at regular intervals for continuous spectra. Adding up the flux at each wavelength gives a total flux produced by a source in the visible spectrum.

Some sources are easier to do this with than others. A standard incandescent lamp produces a continuous spectrum in the visible, and various intervals must be used to determine the luminous flux. For sources like a mercury vapor lamp, however, it is slightly easier. Mercury emits light primarily in a line spectrum. It emits radiant flux at six primary wavelengths. This makes it easier to determine the luminous flux of this lamp versus the incandescent.

Generally, it is not necessary to determine the luminous flux for yourself. It is commonly given for a lamp based on laboratory testing during manufacture. For instance, the luminous flux for a 100 W incandescent lamp is approximately 1700 lm. We can use this information to extrapolate to similar lamps. Thus the average luminous efficacy for an incandescent lamp is about 17 lm W^{-1}. We can now use this as an approximation for similar incandescent sources at various wattages.

Irradiance (Flux Density)

$1 \text{ W m}^{-2} = 10^{-4} \text{ W cm}^{-2} = 6.83 \times 10^2 \text{ lux (lx) at 555 nm}$
$1 \text{ lm m}^{-2} = 1 \text{ lx} = 10^{-4} \text{ lm cm}^{-2}$
$1 \text{ photon at 400 nm} = 4.96 \times 10^{-19} \text{ J}$
$1 \text{ W m}^{-2} \text{ at 400 nm} = 1 \text{ J m}^{-2} \text{ sec}^{-1} = 3.3 \text{ μmol m}^{-2} \text{ s}^{-1} = 3.3 \text{ μEinsten m}^{-2} \text{ sec}^{-1}$
$1 \text{ photon at 500 nm} = 3.97 \times 10^{-19} \text{ J}$
$1 \text{ W m}^{-2} \text{ at 500 nm} = 1 \text{ J m}^{-2} \text{ sec}^{-1} = 4.2 \text{ μmol m}^{-2} \text{ sec}^{-1} = 4.2 \text{ μEinsten m}^{-2} \text{ sec}^{-1}$
$1 \text{ photon at 600 nm} = 3.31 \times 10^{-19} \text{ J}$
$1 \text{ W m}^{-2} \text{ at 600 nm} = 1 \text{ J m}^{-2} \text{ sec}^{-1} = 5.0 \text{ μmol m}^{-2} \text{ sec}^{-1} = 5.0 \text{ μEinsten m}^{-2} \text{ sec}^{-1}$
$1 \text{ photon at 700 nm} = 2.83 \times 10^{-19} \text{ J}$
$1 \text{ W m}^{-2} \text{ at 700 nm} = 1 \text{ J m}^{-2} \text{ sec}^{-1} = 5.9 \text{ μmol m}^{-2} \text{ sec}^{-1} = 5.9 \text{ μEinsten m}^{-2} \text{ sec}^{-1}$

Radiance

$1 \text{ W m}^{-2} \text{ sr}^{-1} = 6.83 \times 10^2 \text{ lm m}^{-2} \text{ sr}^{-1} \text{ at 555 nm} = 683 \text{ cd cm}^{-2} \text{ at 555 nm}$

Radiant Intensity

$1 \text{ W sr}^{-1} = 12.566 \text{ W (isotropic)} = 4\pi \text{ W} = 683 \text{ cd at 555 nm}$

Luminous Intensity

$1 \text{ lm sr} = 1 \text{ cd} = 4\pi \text{ lm (isotropic)} = 1.464 \times 10^{-3} \text{ W sr}^{-1} \text{ at 555 nm}$

Luminance

$1 \text{ lm m}^{-2} \text{ sr}^{-1} = 1 \text{ cd m}^{-2} = 10^{-4} \text{ lm cm}^{-2} \text{ sr}^{-1} = 10^{-4} \text{ cd cm}^{-2}$

Geometries

Converting between geometry-based measurement units is difficult, and should only be attempted when it is impossible to measure in the actual desired units. You must be aware of what each of the measurement geometries implicitly assumes before you can convert. The example below shows the conversion between lux and lumens.

If you measure 22.0 lux from a light bulb at a distance of 3.162 m, how much light, in lumens, is the bulb producing? Assume that the lamp is an isotropic point source, with the exception that the base blocks a 30° solid angle. Using Equation (5.15), the irradiance at 1.0 m is $E_{1.0 \text{ m}} = (3.162/1.0)^2 * 22.0 = 220 \text{ lm m}^{-2}$. On the basis of steradian definition [Equation (5.4)], we know that 1 sr cuts off a spherical surface area of 1 m² at a distance of 1 m from the source. Therefore, 220 lm m^{-2} corresponds to 220 lm sr^{-1}. The solid angle of the lamp is equal to $2\pi \text{ hr}^{-1}$ [Equation (5.4)], where h is the height of the spherical calotte, and corresponds to $2\pi * [1 - \cos(360 - 30/2)] = 12.35$ sr, while the shadowed solid angle corresponds to 0.21 sr. Therefore, the total lumen output will be 220 lm $* \text{ sr}^{-1}$ 12.35 sr = 2717 lm, (Figure 5.13).

PAR Detectors

Photosynthetic irradiance is the radiant flux density of PAR and is expressed as the radiant energy (400–700 nm) incident on a unit of surface per unit time. A PAR detector is typically an irradiance detector that is equally sensitive to light between 400 and 700 nm and insensitive to light outside this region. Irradiance is now internationally expressed in moles of photons per unit area and per unit time as $\text{μmol m}^{-2} \text{ sec}^{-1}$ (formerly $\text{μEinsten m}^{-2} \text{ sec}^{-1}$), where 1 μmol (μEinsten) corresponds to 1 micromole of photons, that is, 6.02×10^{17} photons, at a given wavelength. Modern instruments measure in situ irradiance flux densities directly in $\text{μmol m}^{-2} \text{ sec}^{-1}$.

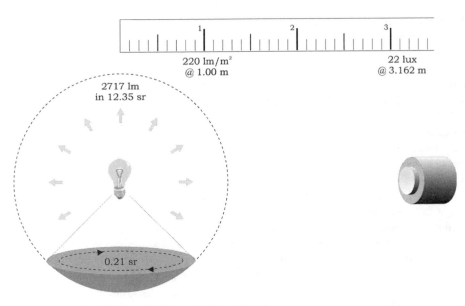

FIGURE 5.13 Example of conversion between lux and lumen.

In terms of collector geometry, a PAR detector is usually equipped with either a planar (cosine) or a spherical (scalar) collector. A planar or 2π collector measures PAR in a $180°$ arc while a spherical or 4π collector measures the light impinging from all directions on a single point (Figure 5.14).

The irradiance impinging on the surface of the water or sediment is commonly measured with a planar detector. Spherical detectors provide a measurement of the amount of light available for photosynthesis by phytoplankton suspended in the water column. A leaf resembles a small planar collector whereas algae resemble small spherical collectors.

Direct solar radiation reaching the water surface varies with the angular height of the radiation and therefore, with the time of day, season, and latitude. The quantity and quality of light also vary

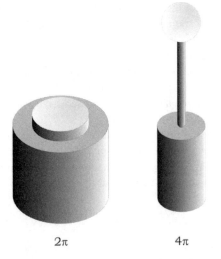

FIGURE 5.14 Examples of PAR detectors equipped with a planar 2π collector or with a spherical 4π collector.

with the molecular transparency of the atmosphere and the distance the light must travel through it; therefore, it varies with altitude and meteorological conditions. In water, upwelling (bottom reflected radiance) and downwelling (depth attenuated radiance) irradiance are usually quantified using paired cosine collectors mounted on a frame that is lowered in the water column. Spherical collectors are also mounted on a frame and lowered in the water to obtain light profiles. For photophysiological studies, spherical collectors provide a realistic measure of the total amount of PAR available for photosynthesis in a water body.

Accurate measurement of photosynthetic radiation is essential for controlled environmental research. The most common method of measuring PAR gives equal value, that is, equal energy content, to all photons with wavelength between 400 and 700 nm and is referred to as the photosynthetic photon flux (PPF) measured in $\mu mol\ m^{-2}\ sec^{-1}$. The ideal PPF sensor would respond equally to all photons between 400 and 700 nm. However, photosynthesis is driven also by photons with wavelength below 400 and above 700 nm, and photons of different wavelength induce inequal amount of photosynthesis. For these reasons, an accurate measurement of PAR should follow the relative quantum efficiency (RQE) curve, which weights the photosynthetic value of all photons with wavelength from 360 to 760 nm. The RQE resembles the absorption spectrum of the absorbing algal structures. A sensor that responds according to this curve measures yield photon flux (YPF) in micromoles per square meter per second, the same unit of PPF.

Detector designed to measure YPF or PPF are commercially available. Both types consist of an optical diffuser collector (planar or spherical) and multiple-spectral filters in front of a broad-spectrum photovoltaic detector.

Planar detectors measure only unreflected sunlight. They should possess cosine corrected response; as the light impinging on the measurement plane is proportional to the cosine of the angle at which the light is incident, the response of the detector must also be proportional to the cosine of the incidence angle [Equation (5.22)]. An integrating sphere measures both unreflected and all the scattered sunlight; it needs no corrections because it is the ideal cosine detector, responding equally to light coming from all directions.

The peak responsivity and bandwidth of a detector can be controlled through the use of filters. Filters can suppress out-of-band light but cannot boost signal, and restricts the spatial responsivity of the detector. Filters operate by absorption or interference. Colored glass filters are doped with materials that selectively absorb light by wavelength, and obey Lambert–Beer law. The peak transmission is inherent to the additives, while bandwidth is dependent on thickness. Interference filters, which select very narrow bandwidths, rely on thin layers of dielectric to cause constructive interference between the wavefronts of the incident white light. So by varying filter thickness, we can selectively modify the spectral responsivity of a detector to match a particular function. Pass band filters are often used to select a specific wavelength radiation. Any of these filter types can be combined to form a composite filter that matches a particular photobiological process. Multiple filters cemented in layers give a net transmission equal to the product of the individual transmissions. The detector's overall spectral sensitivity is equal to the product of the responsivity of the detector and the transmission of the filters. Given a desired overall sensitivity and known detector responsivity, we can solve for the ideal filter transmission curve.

Different measurement geometries necessitate specialized input optics. Radiance and luminance measurements require a narrow viewing angle ($<4°$) in order to satisfy the conditions underlying the measurement units. Power measurements, on the other hand, require a uniform response to radiation regardless of input angle to capture all the light.

PHOTOSYNTHESIS–IRRADIANCE RESPONSE CURVE (*P* VERSUS *E* CURVE)

The P versus E curve, or light-response curve, shows how photosynthetic rate varies with increasing irradiance. The rate of photosynthesis can be measured as net O_2 evolution. This curve describes a complex multi-step process that depends on several limiting factors. Three regions are present in the curve: a light-limited region, a light-saturated region, and a light-inhibited region. The basic description of the $P-E$ curves requires a non-linear mathematical function to account for the light saturation effects. Quite a few such functions have been employed with varying degrees of success, such as rectangular hyperbola, quadratic, exponential, and hyperbolic tangent.

In our opinion, the rectangular hyperbolic function such as the Michaelis-Menten formulation that describes the nutrient uptake kinetics is quite suitable to describe the dynamic relationship between phosynthetic rate and irradiance, though the Michaelis-Menten formulation shows a much less sharper transition from the limited to the saturated region. The $P-E$ curve (without the light-inhibited) region can be described with the following rectangular hyperbolic function:

$$P_E = \frac{P_{\max} * E_\lambda}{K_m + E_\lambda} \tag{5.30}$$

where P_E is the photosynthetic rate at any irradiance E, E_λ is the spectral irradiance (in $\mu mol\ m^{-2}\ sec^{-1}$) and K_m is the half saturation costant when $P_E = P_{\max}/2$ (Figure 5.15a).

In the light-limited region, that is, at low irradiance levels, the rate of photon absorption determines the rate of steady-state electron transport from H_2O to CO_2. In the light-limited region, the available light is insufficient to support the maximum potential rate of the light-dependent reactions, and thus limits the overall rate of photosynthesis. In this region, the rate of photosynthesis (P_E) can be described as:

$$P_E = E_\lambda * \alpha \tag{5.31}$$

where E_λ is the spectral irradiance and α is a measure of "photosynthetic efficiency," or how efficiently solar energy is converted into chemical energy. α takes into account that the light absorbed by the algal cell is proportional to the functional absorption cross-section (σ_{PSII}) of the PSII (the effective area that a molecule presents to an incoming photon and that is proportional to the probability of absorption) and to the number of photosynthetic units (n):

$$\alpha = \sigma_{PSII} * n \tag{5.32}$$

Equation (5.31) shows that photosynthetic rate is linearly proportional to irradiance at low irradiance levels. Greater is the slope (α) more efficient is the photosynthetic process (Figure 5.15b). Keeping constant the number of photosynthetic units but increasing the functional absorption cross-section the slope will increase. The slope can be normalized to chlorophyll biomass and, if so, a superscript "B" is added to denote this normalization, thus α^B. In this case, the curve dimensions are O_2 evolved per unit chlorophyll and quanta per unit area.

At very low irradiance level, the rate of oxygen consumption will be greater than the rate of oxygen evolution; hence, respiration is greater than photosynthesis and net oxygen evolution will be negative. Therefore, the $P-E$ curve does not pass through the origin. The irradiance value on the x-axis at which photosynthesis balances respiration is called the light compensation point (E_c). Therefore, the Equation (5.30) becomes:

$$P_E = \frac{P_{\max}(E_\lambda - E_c)}{K_m + (E_\lambda - E_c)} \tag{5.33}$$

This phenomenon (chlororespiration) is more pronounced in cyanobacteria, where the photosynthetic and respiratory pathways share common electron carriers (cytochromes). The irradiance levels needed to reach compensation point (E_c) are about 10 $\mu mol\ m^{-2}\ sec^{-1}$ for shallow water, but are much lower in dim habitats.

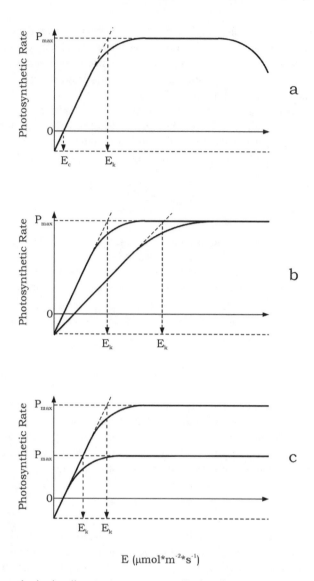

FIGURE 5.15 Photosynthesis–irradiance response curves: E_c, irradiance compensation point; E_k, saturating irradiance; and P_{max}: maximum photosynthetic rate. (a) typical plot; (b) comparison of two curves with different slopes: keeping constant the number of photosynthetic units, but increasing the functional absorption cross-section, the slope increases; and (c) comparison of two curves with different maximum photosynthetic rate: increasing the number of photosynthetic units, P_{max} increases.

As irradiance increases, more ATP and NADPH are produced, and the overall rate of photosynthesis becomes increasingly non-linear, rising towards its maximum or saturation level, P_{max}. This pattern will continue until some other factors becomes limiting. By definition, in the light saturated region, the rate of photon absorption exceeds the rate of steady-state electron transport from H_2O to CO_2. Saturating irradiance, E_k, is defined as the point at which the extrapolated initial slope crosses P_{max}. E_k represents a optimum on the photosynthesis irradiance curve.

Light-saturated photosynthetic rate is independent from the functional absorption cross-section of the photosynthetic apparatus and is only related to the number of photosynthetic unit (n) and their steady-state electron turnover rate through the PSII reaction center $(1/\tau_{PSII})$:

$$P_{max} = n\frac{1}{\tau_{PSII}} \tag{5.34}$$

This equation says that increasing the number of photosynthetic units (but not their size) P_{max}, that is, the saturation level, increases, and that P_{max} cannot be derived from measurements of light absorption (Figure 5.15c). In other words, above the saturation point (E_k), the light-dependent reaction are producing more ATP and NADPH than can be used by the light-independent reaction for CO_2 fixation, that is, increasing irradiance no longer causes any increase in photosynthetic rate. Above E_k and under normal condition, availability of CO_2 is the limiting factor, because the concentration of CO_2 in the atmosphere is very low (0.035% v/v).

Saturating irradiances show some correlation with habitat, but generally, they are low compared with full sun. Intertidal species require 400–600 μmol m^{-2} sec^{-1} (ca. 10% of the full sun irradiance), upper and midsublittoral species 150–250 μmol m^{-2} sec^{-1} and deep sublittoral species less than 100 μmol m^{-2} sec^{-1}. Diatoms under ice saturate at 5 μmol m^{-2} sec^{-1}.

Further increase in irradiance beyond light saturation can lead to a reduction in photosynthetic rate from the maximum saturation level. This reduction, which is dependent on both the irradiance and the duration of the exposure, is often termed photoinhibition. Photoinhibition can be thought as a modification of P_{max} either by a reduction in the number of photosynthetic units or by an increase in the maximum turnover rate [Equation (5.34)]; thus photoinhibition leads a reduction in the photochemical efficiency of PSII, through a reduction in the population of functional (O_2 evolving) reaction centers. Increasing irradiance levels increase the probability that more than one photon, two for example, strike the same reaction centers at the same time. The added energies of two blue photons, for example, could be very harmful as the resulting energy will correspond to a UV photon and could damage the chromophores.

PHOTOACCLIMATION

Photoacclimation is a complex light-response that changes cellular activities on many time scales. The aquatic environment presents a highly variable irradiance (E) field with changes occurring over a wide range of time scales. For example, changes in E on short time scales can result from focusing and defocusing of radiation by waves at the surface. Longer time scale changes can result from variable cloud cover or turbulent motion that transports phytoplankton across the exponential E gradient of the surface mixed layer. In a well-mixed surface layer, phytoplankton experiences long periods of low E interspersed by short periods of saturating or even supersaturating E. The diurnal solar cycle causes changes in E on even longer time scales. To cope with the highly variable radiation environment, phytoplankton has developed numerous strategies to optimize photosynthesis, while minimizing susceptibility to photodamage. Photosynthetic acclimation to E over time scales of hours to days proceeds through changes in cellular pigmentation or structural characteristics, for example, size and number of photosynthetic units. On shorter time scales, cells adjust photon utilization efficiencies by changing the distribution of harvested energy between photosystems (state transitions) or by dissipating excess energy through non-photochemical processes, for example, xanthophyll cycle or photoinhibition.

Therefore, photoacclimation involves change in macromolecular composition in photosynthetical apparatus. It is relatively easy to observe acclimation in unicellular algae and seeweeds, where chlorophylls per cell or per unit surface can increase five- to ten-fold as irradiance decreases. The response is not a linear function of irradiance; rather at extremely low light levels, cells often become a bit chlorotic, and on exposure to slightly higher (but still low) irradiance, chlorophyll reaches a maximum. Increase in irradiance leads to a decrease in the cellular complement of chlorophylls until a minimum value is reached. The absolute irradiance levels that induce these effects are species specific and chlorotic response is not universal. The changes in pigmentation resulting from photoacclimation have profound consequences for light absorption properties of the cells. First, cells acclimated to high irradiance levels generally have high carotenoids concentration relative to chlorophyll *a*. Carotenoids such as β-carotene and zeaxanthin do not transfer excitation energy to the reaction centers and consequently act to screen the cell from excess light. Some xanthophylls such as lutein, transfer excitation energy but with reduced efficiency, and therefore effectively reduce the functional absorption cross-section of the associated photosystem. Because these carotenoids absorb light without a concomitant increase in the functional cross-section of PSII, organisms acclimated to high irradiance levels often have lower maximum quantum yield of photosynthetic O_2 evolution. Second, when cells acclimate to low irradiance levels, the subsequent increase in pigmentation is associated with a decrease in functional optical absorption section normalized to chlorophyll *a*. This effect is due primarily to the self-shading of the chromophores between layers of thylakoids membranes, and is an inverse function of the number of membranes in the chloroplast, that is, the more the membranes, the lower the optical cross-section. Thus, as cells accumulate chlorophyll, each chlorophyll molecule becomes less effective in light absorption; a doubling of cellular chlorophyll does not produce a doubling in the rate of light absorption. The reduction in the chlorophyll-specific optical absorption cross-section can be visualized considering the fate of photons incident on two stacks of thylakoid membranes: one is a thin stack from a cell acclimated to high irradiance levels and the second is a thick stack from a cell acclimated to low irradiance levels. The probability of a photon passing through a thick stack of membranes without being absorbed is small compared with a thin stack of membranes. This so-called "package effect" reduces the effectiveness of increased pigmentation in harvesting light and has important implications for the capital costs of light harvesting, that is, the investments in the physical structures of the organism required for the metabolic processes. The diminution in the optical absorption cross-section with increased chlorophyll is also a function of cell size: the larger the cell the more important is this effect. At some point a cell is, for most practical purposes, optically black and further increases in pigment levels confer no advantage in light absorption.

There are two basic photoacclimation responses in algae. In one, acclimation is accomplished primarily by changes in the number of photosynthetic reaction centers, while the effective absorption cross-section of the reaction centers remains relatively constant. The second is characterized by relatively large changes in the functional size of the antennae serving the reaction centers, while the number of reaction centers remains relatively constant. Complementary changes in either of the responses produce the same effect on the initial slope of the photosynthesis–irradiance curve. As the functional size of the antennae serving PSII, and not the number of rection centers, determines the light-saturation parameter, organisms that vary the cross-section would tend to have more control over this parameter, as long as the turnover time remains constant.

SUGGESTED READING

Adir, N., Zer, H., Shochat, S., and Ohad, I., Photoinhibition, a historical perspective, *Photosynthetic Research*, 76, 343–370, 2003.

Barnes, C., Tibbitts, T., Sager, J., Deitzer, G., Bubenheim, D., Koerner, G., and Bugbee, B., Accuracy of quantum sensor measuring yield photon flux and photosynthetic photon flux, *HortScience*, 28, 1197–1200, 1993.

Behrenfeld, M. J., Prasil, O., Babin, M., and Bruyant, F., In search of a physiological basis for covariations in light-limited and light-saturated photosynthesis, *Journal of Phycology*, 40, 4–25, 2004.

Blanchard, G. F., Guarini, J.-M., Dang, C., and Richard, P., Characterizing and quantifying photoinhibition in intertidal microphytobenthos, *Journal of Phycology*, 40, 692–696, 2004.

Buhrer, H., Light within algal cultures; implications from light intensity within a lens, *Aquatic Science*, 62, 91–103, 2000.

Dusenbery, D. B., *Sensory Ecology — How an Organism Acquires and Responds to Information*, W.H. Freeman and Co., New York, 1992.

Falkowski P. G., and Raven, J. A., *Aquatic Photosynthesis*, Blackwell Science, 1997.

Havelkova-Dousova, H., Prasil, O., and Behrenfeld, M. J., Photoacclimation of *Dunaliella tertiolecta* (Chlorophyceae) under fluctuating irradiance, *Photosynthetica*, 42, 273–281, 2004.

Jones, H. G., Archer, N., Rotenberg, E., and Casa, R., Radiation measurement for plant ecophysiology, *Journal of Experimental Botany*, 54, 879–889, 2003.

Levi, L., *Applied Optics — A Guide to Optical System Design, Volume I and Volume II*, John Wiley and Sons, Inc., New York, 1980.

Lobban, C. S., and Harrison, P. J., *Seaweed Ecology and Physiology*, Cambridge University Press, 1994.

MacIntyre, H. L., Kana, T. M., Anning, T., and Geider, R. J., Photoacclimation of photosynthesis irradiance response curves and photosynthetic pigments in microalgae and cyanobacteria, *Journal of Phycology*, 38, 17–38, 2002.

Masuda, T., Polle, J. E. W., and Melis, A., Biosynthesis and distribution of chlorophyll among the photosystems during recovery of *Dunaliella salina* from irradiance stress, *Plant Physiology*, 128, 603–614, 2002.

Schaller, K., Ruth, D., and Uhl, R., How *Chlamydomonas* keeps track of the light once it has reached the right phototactic orientation, *Biophysic Journal*, 73, 1562–1572, 1997.

Singsaas, E. L., Ort, D. R., and De Lucia, E. H., Variation in measured values of photosynthetic quantum yield in ecophysiological studies, *Oecologia*, 128, 15–23, 2001.

http://hyperphysics.phy-astr.gsu.edu/hbase/hframe.html
http://www.4colorvision.com/files/photopiceffic.htm
http://www.4colorvision.com/files/scotopiceffic.htm
http://www.biospherical.com/Index.htm
http://www.cvrl.org/
http://www.fb.u-tokai.ac.jp
http://www.helios32.com/Measuring%20Light.pdf
http://www.helios32.com/Thinking%20Photometrically%20II.pdf
http://www.intl-light.com/ildocs/handbook.pdf
http://www.konicaminoltaeurope.com/products/industrial/en/support/brochure/download/pdfs/
  the_language_of_light.pdf
http://www.licor.com/env/Products/Sensors/rad.jsp
http://www.optics.arizona.edu/palmer/OPTI400/Lectures/basicradiometry.pdf
http://www.plantphys.net/index.php

6 Algal Culturing

COLLECTION, STORAGE, AND PRESERVATION

As already pointed out in the previous chapters, algae grow in almost every habitat in every part of the world. They can be found on very different natural substrates, from animals (snails, crabs, sloths, and turtles are algal hosts) to plants (tree trunks, branches and leaves, water plants, and macroalgae), from springs and rivers to hypersaline lagoons and salt lakes. They also colonize artificial habitats, such as dams and reservoirs, fountains and pools, but cans, bottles, plant pots, or dishes allow algae to extend their natural range. The ubiquity of these organisms together with the plasticity of their metabolic requirements make many algal species easily available for investigation, collection, or simple observation.

Floating microalgae can be collected with a mesh net (e.g., with $25-30$ μm pores) or, if in sufficient quantity (i.e., coloring the water), by simply scooping a jar through the water. A small amount of the bottom sediments will also provide many of the algal species that live in or on these sediments. Some algae live attached to other types of substrate, such as dead leaves, twigs, and any underwater plants, which may be growing in the water. Macroalgae and the attached microalgae can be collected by hand or with a knife, including part or all of the substrate (rock, plant, wood, etc.) if possible. Algae growing on soil are difficult to collect and study, many requiring culturing before sufficient and suitable material is available for identification.

Any sample should be labeled with standard information such as the locality, date of collection, and as many of the following features as possible: whether the water is saline, brackish or fresh; whether the collection site is terrestrial, a river, a stream, or a lake; whether the alga is submerged during water level fluctuations or floods; whether the water is muddy or polluted; whether the alga is free floating or attached, and if the latter, the type of substrate to which it is attached; and the color, texture and size of the alga. Algae can be stored initially in a glass jar, plastic bottle or bag, or in a vial with some water from the collecting site. The container should be left open or only half filled with liquid and wide shallow containers are better than narrow deep jars. If refrigerated or kept on ice soon after collecting most algae can be kept alive for short periods (a day or two). If relatively sparse in the sample, some algae can continue to grow in an open dish stored in a cool place with reduced light. For long-term storage, specimens can be preserved in liquid, dried, or made into a permanent microscope mount. Even with ideal preservation, examination of fresh material is sometimes essential for an accurate determination. Motile algae particularly must be examined while flagella and other delicate structures remain intact, because any kind of preservation procedure causes the detachment of the flagella.

Commercial formalin (40% formaldehyde in water), diluted between 1/10 and 1/20 with the collecting solution, is the most commonly used fixative. As formaldehyde is considered carcinogenic, any contact with skin, eyes, and air passages should be avoided. This compound mixed with other chemicals such as glacial acetic acid and alcohol (FAA 1:1:8 by volume: 40% formaldehyde 1:glacial acetic acid 1:95% alcohol 8) gives better preservation results for some of the more fragile algae, whereas the standard alcohol and water mix (e.g., 70% ethyl alcohol) will ruin all but the larger algae. However, FAA may cause thin-walled cells to burst. Color is an important taxonomic characteristic, especially for cyanobacteria, and formalin is a good preservative for green algae, cyanobacteria and dinoflagellates because cell color remains intact if samples are stored in the dark. Algae can be kept in diluted formalin for a number of years, but the solution is usually replaced by 70% ethyl alcohol with 5% glycerin (the latter to prevent accidental drying out).

Lugol's solution is the preferred preservative commonly used for short-term (e.g., a few months, but possibly a year or more) storage of microalgae. It is excellent for preserving chrysophytes but it makes the identification of dinoflagellates difficult, if not impossible. Samples can be preserved and kept (in dark bottles away from light) for as long as 1 yr in Lugol's solution (0.05–1% by volume). The solution is prepared by dissolving 20 g of potassium iodide and 10 g of iodine crystals in 180 ml of distilled water and by adding 20 ml glacial acetic acid. Note that Lugol's solution has a shelf life that is affected by light of about 6 months.

Dried herbarium specimens can be prepared by "floating out" similar to aquatic flowering plants. Ideally, fresh specimens should be fixed prior to drying. Most algae will adhere to absorbent herbarium paper. Smaller, more fragile specimens or tangled, mat-forming algae may be dried onto mica or cellophane. After "floating out," most freshwater algae should not be pressed but simply left to air dry in a warm dry room. If pressed, they should be covered with pieces of waxed paper, plastic or muslin cloth so that the specimen does not stick to the drying paper in the press.

To examine a dried herbarium specimen, a few drops of water are added to the specimen to make it swell and lift slightly from the paper. This makes it possible to remove a small portion of the specimen with forceps or a razor blade.

Observations (preferably including drawings or photographs) based on living material are essential for the identification of some genera and a valuable adjunct to more leisurely observations on preserved material for others. The simplest method is to place a drop of the water including the alga onto a microscope slide and carefully lower a coverslip onto it. It is always tempting to put a large amount of the alga onto the slide but smaller fragments are much easier to view under a microscope. Microalgae may be better observed using the "hanging drop method," that is, a few drops of the sample liquid are placed on a coverslip which is turned over onto a ring of paraffin wax, liquid paraffin or a "slide ring."

A permanent slide can be prepared with staining such as aniline blue (1% aqueous solution, pH 2.0–2.5), toluidine blue (0.05% aqueous solution, pH 2.0–2.5), and potassium permanganate, $KMnO_4$ (2% aqueous solution), which are useful stains for macroalgae (different stains suit different species) and India Ink, which is a good stain for highlighting mucilage and some flagella-like structures. Staining time ranges from 30 sec to 5 min (depending on the material), after which the sample is rinsed in water. Mounting is achieved by adding a drop or two of glycerine solution (75% glycerine, 25% water) to a small piece of the sample placed on a microscope slide then carefully lowering the coverslip. Sealing with nail polish is essential. This method is unsuitable for most unicellular algae which should be examined fresh or in temporary mounts of liquid-preserved material.

Magnifications of between 40 and 1000 times are required for the identification of all but a few algal genera. A compound microscope with $10\times - 12\times$ eyepiece and $4\times - 10\times - 40\times$ objectives is therefore an essential piece of equipment for anyone wishing to discover the world of algal diversity. An oil immersion $100\times$ objective would be a useful addition, particularly when aiming at identifying to species level. Phase-contrast or interference (e.g., Nomarski) microscopy can improve the contrast for bleached or small specimens. A dissecting microscope providing $20\times$, $40\times$ up to $60\times$ magnifications is a useful aid but is secondary to a compound microscope. A camera lucida attachment is helpful for producing accurate drawings while an eyepiece micrometer is important for size determinations. Formulas for calculating biomass for various phytoplankton shapes using geometric forms and measurements and shape code for each taxa exists in the literature, and are routinely used in the procedure for phytoplankton analysis that require biovolume calculations (Table 6.1). Moreover, software packages for image recognition and analysis are available, which can process phytoplankton images acquired by means of a TV camera mounted onto a compound microscope. Scanning and transmission electron microscopes are beyond the reach of all but specialist institutions; however, they are an essential tool for identifying some of the very small algae and investigating the details of their ultrastructure.

TABLE 6.1
Shape Codes and Corresponding Dimensions Required for Calculating Biovolumes of Various Phytoplankton Species

Shape	Code	Length	Width	Depth	Diameter
			Dimensions Required		
Cone	CON	L	W		
Cylinder	CYL	L	W	DP	D
Dumbell box	DBB	L	W	DP	
Dumbell	DBL	L	W	DP	
Diamond box	DMB	L	W	DP	
Fusiform	FUS	L	W		
Ovoid box	OVB	L	W	DP	
Ovoid	OVO	L	W		
Rectangular box	RTB	L	W	DP	
Sphere	SPH				D
Teardrop	TRP	L	W		

CULTURE TYPES

A culture can be defined as an artificial environment in which the algae grow. In theory, culture conditions should resemble the alga's natural environment as far as possible; in reality many significant differences exist, most of which are deliberately imposed. In fact, following isolation from the natural environment, algal strains are maintained under largely artificial conditions of media composition, light, and temperature. The imposition of an artificial environment on a cell population previously surviving under complex, fluctuating conditions and following a seasonal life cycle inevitably causes a period of physiological adaptation or selection, during which population growth will not occur or is very slow.

While contaminated algal cultures have previously been satisfactory for certain application and experiments, modern experimental methods and application demand that contaminants are not generally present, and that the taxonomy and growth characteristics of strains are defined. Hence, for most purposes, algal cultures are maintained as unialgal, contaminant-free or axenic stocks. "Unialgal" cultures contain only one kind of alga, usually a clonal population (but which may contain bacteria, fungi, or protozoa), whereas "axenic" cultures should contain only one alga and no bacteria, fungi, or protozoa.

To obtain a unialgal culture one species must be isolated from all the rest; three major techniques borrowed from microbiology are available for obtaining unialgal isolates: streaking and successive plating on agar media, serial dilution, and single-cell isolations using capillary pipettes. Streaking is useful for single-celled, colonial, or filamentous algae that will grow on an agar surface. Filaments can be grabbed with a slightly curved pipette tip and dragged through soft agar (less than 1%) to remove contaminants. It is best to begin with young branches or filament tips that have not yet been extensively epiphytized.

Many flagellates, however, as well as other types of algae must be isolated by single-organism isolations or serial-dilution techniques. A particularly effective means of obtaining unialgal cultures is isolation of zoospores immediately after they have been released from parental cell walls, but before they stop swimming and attached to a surface. Recently released zoospores are devoid of contaminants, unlike the surfaces of most algal cells, but catching zoospores requires a steady hand and experience.

Sterile cultures of microalgae may be obtained from specialized culture collections. Alternatively, axenic cultures can be obtained by treating isolated algae to an extensive washing procedure, or with one or more antibiotics. Resistant stages such as zygotes or akinetes can be treated with bleach to kill epiphytes, then planted on agar for germination. It is usually necessary to try several different concentrations of bleach and times of exposure to find a treatment that will kill epiphytes without harming the alga. When diatoms represent the contaminating species, addition of low concentrations (5 mg l^{-1}) of germanium dioxide, GeO$_2$, to a culture medium can inhibit diatom growth, because it disrupts silica deposition.

"Cleaning" previously contaminated cultures is a skilful and time-consuming process, and could take several years in sizeable collections. Extensive measures must be taken to keep pure uni-algal cultures chemically and biologically clean. Chemical contamination may have unquantifiable, often deleterious, and therefore undesirable effects on algal growth. Biological contamination of pure algal cultures by other eukaryotes and prokaryotic organisms in most cases invalidates experimental work, and may lead to the extinction of the desired algal species in culture through out-competition or grazing. In practice, it is very difficult to obtain bacteria-free (axenic) cultures, and although measures should be taken to minimize bacterial numbers, a degree of bacterial contamination is often acceptable.

If biological contaminants appear in a culture, the best remedy is to isolate a single cell from the culture with a micropipette, and try to establish a new, clean clonal culture. Alternatively the culture can be streaked on an agar plate in the hope of attaining a colony free of contaminants. Neither of these methods works well, however, for eliminating bacteria that attach firmly to the surface of microalgae. Placing a test-tube of microalgal culture in a low-intensity 90 kilocycles sec^{-1} ultra-sonic water bath for varying lengths of time (a few seconds to tens of minutes) can sometimes physically separate bacteria without killing the algae, making it easier to obtain an axenic culture by micropipette isolation. Often, however, to achieve an axenic culture, antibiotics must be added to the growth medium to discourage growth of contaminating cyanobacteria and other bacteria. Best results appear to occur when an actively growing culture of algae is exposed to a mixture of penicillin, streptomycin, and gentamycin for around 24 h. This drastically reduces the growth of bacteria while allowing the microalgae to continue to grow, increasing the chances of obtaining an axenic cell when using micropipette or agar streaking isolation. Different algal species tolerate different concentrations of antibiotics, so a range of concentrations should be used (generally 50–500‰ w/v). Other antibiotics that can be used include chloramphenicol, tetracycline, and bacitracin. Antibiotic solutions should be made with distilled water and filter-sterilized (0.2 μm filter units) into sterile tubes, and should be stored frozen until use. Another approach is to add a range of antibiotic concentrations to a number of subcultures and then select the culture that has surviving algal cells but no surviving bacteria or other contaminants. Sterility of cultures should be checked by microscopic examination (phase contrast) and by adding a small amount of sterile bacterial culture medium (e.g., 0.1% peptone) to a microalgal culture and observing regularly for bacterial growth. Absence of bacterial growth does not, however, ensure that the microalgal culture is axenic, because the majority of bacteria do not respond to standard enrichments. In reality there is no way of demonstrating that a microalgal culture is completely axenic. In practice, therefore, axenic usually means "without demonstrable unwanted prokaryotes or eukaryotes." Some micro-algal cultures may die when made axenic, probably due to the termination of obligate symbiotic relationships with bacteria.

The collection of algal strains should be carefully protected against contamination during handling and poor temperature regulation. To reduce risks, two series of stocks are often retained, one which supplies the starter cultures for the production system and the other which is only subjected to the handling necessary for maintenance. Stock cultures are kept in test-tubes at a light intensity of about 1.5 W m^2 and a temperature of 16–19°C. Constant illumination is suitable for the maintenance of flagellates, but may result in decreased cell size in diatom stock cultures. Stock cultures are maintained for about a month and then transferred to create a new culture line.

CULTURE PARAMETERS

A culture has three distinct components: a culture medium contained in a suitable vessel; the algal cells growing in the medium; air, to allow exchange of carbon dioxide between medium and atmosphere.

For an entirely autotrophic alga, all that is needed for growth is light, CO_2, water, nutrients, and trace elements. By means of photosynthesis the alga will be able to synthesize all the biochemical compounds necessary for growth. Only a minority of algae is, however, entirely autotrophic; many are unable to synthesize certain biochemical compounds (certain vitamins, e.g.) and will require these to be present in the medium (obligate mixotropy condition, see Chapter 1).

The most important parameters regulating algal growth are nutrient quantity and quality, light, pH, turbulence, salinity, and temperature. The most optimal parameters as well as the tolerated ranges are species specific and the various factors may be interdependent and a parameter that is optimal for one set of conditions is not necessarily optimal for another.

TEMPERATURE

The temperature at which cultures are maintained should ideally be as close as possible to the temperature at which the organisms were collected; polar organisms ($<10°C$); temperate ($10-25°C$); tropical ($>20°C$). Most commonly cultured species of microalgae tolerate temperatures between 16 and 27°C, although this may vary with the composition of the culture medium, the species, and strain cultured. An intermediate value of $18-20°C$ is most often employed. Temperature-controlled incubators usually use constant temperature (transfers to different temperatures should be conducted in steps of 2°C per week), although some models permit temperature cycling. Temperatures lower than 16°C will slow down growth, whereas those higher than 35°C are lethal for a number of species.

LIGHT

As for plants, light is the source of energy which drives photosynthetic reactions in algae and in this regard intensity, spectral quality, and photoperiod need to be considered. Light intensity plays an important role, but the requirements greatly vary with the culture depth and the density of the algal culture: at higher depths and cell concentrations the light intensity must be increased to penetrate through the culture. Too high light intensity (e.g., direct sunlight, small container close to artificial light) may result in photoinhibition. Most often employed light intensities range between 100 and 200 $\mu E\ sec^{-1}\ m^{-2}$, which corresponds to about 5–10% of full daylight (2000 $\mu E\ sec^{-1}\ m^{-2}$). Moreover, overheating due to both natural and artificial illumination should be avoided. Light may be natural or supplied by fluorescent tubes emitting either in the blue or the red light spectrum, as these are the most active portions of the light spectrum for photosynthesis. Light intensity and quality can be manipulated with filters. Many microalgal species do not grow well under constant illumination, although cultivated phytoplankton develop normally under constant illumination, and hence a light/dark (LD) cycle is used (maximum 16:8 LD, usually 14:10 or 12:12).

pH

The pH range for most cultured algal species is between 7 and 9, with the optimum range being 8.2–8.7, though there are species that dwell in more acid/basic environments. Complete culture collapse due to the disruption of many cellular processes can result from a failure to maintain an acceptable pH. The latter is accomplished by aerating the culture. In the case of high-density algal culture, the addition of carbon dioxide allows to correct for increased pH, which may reach limiting values of up to pH 9 during algal growth.

SALINITY

Marine algae are extremely tolerant to changes in salinity. Most species grow best at a salinity that is slightly lower than that of their native habitat, which is obtained by diluting sea water with tap water. Salinities of $20-24$ g l^{-1} are found to be optimal.

MIXING

Mixing is necessary to prevent sedimentation of the algae, to ensure that all cells of the population are equally exposed to the light and nutrients, to avoid thermal stratification (e.g., in outdoor cultures), and to improve gas exchange between the culture medium and the air. The latter is of primary importance as the air contains the carbon source for photosynthesis in the form of carbon dioxide. For very dense cultures, the CO_2 originating from the air (containing 0.03% CO_2) bubbled through the culture is limiting the algal growth and pure carbon dioxide may be supplemented to the air supply (e.g., at a rate of 1% of the volume of air). CO_2 addition furthermore buffers the water against pH changes as a result of the CO_2/HCO_3^- balance.

Mixing of microalgal cultures may be necessary under certain circumstances: when cells must be kept in suspension in order to grow (particularly important for heterotrophic dinoflagellates); in concentrated cultures to prevent nutrient limitation effects due to stacking of cells and to increase gas diffusion. It should be noted that in the ocean cells seldom experience turbulence, and hence mixing should be gentle. Depending on the scale of the culture system, mixing is achieved by stirring daily by hand (test-tubes, erlenmeyers), aerating (bags, tanks), or using paddle wheels and jet pumps (ponds). Not all algal species can tolerate vigorous mixing. The following methods may be used: bubbling with air (may damage cells); plankton wheel or roller table (about 1 r.p.m.); and gentle manual swirling. Most cultures do well without mixing, particularly when not too concentrated, but when possible, gentle manual swirling (once each day) is recommended.

CULTURE VESSELS

Culture vessels should have the following properties: non-toxic (chemically inert); reasonably transparent to light; easily cleaned and sterilized; and provide a large surface to volume ratio (depending on the organism).

Certain materials which could potentially be used for culture vessels may leach chemicals which have a deleterious effect on algal growth into the medium. The use of chemically inert materials is particularly important when culturing oceanic plankton and during isolation. Recommended materials for culture vessels and media preparation include Teflon, polycarbonate, polystyrene, and borosilicate glass.

Culture vessels are usually borosilicate glass conical flasks (narrow or wide mouth Erlenmeyer flasks) of various volumes (from tens of milliliters to $3-5$ l) or test-tubes for liquid culture, and test-tubes for agar cultures. Borosilicate glass flasks and tubes, which have been shown to inhibit growth of some species, can be replaced by more expensive transparent polycarbonate vessels, which offer excellent clarity and good physical strength. Like borosilicate, polycarbonate is autoclavable, but is expensive and becomes cloudy and cracks with repeated autoclaving, undergoing some loss of mechanical strength. Teflon is very expensive and it is used only for media preparation, and polystyrene, the cheaper alternative to Teflon and polycarbonate, is not autoclavable. Polystyrene tissue culture flasks can be purchased as single-use sterile units (Iwaki, Nunc, Corning) and used for transport purposes.

The vessels are capped by non-adsorbent cotton-wool plugs, which will allow gas transfer but prevent entry of microbial contaminants. A more efficient and costly way of capping is to use foam plugs or silicone bubble stoppers (bungs), which also allow efficient gas transfer: they are re-usable and autoclavable. Glass, polypropylene, or metal covers and caps are also used for flasks and tubes.

Materials which should generally be avoided during microalgal culturing include all types of rubber and PVC.

MEDIA CHOICE AND PREPARATION

The correct maintenance of algal strains is dependent on the choice of growth media and culture parameters. Two approaches are possible for selection of media composition:

- In theory it is best to work on the principle that if the alga does not need the addition of any particular chemical substance to the culture media (i.e., if it has no observable positive effect on growth rate), do not add it.
- In practice it is often easier to follow well-known (and presumably, therefore, well tried) media recipes, and safer to add substances "just in case" (providing they have no observable detrimental effect on algal growth).

When choosing a culture medium, the natural habitat of the species in question should be considered in order to determine its environmental requirements. It is important to know whether the environment is eutrophic, hence nutrient rich, or oligotrophic, hence nutrient poor, and whether the algae belong to a r-selected or a k-selected species. r-selected species are characterized by a rapid growth rate, autotrophic metabolism, and a wide environmental plasticity, whereas k-selected species shows a slow growth rate, mixotrophic or photoheterotrophic metabolism, and a low environmental tolerance.

The media recipes currently available are not always adequate for many species, and the exact choice for a particular species therefore is dependent on trial and error. It must be remembered that in culturing in general there are (within limits) no right and wrong methods; culture media have only developed trying out various additions, usually based on theoretical considerations. Refinement of media composition for laboratory-maintained algal cultures have been the object of research for several decades, resulting in many different media recipes being reported in the literature and being used in different laboratories.

Media can be classified as being defined or undefined. Defined media, which are often essential for nutritional studies, have constituents that are all known and can be assigned a chemical formula. Undefined media, on the other hand, contain one or more natural or complex ingredients, for example, agar or liver extract and seawater, the composition of which is unknown and may vary. Defined and undefined media may be further subdivided into freshwater or marine media.

In choosing or formulating a medium, it may be important to decide whether it is likely to promote heavy bacterial growth. Richly organic media should be avoided unless the algae being cultivated are axenic. For contaminated cultures, mineral media should be used. These may contain small amount of organic constituents, such as vitamins or humic acids and, in either case, provide insufficient carbon for contaminating organisms to outgrow the algae.

Culture collections have attempted to rationalize the number of media recipes, and to standardize recipes for algal strain maintenance. In particular, the use of undefined biphasic media (soil/water mixture) is declining, due to lack of reproducibility in media batches and occasional contamination of the media from soil samples.

Media may be prepared by combining concentrated stock solutions, which are not combined before use, to avoid precipitation and contamination. Reagent grade chemicals and bidistilled (or purer) water should be used to make stock solutions of enrichments. Gentle heating and/or magnetic stirring of stock solutions can be used to ensure complete dissolution. When preparing a stock solution containing a mixture of compounds, each compound has to be dissolved individually in a minimal volume of water before mixing, then combined with the other and the volume diluted to the needed amount.

Another practical way of preparing artificial defined media is to mix the ingredients together and dry them prior to long-term storage. Constituents are added sequentially to distilled water, smallest quantities first, for a solution and finally a stiff slurry. Prior to the addition of the major salts, the pH is adjusted between 4 and 5. The mixture is then transferred to a clean desiccator of suitable size and the final additions are made. The well-stirred slurry is vacuum-dried, and the dry mixture is then stored with calcium chloride as a desiccant. It can be stored for some years if kept dry.

Solid media are usually prepared using 1.0–1.5% w/v agar, tubes being rested at an angle of 30° during agar gelation to form a slope that increases the surface area available for growth.

As described earlier, media are sterilized either by autoclaving at 126°C (20 min, 1 atm pressure), or by filtration, using 0.2 μm pore nitrate cellulose sterile filters. Generally, subculturing is performed using aseptic microbiological techniques; laminar flow cabinets, equipped with Bunsen flame, microbiological loops, and glass or plastic sterile pipettes are required.

FRESHWATER MEDIA

Freshwater media are generally selected because they possess characteristics similar to the natural environment or they differentially select for a specific algal component of the habitat. Media of an artificial nature, with known chemical composition, are often employed as additives to natural media with an unknown chemical composition, such as lake water, to enrich them. They are often used to simulate diverse nutritional or physical requirements of a particular species or groups of species, especially when the exact nutritional requirements are unknown.

Media are generally prepared from premixed stock solutions. Aliquots from these stocks are measured and added to a given volume of water. Some, however, must be prepared by weighting or measuring the desired components and adding them directly to a given volume of liquid. Accuracy in measuring liquid aliquots from stock solution or water, and weighing of chemicals is essential. Improper procedures may result in precipitation of one or more of the components of the medium, such as nitrates and phosphates, or a failure of some of the constituents to go into solutions.

Stock solution can be prepared and stored at low temperature in tightly sealed glassware, because evaporation may alter initial concentrations.

Water generally employed for freshwater media should belong to one of the following types: copper-distilled water; single glass-distilled water; double glass-distilled water; membrane filtered water; and deionized water. In most laboratories single or double-distilled water is routinely used, which can be deionized by passing it through a prepacked deionizing column.

As for the marine media, also freshwater media can be "defined" and "undefined." Defined medium such as Beijerinck or Bold Basal Medium have been proved successful for many algal classes. Most of these defined media can be used for additional algal groups by adding a variety of other components or modifying the amounts of certain reagents. These "undefined" media often have the advantage of supporting growth of large number of different algal species, but, when highly organic, they have also the disadvantage of encouraging more bacterial growth than strictly inorganic media.

Some of the most commonly used freshwater media, defined and undefined, are listed in Table 6.2–Table 6.9.

TABLE 6.2
BG11 Medium Composition

Reagents	Per Liter
NaNO$_3^a$	1.5 g
K$_2$HPO$_4$ * 3H$_2$O	0.004 g
MgSO$_4$ * 7H$_2$O	0.075 g
CaCl$_2$ * 2H$_2$O	0.027 g
Citric acid (C$_6$H$_8$O$_7$)	0.006 g
Ammonium ferric citrate (C$_6$H$_8$O$_7$ * nFe * nNH$_3$)	0.006 g
EDTANa$_2$Mg	0.001 g
Na$_2$CO$_3$	0.02 g
Microelement stock solution	1 ml
Microelement Stock Solution	
H$_3$BO$_3$	2.860 g
MnCl$_2$ * 4H$_2$O	1.810 g
ZnSO$_4$ * 7H$_2$O	0.222 g
Na$_2$MoO$_4$ * 2H$_2$O	0.390 g
CuSO$_4$ * 5H$_2$O	0.079 g
Co(NO$_3$)$_2$ * 6H$_2$O	0.0494 g

pH = 7.4
^{a}To be omitted for N$_2$-fixing cyanobacteria

TABLE 6.3
Diatom Medium Composition

Reagents	Per Liter
Ca(NO$_3$)$_2$ * 4H$_2$O	20 mg
KH$_2$PO$_4$	12.4 mg
MgSO$_4$ * 7H$_2$O	25 mg
NaHCO$_3$	15.9 mg
EDTA FeNa	2.25 mg
EDTA Na$_2$	2.25 mg
H$_3$BO$_3$	2.48 mg
MnCl$_2$ * 4H$_2$O	1.39 mg
(NH$_4$)$_6$Mo$_7$O$_{24}$ * 4H$_2$O	1.0 mg
Biotin (Vitamin H)	0.04 mg
Thiamine HCl (Vitamin B$_1$)	0.04 mg
Cyanocobalamin (Vitamin B$_{12}$)	0.04 mg
Na$_2$SiO$_3$ * 9H$_2$O	57 mg

pH = 6.9

TABLE 6.4
DY-III Medium Composition

Reagents	Per Liter
$MgSO_4 * 7H_2O$	50 mg
KCl	3 mg
NH_4Cl	2.68 mg
$NaNO_3$	20 mg
H_3BO_3	0.8 mg
Na_2 EDTA $* 2H_2O$	8 mg
$Na_2SiO_3 * 9H_2O$	14 mg
$FeCl_3 * 6H_2O$	1 mg
$CaCl_2$	75 mg
MES buffer	200 mg
Microelement stock solution	1 ml
Vitamin solution	1 ml
Microelement Stock Solution	
$MnCl_2 * 4H_2O$	200 mg
$ZnSO_4 * 7H_2O$	40 mg
$CoCl_2 * 6H_2O$	8 mg
$Na_2MoO_4 * 6H_2O$	20 mg
$Na_3VO_4 * nH_2O$	2 mg
H_2SeO_3	2 mg
Vitamin Solution	
Biotin (Vitamin H)	0.5 μg
Thiamine HCl (Vitamin B_1)	100 mg
Cyanocobalamin (Vitamin B_{12})	0.5 mg

pH = 6.7

TABLE 6.5
Aaronson Medium Composition

Reagents	Per Liter
NH_4Cl	0.5 g
KH_2PO_4	0.3 g
$MgSO_4 * 7H_2O$	1.0 g
$MgCO_3$	0.4 g
$C_6H_9NO_6$ (Nitrilotriacetic acid)	0.2 g
$CaCO_3$	0.05 g
$C_5H_9NO_4$ (Glutamic acid)	3.0 g
Glucose	10.0 g
Arginine	0.4 g
Hystidine monochloride	0.4 g
Biotin (Vitamin H)	0.00001 g
Thiamine HCl (Vitamin B_1)	0.001 g
Microelement stock solution	10 ml

Microelement Stock Solution	
$Fe(NH_4)_2 (SO_4)_2 * 6H_2O$	0.5 g
H_3BO_3	0.025 g
$MnSO4 * H_2O$	0.125 g
$CoSO_4 * 7H_2O$	0.025 g
$CuSO_4 * 5H_2O$	0.020 g
$ZnSO_4 * H_2O$	0.25 g
$(NH_4)_6Mo_7O_{24} * 4H_2O$	0.0125 g
$Na_3VO_4 * 16H_2O$	0.025 g

pH = 5.0

TABLE 6.6
Cramer and Myers Medium Composition

Reagents	Per Liter
$(NH_4)_2HPO_4$	1 g
KH_2PO_4	1 g
$MgSO_4 * 7H_2O$	0.2 g
$CaCl_2$	0.02 g
Sodium Citrate ($Na_3C_6H_5O_7 * 2H_2O$)	0.8 g
$Fe_2(SO_4)_3 * nH_2O$	3 mg
$MnCl_2 * 4H_2O$	1.8 mg
$CoSO_4 * 7H_2O$	1.5 mg
$ZnSO_4 * 7H_2O$	0.4 mg
$Na_2MoO_4 * 2H_2O$	0.2 mg
$CuSO_4 * 5H_2O$	0.02 mg
H_3BO_3	2.48 mg
Thiamine HCl (Vitamin B_1)	0.5 mg
Cyanocobalamin (Vitamin B_{12})	0.02 mg

Carbon Sources (Choose One)

Glucose[•]	10 g
Succinic acid ($C_4H_6O_4$)[•]	5 g
Glutamic acid ($C_5H_9NO_4$)[•]	5 g
Sodium Acetate ($C_2H_3O_2Na * 5H_2O$)[••]	1.64 g

[•] pH = 3.4
[••] pH = 6.8

TABLE 6.7
Beijerinck Medium Composition

Reagents	Per Liter
Stock I	100 ml
Stock II	40 ml
Stock III	60 ml
Micronutrients	1 ml
Stock I	
NH_4NO_3	1.5 g
K_2HPO_4	0.2 g
$MgSO_4 * 7H_2O$	0.2 g
$CaCl_2 * 2H_2O$	0.1 g
Stock II	
KH_2PO_4	9.07 g
Stock III	
K_2HPO_4	11.61 g
Micronutrients	
H_3BO_3	10 g
$MnCl_2 * 4H_2O$	5 g
EDTA	50 g
$CuSO_4 * 5H_2O$	1.5 g
$ZnSO_4 * H_2O$	22 g
$CoCl_2 * 6H_2O$	1.5 g
$FeSO_4 * 7H_2O$	5 g
$(NH_4)_6Mo_7O_{24} * 4H_2O$	1 g

pH = 6.8

TABLE 6.8
Bold Basal Medium Composition

Reagents	Per Liter
KH_2PO_4	175 mg
$CaCl_2 * 2H_2O$	25 mg
$MgSO_4 * 7H_2O$	75 mg
$NaNO_3$	250 mg
K_2HPO_4	75 mg
NaCl	25 mg
H_3BO_3	11.42 mg
Microelement stock solution	1 ml
Solution 1	1 ml
Solution 2	1 ml
Microelement Stock Solution	
$ZnSO_4 * 7H_2O$	8.82 g
$MnCl_2 * 4H_2O$	1.44 g
MoO_3	0.71 g
$CuSO_4 * 5H_2O$	1.57 g
$Co(NO_3)_2 * 6H_2O$	0.49 g
Solution 1	
Na_2EDTA	50 g
KOH	3.1 g
Solution 2	
$FeSO_4$	4.98 g
H_2SO_4 (Conc.)	1 ml
pH = 6.8	

TABLE 6.9
Mes-Volvox Medium Composition

Reagents	Per Liter
MES buffer	1.95 g
$Ca(NO_3)_2 * 4H_2O$	117.8 mg
Na_2-glycerophosphate $* 5H_2O$	60 mg
$MgSO_4 * 7H_2O$	40 mg
KCl	50 mg
NH_4Cl	26.7 mg
Biotin (Vitamin H)	0.0025 mg
Cyanocobalamin (Vitamin B_{12})	0.0015 mg
Microelement stock solution	6 ml
Microelement Stock Solution	
Na_2EDTA	750 mg
$FeCl_3 * 6H_2O$	97 mg
$MnCl_2 * 4H_2O$	41 mg
$ZnCl_2$	5 mg
$CoCl_2 * 6H_2O$	2 mg
$Na_2MoO_4 * 2H_2O$	4 mg

pH = 6.7

TABLE 6.10
Average Concentration of Typical Seawater Constituents

Element	Average Molar Concentration (Range)
Group I	
Na^+	4.7×10^{-1}
K^+	1.02×10^{-2}
Mg^{2+}	5.3×10^{-2}
Ca^{2+}	1.03×10^{-2}
Cl^-	5.5×10^{-1}
SO_4^{2-}	2.8×10^{-2}
HCO_3^-	2.3×10^{-3}
BO_3^{3-}	4.2×10^{-4}
Group II	
Br^-	8.4×10^{-4}
F^-	6.8×10^{-5}
IO_3^-	4.4×10^{-7}
Li^+	2.5×10^{-5}
Rb^+	1.4×10^{-6}
Sr^{2+}	8.7×10^{-5}
Ba^{2+}	1×10^{-7}
MoO_4^{2-}	1.1×10^{-7}
VO_4^{3-}	2.3×10^{-8}
CrO_4^{2-}	4×10^{-9}
AsO_4^{3-}	2.3×10^{-8}
SeO_4^{2-}	1.7×10^{-9}
Group III	
NO_3^-	3×10^{-5} (10^{-8} to 4.5×10^{-5})
PO_4^{3-}	2.3×10^{-6} (10^{-7} to 3.5×10^{-6})
Fe^{3+}	1×10^{-9} (10^{-10} to 10^{-7})
Zn^{2+}	6×10^{-9} (5×10^{-11} to 10^{-7})
Mn^{2+}	5×10^{-10} (2×10^{-10} to 10^{-6})
Cu^{2+}	4×10^{-9} (5×10^{-10} to 6×10^{-9})
Co^{2+}	2×10^{-11} (10^{-11} to 10^{-10})
SiO_4^{4-}	1×10^{-4} (10^{-7} to 1.8×10^{-4})
Ni^{2+}	8×10^{-9} (2×10^{-9} to 1.2×10^{-8})

MARINE MEDIA

Seawater is an ideal medium for growth of marine species, but it is an intrinsically complex medium, containing over 50 known elements in addition to a large but variable number of organic compounds. Usually it is necessary to enrich seawater with nutrients such as nitrogen, phosphorus, and iron. Synthetic formulations have been designed primarily to provide simplified, defined media. Marine species generally have fairly wide tolerances, and difficulties attributed to media can frequently be related to problems of isolation, conditions of manipulations and incubation, and physiological state of the organism. A single medium will generally serve most needs of an investigator. Many media are only major variations of some widely applicable, and often equally effective media. Whatever the choice, a medium should be as simple as possible in composition and preparation.

Media for the culture of marine phytoplankton consist of a seawater base (natural or artificial) which may be supplemented by various substances essential for microalgal growth, including nutrients, trace metals and chelators, vitamins, soil extract, and buffer compounds.

The salinity of the seawater base should first be checked (30–35‰ for marine phytoplankton), and any necessary adjustments (addition of fresh water/evaporation) made before addition of enrichments.

Seawater, stock solutions of enrichments, and the final media must be sterile in order to prevent (or more realistically minimize) biological contamination of unialgal cultures. Autoclaving is a process which has many effects on seawater and its constituents, potentially altering or destroying inhibitory organic compounds, as well as beneficial organic molecules. Because of the steam atmosphere in an autoclave, CO_2 is driven out of the seawater and the pH is raised to about 10, a level which can cause precipitation of the iron and phosphate added in the medium. Some of this precipitate may disappear upon re-equilibriation of CO_2 on cooling, but both the reduced iron and phosphate levels and the direct physical effect of the precipitate may limit algal growth. The presence of ethylenediaminetetraacetic acid (EDTA) and the use of organic phosphate may reduce precipitation effects. Addition of 5% or more of distilled water may also help to reduce precipitation (but may affect final salinity). The best solution, however, if media are autoclaved, is to sterilize iron and phosphate (or even all media additions) separately and add them aseptically afterwards.

Some marine microalgae grow well on solid substrate. A 3% high grade agar can be used for the solid substrate. The agar and culture medium should not be autoclaved together, because toxic breakdown products can be generated. The best procedure is to autoclave 30% agar in deionized water in one container and nine times as much seawater base in another. After removing from the autoclave, sterile nutrients are added aseptically to the water, which is then mixed with the molten agar. After mixing, the warm fluid is poured into sterile Petri dishes, where it solidifies when it cools. The plate is inoculated by placing a drop of water containing the algae on the surface of the agar, and streaking with a sterile implement. The plates are then maintained under standard culture conditions.

Seawater Base

The quality of water used in media preparation is very important. Natural seawater can be collected near shore, but its salinity and quality is often quite variable and unpredictable, particularly in temperate and polar regions (due to anthropogenic pollution, toxic metabolites released by algal blooms in coastal waters). The quality of coastal water may be improved by ageing for a few months at 4°C (allowing bacteria degradation), by autoclaving (heat may denature inhibitory substances), or by filtering through acid-washed charcoal (which absorbs toxic organic compounds). Most coastal waters contain significant quantities of inorganic and organic particulate matter, and therefore must be filtered before use (e.g., Whatman no. 1 filter paper).

The low biomass and continual depletion of many trace elements from the surface waters of the open ocean by biogeochemical processes makes this water much cleaner, and therefore preferable for culturing purposes. Seawater can be stored in polyethylene carboys, and should be stored in cool dark conditions.

Artificial seawater, made by mixing various salts with deionized water, has the advantage of being entirely defined from the chemical point of view, but it is very laborious to prepare, and often does not support satisfactory algal growth. Trace contaminants in the salts used are at rather high concentrations in artificial seawater because so much salt must be added to achieve the salinity of full strength seawater. Commercial preparations are available, which consists of synthetic mixes of the major salts present in natural sea water, such as Tropic Marine Sea Salts produced in Germany for Quality Marine (USA) and Instant Ocean Sea Salts by Aquarium System (USA).

Nutrients, Trace Metals, and Chelators

The term 'nutrient' is colloquially applied to a chemical required in relatively large quantities, but can be used for any element or compound necessary for algal growth.

The average concentrations of constituents of potential biological importance found in typical seawater are summarized in Table 6.10. These nutrients can be divided into three groups, I, II, and III; with decreasing concentration:

- *Group I.* Concentrations of these constituents exhibit essentially no variation in seawater, and high algal biomass cannot deplete them in culture media. These constituents do not, therefore, have to be added to culture media using natural seawater, but do need to be added to deionized water when making artificial seawater media.
- *Group II.* Also have quite constant concentrations in seawater, or vary by a factor less than 5. Because microalgal biomass cannot deplete their concentrations significantly, they also do not need to be added to natural seawater media. Standard artificial media (and some natural seawater media) add molybdenum (as molybdate), an essential nutrient for algae, selenium (as selenite), which has been demonstrated to be needed by some algae, as well as strontium, bromide, and flouride, all of which occur at relatively high concentrations in seawater, but none of which have been shown to be essential for microalgal growth.
- *Group III.* All known to be needed by microalgae (silicon is needed only by diatoms and some chrysophytes, and nickel is only known to be needed to form urease when algae are using urea as a nitrogen source). These nutrients are generally present at low concentrations in natural seawater, and because microalgae take up substantial amounts, concentrations vary widely (generally by a factor of 10 to 1000). All of these nutrients (except silicon and nickel in some circumstances) generally need to be added to culture media in order to generate significant microalgal biomass.

Nitrate is the nitrogen source most often used in culture media, but ammonium can also be used, and indeed is the preferential form for many algae because it does not have to be reduced prior to amino acid synthesis, the point of primary intracellular nitrogen assimilation into the organic linkage. Ammonium concentrations greater than 25 μM are, however, often reported to be toxic to phytoplankton, so concentrations should be kept low.

Inorganic (ortho)phosphate, the phosphorus form preferentially used by microalgae, is most often added to culture media, but organic (glycero)phosphate is sometimes used, particularly when precipitation of phosphate is anticipated (when nutrients are autoclaved in the culture media rather than separately, e.g.). Most microalgae are capable of producing cell surface phosphatases, which allow them to utilize this and other forms of organic phosphate as a source of phosphorus.

The trace metals that are essential for microalgal growth are incorporated into essential organic molecules, particularly a variety of coenzyme factors that enter into photosynthetic reactions. Of these metals, the concentrations (or more accurately the biologically available concentrations) of Fe, Mn, Zn, Cu and Co (and sometimes Mo and Se) in natural waters may be limiting to algal growth. Little is known about the complex relationships between chemical speciation of metals and biological availability. It is thought that molecules that complex with metals (chelators) influence the availability of these elements. Chelators act as trace metal buffers, maintaining constant concentrations of free ionic metal. It is the free ionic metal, not the chelated metal, which influences microalgae, either as a nutrient or as a toxin. Without proper chelation some metals (such as Cu) are often present at toxic concentrations, and others (such as Fe) tend to precipitate and become unavailable to phytoplankton. In natural seawater, dissolved organic molecules (generally present at concentrations of $1-10$ mg l^{-1}) act as chelators. The most widely used chelator in culture media additions is EDTA, which must be present at high concentrations because most complexes with Ca and Mg, present in large amounts in seawater. EDTA may have an additional benefit of reducing precipitation during autoclaving. High concentrations have, however, occasionally been reported to be toxic to microalgae. As an alternative the organic chelator citrate is sometimes utilized, having the advantage of being less influenced by Ca and Mg. The ratio of chelator:metal in culture medium ranges from 1:1 in f/2 to 10:1 in K medium. High ratios may result in metal deficiencies for coastal phytoplankton (i.e., too much metal is complexed), and many media therefore use intermediate ratios.

In today's aerobic ocean, iron is present in the oxidized form as various ferric hydroxides and thus is rather insoluble in seawater. While concentrations of nitrogen, phosphorus, zinc, and manganese in deep water are similar to plankton elemental composition, there is proportionally 20 times less iron in deep water than is apparently needed, leading to the suggestion that iron may be the ultimate geochemically limiting nutrient to phytoplankton in the ocean. Very little is known about iron in seawater or phytoplankton uptake mechanisms due to the complex chemistry of the element. Iron availability for microalgal uptake seems to be largely dependent on levels of chelation. It is highly recommended that iron be added as the chemically prepared chelated iron salt of EDTA rather than as iron chloride or other iron salts; the formation of iron chelates is relatively slow, and iron hydoxides will form first in seawater, leading to precipitation of much of the iron in the culture medium.

Apparently, as a result of the extreme scarcity of copper in anaerobic waters, copper did not begin to be utilized by organisms until the earth became aerobic and copper increased in abundance. Consequently copper does not seem to be an obligate requirement, algae either not needing it, or needing so little that free ionic copper concentrations in natural seawater are sufficient to maintain maximum growth rates. Copper may indeed be toxic, particularly to more primitive algae, and hence copper, if added to culture media at all, should be kept at low concentrations.

The provision of manganese, zinc, and cobalt in culture medium should not be problematical because even fairly high concentrations are not thought to be toxic to algae.

Vitamins

Roughly all microalgal species tested have been shown to have a requirement for vitamin B_{12}, which appears to be important in transferring methyl groups and methylating toxic elements such as arsenic, mercury, tin, thallium, platinum, gold, and tellurium, around 20% need thiamine, and less than 5% need biotin.

It is recommended that these vitamins are routinely added to seawater media. No other vitamins have ever been demonstrated to be required by any photosynthetic microalgae.

Soil Extract

Soil extract has historically been an important component of culture media. It is prepared by heating, boiling, or autoclaving a $5-30\%$ slurry of soil in fresh water or seawater and subsequently filtering out the soil. The solution provides macronutrients, micronutrients, vitamins, and trace metal chelators in undefined quantities, each batch being different, and hence having unpredictable effects on microalgae. With increasing understanding of the importance of various constituents of culture media, soil extract is less frequently used.

Buffers

The control of pH in culture media is important because certain algae grow only within narrowly defined pH ranges in order to prevent the formation of precipitates. Except under unusual conditions, the pH of natural seawater is around 8. Because of the large buffering capacity of natural seawater due to its bicarbonate buffering system (refer to Chapter 4) it is quite easy to maintain the pH of marine culture media. The buffer system is overwhelmed only during autoclaving, when high temperatures drive CO_2 out of solution and hence cause a shift in the bicarbonate buffer system and an increase in pH, or in very dense cultures of microalgae, when enough CO_2 is taken up to produce a similar effect.

As culture medium cools after autoclaving, CO_2 reenters solution from the atmosphere, but certain measures must be taken if normal pH is not fully restored:

- The pH of seawater may be lowered prior to autoclaving (adjustment to pH 7–7.5 with 1 M HCl) to compensate for subsequent increases.
- Certain media recipes include additions of extra buffer, either as bicarbonate, Tris (Tris-hydroxymethyl-aminomethane), or glycylglycine to supplement the natural buffering system. Tris may also act as a Cu buffer, but has occasionally been cited for its toxic properties to microalgae such as *Haematococcus* sp. Glycylglycine is rapidly metabolized by bacteria and hence can only be used with axenic cultures. These additions are generally not necessary if media are filter sterilized, unless very high cell densities are expected.
- The problem of CO_2 depletion in dense cultures may be reduced by having a large surface area of media exposed to the atmosphere relative to the volume of the culture, or by bubbling with either air (CO_2 concentration ca. 0.03%) or air with increased CO_2 concentrations (0.5–5%). Unless there is a large amount of biomass taking up the CO_2, the higher concentrations could actually cause a significant decline in pH. When bubbling is employed, the gas must first pass through an inline 0.2 μm filter unit (e.g., Millipore Millex GS) to maintain sterile conditions. For many microalgal species, aeration is not an option because the physical disturbance may inhibit growth or kill cells.

Some of the most commonly used marine media, defined and undefined, are listed in Table 6.11–Table 6.17. Table 6.18 indicates the algal classes that have been successfully cultured in the media included in this chapter.

For a full range of possible media refer to the catalog of strains from culture collections present all over the world such as:

- SAG (Experimentelle Phykologie und Sammlung von Algenkulturen, University of Gottigen, http://www.epsag.uni-goettingen.de/)
- UTEX (Culture Collection of Algae at the University of Texas at Austin, http://www.bio.utexas.edu/research/utex/)
- CCAP (Culture Collection of Algae and Protozoa, Argyll, Scotland, http://www.ife.ac.uk/ccap/)
- UTCC (University of Toronto Culture Collection of Algae and Cyanobacteria, http://www.botany.utoronto.ca/utcc/)
- CCMP (The Provasoli-Guillard National Center for Culture of Marine Phytoplankton, Maine, http://ccmp.bigelow.org/)

TABLE 6.11
Walne's Medium Composition

Reagents	Per Liter Seawater
Solution A	1.0 ml
Solution C	0.1 ml
Solution D (to add for diatoms)	2.0 ml
Solution A	
$FeCl_3 * 6H_2O$	1.3 g
$MnCl_2 * 4H_2O$	0.4 g
H_3BO_3	33.6 g
Na_2EDTA	45.0 g
$NaH_2PO_4 * 2H_2O$	20.0 g
$NaNO_3$	100.0 g
Solution B	1 ml
Solution B per 100 ml	
$ZnCl_2$	2.1 g
$CoCl_2 * 6H_2O$	2.0 g
$(NH_4)_6Mo_7O_{24} * 4H_2O$	0.9 g
$CuSO_4 * 5H_2O$	2.0 g
Concentrated HCl	10 ml
Solution C per 200 ml	
Thiamine HCl (Vitamin B_1)	0.2 g
Cyanocobalamin (Vitamin B_{12})	10 mg
Solution D (per liter)	
$Na_2SiO_3 * 5H_2O$	40.0 g

TABLE 6.12
ASN-III Medium Composition

Reagents	Per Liter Seawater
NaCl	25 g
$MgCl_2 * 6H_2O$	2 g
$MgSO_4 * 7H_2O$	3.5 g
$NaNO_3$	0.75 g
$K_2HPO_4 * 3H_2O$	0.02 g
KCl	0.5 g
$CaCl_2 * 2H_2O$	0.5 g
Citric acid ($C_6H_8O_7$)	0.003 g
Ammonium ferric citrate ($C_6H_8O_7 * n$Fe $* n$NH$_3$)	0.003 g
$EDTANa_2Mg$	0.0005 g
Na_2CO_3	0.02 g
Microelement stock solution	1 ml
Microelement Stock Solution	
H_3BO_3	2.860 g
$MnCl_2 * 4H_2O$	1.810 g
$ZnSO_4 * 7H_2O$	0.222 g
$Na_2MoO_4 * 2H_2O$	0.390 g
$CuSO_4 * 5H_2O$	0.079 g
$Co(NO_3)_2 * 6H_2O$	0.0494 g
pH = 7.5	

TABLE 6.13
CHU-11 Medium Composition

Reagents	Per Liter Seawater
$MgSO_4 * 7H_2O$	0.075 g
$CaCl_2 * 2H_2O$	0.036 g
$NaNO_3$	1.5 g
$K_2HPO_4 * 3H_2O$	0.04 g
Na_2CO_3	0.02 g
$Na_2SiO_3 * 9H_2O$	0.58 g
Citric acid ($C_6H_8O_7$)	0.006 g
Ammonium ferric citrate ($C_6H_8O_7 * n$Fe $* n$NH$_3$)	0.006 g
$EDTANa_2Mg$	0.001 g
Microelement stock solution	1 ml
Microelement Stock Solution	
H_3BO_3	0.5 g
$MnCl_2 * 4H_2O$	2 g
$Zn(NO_3)_2 * 6H_2O$	0.5 g
$Na_2MoO_4 * 2H_2O$	0.025 g
$CuCl_2 * 2H_2O$	0.025 g
$Co(NO_3)_2 * 6H_2O$	0.025 g
$VOSO_4 * 6H_2O$	0.025 g
HCl 1 *M*	3 ml
pH = 7.5	

TABLE 6.14
PCR-S11 Medium Composition

Reagents	Per Liter Seawater
$(NH_4)_2SO_4$	2.68 mg
Na_2 EDTA/$FeCl_3$	8 mg
Na_2PO_4 (pH 7.5)	14 mg
HEPES-NaOH (pH 7.5)	200 mg
Microelement stock solution	0.5 ml
Microelement Stock Solution	
H_3BO_3	18.549 mg
$MnSO_4 * H_2O$	10.140 mg
$ZnSO_4 * 7H_2O$	1.725 mg
$(NH_4)_6Mo_7O_{24} * 4H_2O$	0.494 mg
$CuSO_4 * 5H_2O$	0.749 mg
$Co(NO_3)_2 * 6H_2O$	0.873 mg
$VSO_5 * 5H_2O$	0.098 mg
$Na_2WO_4 * 2H_2O$	0.198 mg
KBr	0.714 mg
KI	0.498 mg
$Cd(NO_3)_2 * 6H_2O$	0.925 mg
$NiCl_2 * 6H_2O$	0.713 mg
$Cr(NO_3)_3 * 9H_2O$	0.240 mg
SeO_2	0.333 mg
$KAl(SO_4)_2 * 12H_2O$	2.846 mg
Cyanocobalamin (Vitamin B_{12})	10 μg
pH = 7.5	

TABLE 6.15
f/2 Medium Composition

Reagents	Per Liter Seawater
$NaNO_3$	0.075 g
$NaH_2PO_4 * H_2O$	0.005 g
Microelement stock solution	1 ml
Vitamin solution	1 ml
Microelement Stock Solution	
$FeCl_3 * 6H_2O$	3.150 g
Na_2 EDTA	4.160 g
$MnCl_2 * 4H_2O$	0.180 g
$CoCl_2 * 6H_2O$	0.010 g
$CuSO_4 * 5H_2O$	0.010 g
$ZnSO_4 * H_2O$	0.022 g
$Na_2MoO_4 * 2H_2O$	0.006 g
Vitamin Solution	
Biotin (Vitamin H)	0.5 mg
Thiamine HCl (Vitamin B_1)	100 mg
Cyanocobalamin (Vitamin B_{12})	0.5 mg
pH = adjust to 8.0 with 1 *M* NAOH or HCl	

TABLE 6.16
K Medium Composition

Reagents	Per Liter Seawater
$NaNO_3$	75 mg
Na_2-glycerophosphate $* 5H_2O$	2.16 mg
NH_4Cl	2.67 mg
Tris-base (pH 7.2)	121.1 mg
Microelement stock solution	1 ml
Vitamin solution	1 ml
Microelement Stock Solution	
Fe Na EDTA	4.3 g
$Na_2EDTA * 2H_2O$	37.22 g
$CuSO_4 * 5H_2O$	4.9 mg
$MnCl_2 * 4H_2O$	178.2 mg
$ZnSO_4 * 7H_2O$	22 mg
$Na_2MoO_4 * 2H_2O$	6.3 g
H_2SeO_3	1.29 µg
$FeCl_3 * 6H_2O$	3,15 g
$CoCl_2 * 6H_2O$	11.9 mg
Vitamin Solution	
Biotin (Vitamin H)	0.1 mg
Thiamine HCl (Vitamin B_1)	200 mg
Cyanocobalamin (Vitamin B_{12})	1 mg
pH = 7.2	

TABLE 6.17
ESAW Medium Composition

Reagents	Per Liter
NaCl	21.19 g
Na_2SO_4	3.55 g
KCl	0.599 g
Na_2HCO_3	0.174 g
KBr	0.0863 g
H_3BO_3	0.023 g
NaF	0.0028 g
$MgCl_2 * 6H_2O$	9.592 g
$CaCl_2 * 2H_2O$	1.344 g
$SrCl_2$	0.0218 mg
$NaNO_3$	46.7 mg
$NaH_2PO_4 * H_2O$	3.09 mg
$Na_2SiO_3 * 9H_2O$	30 mg
Metal Stock I	1 ml
Metal Stock II	1 ml
Vitamin Solution	1 ml
Metal Stock I	
$Na_2EDTA * 2H_2O$	3.09 g
$FeCl_3 * 6 H_2O$	1.77 g
Metal Stock II	
$Na_2EDTA * 2H_2O$	2.44 g
$ZnSO_4 * 7H_2O$	0.073 g
$CoSO_4 * 7H_2O$	0.016 g
$MnSO_4 * 4H_2O$	0.54 g
$Na_2MoO_4 * 2H_2O$	1.48 mg
Na_2SeO_3	0.173 mg
$NiCl_2 * 6H_2O$	1.49 mg
Vitamin Solution	
Biotin (Vitamin H)	1 mg
Thiamine HCl (Vitamin B_1)	100 mg
Cyanocobalamin (Vitamin B_{12})	2 mg

pH = 8.2

TABLE 6.18
Main Algal Groups Cultured in the Media

Medium	Group Cultured
BG11 medium	Freshwater and soil Cyanophyceae
Diatom medium	Freshwater Bacillariophyceae
DY-III medium	Freshwater Chrysophyceae
Aaronson medium	*Ochromonas* sp.
Cramer and Myers medium	Euglenophyceae
Beijerinck medium	Freshwater Chlorophyceae
Bold Basal medium	Broad spectrum medium for freshwater Chlorophyceae, and Xantophyceae, Chrysophyceae, and Cyanophyceae
MES-Volvox medium	Broad spectrum medium for freshwater algae
Walne's medium	Broad spectrum medium for marine algae (especially designed for mass culture)
ASN-III medium	Marine Cyanophyceae
CHU-11 medium	Marine Cyanophyceae
PCR-S11 medium	Prochlorophyceae
f/2 medium	Broad spectrum medium for coastal algae
K medium	Broad spectrum medium for oligotrophic marine algae
ESAW medium	Broad spectrum medium for coastal and open ocean algae

- CSIRO (CSIRO Collection of living microalgae, Tasmania, http://www.marine.csiro.au/ microalgae/collection.html)
- PCC (Pasteur Culture Collection of Cyanobacteria http://www.pasteur.fr/recherche/ banques/PCC/)

STERILIZATION OF CULTURE MATERIALS

All vessels used for culture purpose should be scrubbed (abrasive brushes not appropriate for most plastics) and soaked with warm detergent (not domestic detergents, which leave a residual film on culture ware, but phosphate-free laboratory detergent), then rinsed extensively with tap water. After soaking in 10% HCl for 1 day–1 week (not routinely necessary, but particularly important for new glass and polycarbonate material), vessels should be rinsed extensively with distilled and finally bidistilled water, and left inverted to dry in a clean, dust-free place, or in an oven.

Sterilization can be defined as a process which ensures total inactivation of microbial life, and should not be confused with disinfection, which is defined as an arbitrary reduction of bacterial numbers. The primary purpose of sterilization is to prevent contamination by unwanted organisms, but it may also serve to eliminate unwanted chemicals. Sterilization can be obtained by means of several methods, the choice depending on the purpose and material used, either empty glassware/ plasticware or medium-containing vessels, but also on the facilities available in a laboratory.

Several methods are available for sterilization of material, some of which can be used also for growth media:

- Gas such as ethylene oxide, EtO, finds the best application on heat sensitive equipment on which steam autoclaving (sterilization with heat) cannot be performed. EtO sterilizes by alkylation, it substitutes for hydrogen atoms on molecules such as proteins and DNA, and, by attaching to these molecules and disrupting them, EtO stops these molecules' normal life-supporting functions. This method is widely used for the sterilization of medical devices, but it is not a routinely available technique for algology laboratories. Moreover, EtO is a potent carcinogen.
- Dry heat: some laboratories use dry heat to sterilize empty vessels, putting the material in an oven for at least 3–4 h at 160°C; however, only higher temperature (200–250°C) guarantees an effective result. Vessels are covered with aluminum foil to maintain sterility on removal from oven. This procedure is suitable only for few materials that stand high temperatures, such as glass, teflon, silicone, metal, and cotton.
- Autoclaving (moist heat) is the most widely used technique for sterilizing culture media and vessels, and is the ultimate guarantee of sterility (including the destruction of viruses). A commercial autoclave is the best, but pressure cookers of various sizes are also suitable. Sterility requires 15 min at 1–2 Bar pressure and a temperature of 121°C in the entire volume of the liquid (i.e., longer times for larger volumes of liquid; approximately 10 min for 100 ml, 20 min for 2 l, 35 min for 5 l). Flasks containing media should not be more than half full, and should be left partially open or plugged with cotton wool or covered with aluminium foil or paper, because for sterility the steam must penetrate the material. Autoclave steam may introduce chemical contaminants; empty glass and polycarbonate vessels should be autoclaved containing a small amount of bidistilled water which is poured out (thus diluting contaminants) under sterile conditions immediately prior to use. Vessels should never be closed, because of the risk of implosion, by using cotton wool bungs, or by leaving screw caps slightly open. Ensure that heating elements are covered with distilled water, and the escape valve should not be closed until a steady stream of steam is observed. After autoclaving, the pressure release valve should not be opened until the temperature has cooled to below 80°C. Autoclave steam may contaminate the media (i.e., with trace metals from the autoclave tubing). Autoclaving also produces

leaching of chemicals from the medium receptacle into the medium (silica from glass bottles, toxic chemicals from plastics). Autoclaving in well-used Teflon or polycarbonate vessels reduces leaching of trace contaminants.

- Pasteurization (heating to 90–95°C for 30 min) of media in Teflon or polycarbonate bottles is a potential alternative, reducing the problems of trace metal contamination and alteration of organic molecules inherent with autoclaving. Pasteurization does not, however, completely sterilize seawater containing media; it kills all eukaryotes and most bacteria, but some bacterial spores probably survive. Heating to 90–95°C for at least 30 min and cooling, repeated on two or more successive days ("tyndallization") may improve sterilization efficiency; it is assumed that vegetative cells are killed by heat and heat resistant spores will germinate in the following cool periods and be killed by subsequent heating.

- Ultraviolet radiation (240–280 nm) is not often used for culture materials, because very high intensities are needed to kill everything in a medium such as seawater (1200 W lamp, 2–4 h for culture media in quartz tubes). Such intense UV light necessarily alters and destroys the organic molecules in seawater and generates many long lived free radicals and other toxic reactive chemical species (Brand, 1986). Seawater exposed to intense UV light must, therefore, be stored for several days prior to use to allow the level of these highly reactive chemical species to decline.

- Sterile filtration is probably the best method of sterilizing certain media, especially seawater-based media, without altering their chemistry, as long as care is taken not to contaminate the seawater with dirty filter apparatus. Sterilization efficiency is, however, to some extent reduced compared with heat sterilization methods. Membrane filters of 0.2 μm porosity are generally considered to yield water free of bacteria, but not viruses. 0.1 μm filters can be used, but the time required for filtration of large volumes of culture media may be excessively long. The filtration unit must be sterile: for small volumes (<50 ml) pre-sterilized single use filter units for syringe filtration (e.g., Millipore Millex GS) can be used; for volumes up to 1 l reusable autoclavable self-assembly filter units (glass or polycarbonate) with 47 mm cellulose ester sterile membrane filters (e.g., Millipore HA) can be used with suction provided by a vacuum pump; for larger volumes an inline system with peristaltic pump and cartridge filters may be the best option. Filter units (particularly disposable plastic systems) and the membrane filters can also leak toxic compounds into the filtrate. The first portion of filtrate (e.g., 5% of the volume to be filtered) should be discarded to alleviate this problem.

- Most stock solutions of culture medium additions can be sterilized separately by autoclaving, although vitamin stock solutions are routinely filtered through 0.2 μm single use filter units (e.g., Millipore Millex GS), because heat sterilization will denature these organic compounds. Filter sterilization of all additions may reduce uncertainties about stability of the chemical compounds and contamination from autoclave steam, but absolute sterilization is not guaranteed. Stock solutions can be stored in ultraclean sterile glass, polycarbonate, or Teflon tubes/bottles. In order to minimize effects of any microbial contaminations, all stock solutions should be stored in a refrigerator at 4°C, except vitamin stocks which are stored frozen at −20°C and thawed immediately prior to use.

CULTURE METHODS

Algae can be produced according to a great variety of methods, from closely-controlled laboratory methods to less predictable methods in outdoor tanks. Indoor culture allows control over illumination, temperature, nutrient level, contamination with predators, and competing algae, whereas outdoor algal systems, though cheaper, make it very difficult to grow specific algal cultures for extended periods. Open cultures such as uncovered ponds and tanks (indoors or outdoors) are more readily contaminated than closed culture vessels such as tubes, flasks, carboys, bags, etc. Axenic cultivation can be also chosen, by using algal cultures free of any foreign organisms such as bacteria, but this cultivation is expensive and difficult, because it requires a strict sterilization of all glassware, culture media, and vessels to avoid contamination. These constraints make it impractical (and very expensive) for commercial operations. On the other hand, non-axenic cultivation, though cheaper and less laborious, are more prone to crash, less predictable, and often of inconsistent quality.

Different types of algal cultures are used worldwide, the most routinely adopted include batch, continuous, and semicontinuous ponds and photobioreactors.

BATCH CULTURES

The most common culture system is the batch culture, due to its simplicity and low cost. This is a closed system, volume-limited, in which there is no input or output of materials, that is, resources are finite. The algal population cell density increases constantly until the exhaustion of some limiting factor, whereas other nutrient components of the culture medium decrease over time. Any products produced by the cells during growth also increase in concentration in the culture medium. Once the resources have been utilized by the cells, the cultures die unless supplied with new medium. In practice this is done by subculturing, that is, transferring a small volume of existing culture to a large volume of fresh culture medium at regular intervals. In this method algal cells are allowed to grow and reproduce in a closed container. A typical batch culture set-up can be a 250 ml *Erlenmeyer* culture flask with a cotton/gauze bung; in some cases, the bung can be fitted with a Pasteur pipette and air is bubbled into the culture to maintain high levels of oxygen and carbon dioxide and provide mixing.

Batch culture systems are highly dynamic. The population shows a typical pattern of growth according to a sigmoid curve (Figure 6.1a), consisting of a succession of six phases, characterized by variations in the growth rate (Figure 6.1b); the six phases are summarized in Table 6.19.

The growth curve, relative to the Phases 3, 4 and 5, without the lag, acceleration and crash phases, can be described with a rectangular hyperbolic function similar to the Michaelis-Menten formulation that describes the nutrient uptake kinetics, and the dynamic relationship between photosynthetic rate and irradiance.

After the inoculum, growth does not necessarily start right away, because most cells may be viable, but not in condition to divide. The interval necessary for the transferred cells to adapt to the new situation and start growing is the first phase of the growth curve, the lag phase. This lag or induction phase is relatively long when an algal culture is transferred from a plate to liquid culture. Cultures inoculated with exponentially growing algae have short lag phases, which can seriously reduce the time required for upscaling. The lag in growth is attributed to the physiological adaptation of the cell metabolism to growth, such as the increase of the levels of enzymes and metabolites involved in cell division and carbon fixation. During this phase the growth rate is zero.

After a short phase of growth acceleration, characterized by a continuously increasing growth rate, up to its maximum value, which is achieved in the following exponential phase, the cell density increases as a function of time t according to the exponential function:

$$N_2 = N_1 \cdot e^{\mu} \tag{6.1}$$

where N_2 and N_1 are the number of cells at two successive times and μ is the growth rate. During this phase, the growth rate reached is kept constant. The growth rate is mainly dependent on algal species and cultivation parameters, such as light intensity, temperature, and nutrient availability.

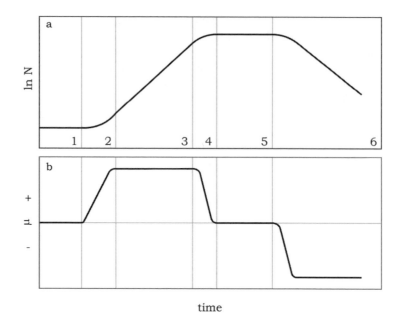

FIGURE 6.1 Growth curve of an algal population under batch culture conditions (a) and corresponding variations of the growth rate (b).

The exponential growth phase normally lasts for a very short period, because cells start to shade each other as their concentration increases. Hence, the culture enters the phase of retardation, and cell growth rate decreases because mainly light, but also nutrients, pH, carbon dioxide, and other physical and chemical factors begin to limit growth. Following this phase, the cell population continues to increase, but the growth rate decreases until it reaches zero, at which point the culture enters the stationary phase, during which the cell concentration remains constant at its maximum value. The final stage of the culture is the death or "crash" phase, characterized by a negative growth rate; during this phase water quality deteriorates, mainly due to catabolite accumulation, and nutrients are depleted to a level incapable to sustain growth. Cell density decreases rapidly and the culture eventually collapses.

TABLE 6.19
Description of the Six Successive Phases of Growth for an Algal Population under Batch Culture Conditions

Phase	Growth	Growth Rate Interpretation	Description
1	Lag	Zero	Physiological adaptation of the inoculum to changing conditions
2	Acceleration	Increasing	Trivial
3	Exponential	Constant	Population growth changes the environment of the cells
4	Retardation	Decreasing	Effects of changing conditions appear
5	Stationary	Zero	One or more nutrients (or light) are exhausted down to the threshold level of the cells
6	Decline	Negative	The duration of stationary phase and the rate of decline are strongly dependent on the kind of organisms

In practice, culture crashes can be caused by a variety of reasons, including the depletion of a nutrient, oxygen deficiency, overheating, pH disturbance, or contamination. The key to the success of algal production is maintaining all cultures in the exponential phase of growth. Also, the nutritional value of the produced algae is inferior once the culture is beyond Phase 4 due to reduced digestibility, deficient composition, and possible production of toxic metabolites.

In batch cultures, cell properties such as size, internal nutrient composition, and metabolic function vary considerably during the above growth phases. This can often make interpretation of the results difficult. During the exponential growth phase, cell properties tend to be constant. However, this phase usually only lasts for a short period of time, and if one wishes to estimate growth rates of the exponential phase of batch cultures, daily sampling appeared to be insufficient to allow a reasonably accurate estimate. Moreover, the accuracy of growth rate determination is highest in artificial, defined media as compared to cells grown in natural surface water media.

A significant advantage of batch culture systems is their operational simplicity. The culture vessels most often consist of an Erlenmeyer flask with a sample to flask volume ratio of about 0.2 in order to prevent carbon dioxide limitation. This volume ratio is only critical if the flasks are shaken by hand once a day during the culturing run. If the flasks are cultured on a rotating shaker table a sample to flask volume ratio of 0.5 is permitted.

Batch culture systems are widely applied because of their simplicity and flexibility, allowing to change species and to remedy defects in the system rapidly. Although often considered as the most reliable method, batch culture is not necessarily the most efficient method. Batch cultures are harvested just prior to the initiation of the stationary phase and must thus always be maintained for a substantial period of time past the maximum specific growth rate. Also, the quality of the harvested cells may be less predictable than that in continuous systems and, for example, vary with the timing of the harvest (time of the day, exact growth phase).

Another disadvantage is the need to prevent contamination during the initial inoculation and early growth period. Because the density of the desired phytoplankton is low and the concentration of nutrients is high, any contaminant with a faster growth rate is capable of outgrowing the culture. Batch cultures also require a lot of labor to harvest, clean, sterilize, refill, and inoculate the containers.

CONTINUOUS CULTURES

In continuous cultures, resources are potentially infinite: cultures are maintained at a chosen point on the growth curve by the regulated addition of fresh culture medium. In practice, a volume of fresh culture medium is added automatically at a rate proportional to the growth rate of the alga, while an equal volume of culture is removed. This method of culturing algae permits the maintenance of cultures very close to the maximum growth rate, because the culture never runs out of nutrients.

Fresh growth medium is stored in the large vessel. Air is pumped into the airspace in this medium vessel. This air pressure will push the medium through a tube which is connected to the culture vessel. By opening and closing the clamp on this medium line one can add medium to the culture vessel. Air is also pumped into the culture vessel. This air passes down a long glass tube to the bottom of the culture and bubbles up. This serves to keep the culture well suspended as well as high in oxygen and CO_2. The air flowing into the culture vessel flows out through an outflow tube. As fresh medium is added to the culture vessel the level of the liquid in the culture vessel rises; when that level reaches the bottom of the outflow tube, old medium and cells flow out of the culture vessel into a waste flask. There is another glass tube in the culture vessel, the sample port. When a sample of cells from the culture vessel is needed the clamp on the sample port can be opened up and medium and cells flow out.

Continuous culture systems have been widely used to culture microbes for industrial and research purposes. The early development of a continuous culture system can be traced back to the 1950s, when the first chemostat, also called bactogen, was developed.

Batch and continuous culture systems differ in that in a continuous culture system, nutrients are supplied to the cell culture at a constant rate, and in order to maintain a constant volume, an equal volume of cell culture is removed. This allows the cell population to reach a "steady state" (i.e., growth and cell division where the growth rate and the total number of cells per milliliter of culture remains constant).

Two categories of continuous cultures can be distinguished *sensu* Fogg and Thake (1987):

- Turbidostat culture, in which fresh medium is delivered only when the cell density of the culture reaches some predetermined point, as measured by the extinction of light passing through the culture. At this point, fresh medium is added to the culture and an equal volume of culture is removed. The diluted culture increases in cell density until the process is repeated.
- Chemostat culture, in which a flow of fresh medium is introduced into the culture at a steady, predetermined rate. The latter adds a limiting vital nutrient (e.g., nitrate) at a fixed rate and in this way the growth rate and not the cell density is kept constant. In a chemostat, the medium addition ultimately determines growth rate and cell density.

In many chemostat continuous culture systems, the nutrient medium is delivered to the culture at a constant rate by a peristaltic pump or solenoid gate system. The rate of media flow can be adjusted, and is often set at approximately 20% of culture volume per day. Air is pumped into the culture vessel through a sterile filter. This bubbling air has three effects: it supplies CO_2 and O_2 to the culture, aids in circulation and agitation of the cultures, and pressurizes the head space of the culture vessel so as to provide the force to "remove" an amount of media (and cells) equal to the volume of inflowing media. The culture may be aseptically sampled by opening the clamp on a sample port. The magnetic stirrer and aeration help to prevent the cells from collecting in the bottom of the culture vessel. A truly continuous culture will have the medium delivered at a constant volume per unit time. However, delivery systems such as peristaltic pumps or solenoid gates are inherently unreliable. In order to deliver exactly the same amounts of medium to several cultures growing at once, a "semicontinuous" approach can be taken.

The rate of flow of medium into a continuous culture system is known as the "dilution rate." When the number of cells in the culture vessel remains constant over time, the dilution rate is said to equal the rate of cell division in the culture, as the cells being removed by the outflow of medium are being replaced by an equal number through cell division in the culture. The principal advantage of continuous culture is that the rate of dilution controls the rate of microbial growth via the concentration of the growth-limiting nutrient in the medium. As long as the dilution rate is lower than the maximum growth rate attainable by the algal species, the cell density will increase to a point at which the cell division rate ("birth rate") exactly balances the cell washout rate ("death rate"). This steady-state cell density is also characterized by a constancy of all metabolic and growth parameters. On the other hand, if the dilution rate exceeds the maximum cell division rate, then cells are removed faster than they are produced and total washout of the entire cell population eventually occurs.

The disadvantages of the continuous system are its relatively high cost and complexity. The requirements for constant illumination and temperature mostly restrict continuous systems to indoors and this is only feasible for relatively small production scales. Continuous cultures have the advantages of producing algae of more predictable quality. Furthermore, they are amenable to technological control and automation, which in turn increases the reliability of the system and reduces the need for labor.

SEMI-CONTINUOUS CULTURES

In a semi-continuous system the fresh medium is delivered to the culture all at once, by simply opening a valve in the medium delivery line. Fresh medium flows into the culture vessel, and spent culture flows out into a collecting vessel. Once the required medium has entered the culture, the valve is closed, and the culture is allowed to grow for 24 h, when the procedure is repeated. The semicontinuous technique prolongs the use of large tank cultures by partial periodic harvesting followed immediately by topping up to the original volume and supplementing with nutrients to achieve the original level of enrichment. The culture is grown up again, partially harvested, etc. Semi-continuous cultures may be indoors or outdoors, but usually their duration is unpredictable. Competitors, predators, or contaminants and metabolites eventually build up, rendering the culture unsuitable for further use. As the culture is not harvested completely, the semi-continuous method yields more algae than the batch method for a given tank size.

COMMERCIAL-SCALE CULTURES

Existing commercial microalgae culture systems range in volume from about 10^2 l to more than 10^9 l. However, aside from the specialized small-scale culture systems (<1000 l) other types of culture systems predominate: large open ponds, circular ponds with a rotating arm to mix the cultures, raceway ponds, or large bags.

There are several considerations as to which culture system to use. Factors to be considered include: the biology of the alga, the cost of land, labor, energy, water, nutrients, climate (if the culture is outdoors), and the type of final product. The various large-scale culture systems also need to be compared on their basic properties such as their light utilization efficiency, ability to control temperature, the hydrodynamic stress placed on the algae, the ability to maintain the culture unialgal or axenic and how easy they are to scale up from laboratory scale to large-scale. The final choice of system is almost always a compromise between all of these considerations to achieve an economically acceptable outcome. Successful further development of the industry requires significant improvements in the design and construction of the photobioreactors as well as a better understanding of the physiology and physical properties of the microalgae to be grown.

A common feature of most of the algal species currently produced commercially (i.e., *Chlorella, Spirulina,* and *Dunaliella*) is that they grow in highly selective environments, which means that they can be grown in open air cultures and still remain relatively free of contamination by other algae and protozoa. Thus, *Chlorella* grows well in nutrient-rich media, *Spirulina* requires a high pH and bicarbonate concentration, and *Dunaliella salina* grows at very high salinity. Those species of algae which do not have this selective advantage must be grown in closed systems. This includes most of the marine algae grown as aquaculture feeds (e.g., *Skeletonema, Chaetoceros, Thalassiosira, Tetraselmis,* and *Isochrysis*) and the dinoflagellate *Crypthecodinium cohnii* grown as a source of long-chain polyunsaturated fatty acids, as well as almost all other species being considered for commercial mass culture. In the particular case of *C. cohnii*, a large scale fermentation plant is operated by Martek Biosciences in Winchester (USA).

Outdoor Ponds

Different types of ponds have been designed and experimented with for microalgae cultivation, which vary in size, shape, materials used for construction, and mixing device. Large outdoor ponds can be unlined, with a natural bottom, or lined with unexpensive materials such as clay, brick, or cement, or expensive plastics such as polyethylene, PVC sheets, glass fiber, or polyurethane. Unlined ponds suffer from silt suspension, percolation, and heavy contamination, and their use is limited to a few algal species, and to particular soil and environmental conditions.

Also natural systems such as euthrophic lakes or small natural basins can be exploited for microalgal production, provided suitable climatic conditions and sufficient nutrients. Examples are the numerous temporary or permanent lakes along the northeast border of Lake Chad, where *Arthrospira* sp. grows almost as monoculture, and is collected for human consumption by the Kanembou people inhabiting those areas. *Arthrospira* sp. naturally blooms also in old volcanic craters filled with alkaline waters in the Myanmar region. Production began at Twin Taung Lake in 1988, and by 1999 increased to 100 tons per year. About 60% is harvested from boats on the surface of the lake, and about 40% is grown in outdoor ponds alongside the lake. During the blooming season in the summer, when the cyanobacterium forms thick mats on the lake, people in boats collect a dense concentration of spirulina in buckets. *Arthrospira* is harvested on parallel inclined filters, washed with fresh water, dewatered, and pressed again. This paste is extruded into noodle like filaments which are dried in the sun on transparent plastic sheets. Dried chips are taken to a pharmaceutical factory in Yangon, pasteurized, pressed into tablets ready to be sold. Another cyanobacterium to be used as a food supplement is *Aphanizomenon flos-aquae*, which since the early 1980s has been harvested from Upper Klamath Lake, Oregon, and sold as a food and health food supplement. In 1998 the market for *A. flos-aquae* as a health food supplement was about $100 million with an annual production greater than 10^6 kg (dry weight). *A. flos-aquae* blooms are often biphasic, with a first peak in late June to early July, and a second peak late in September to mid-October. The harvested biomass is screened and centrifuged to remove small extraneous material. The algal concentrate is then gravity-fed into a vertical centrifuge that applies high centrifugal force to separate cells and colonies, removing about 90% of the remaining water. At this stage the algal product is 6–7% solids. Once concentrated, the product is chilled to 2°C and stored before being pumped to the freezers. The frozen algae is then put into storage boxes and shipped to the freezer facility for storage. When needed, the frozen product is shipped to an external commercial freeze drying facility to be freeze dried into a powder containing 3–5% water content. This final product is processed into consumable products such as capsules or tablets.

Natural ponds that do not necessitate mixing, and need only minimal environmental control, represent other extensive cultivation systems.

The largest natural ponds used for commercial production of microalgae are *Dunaliella salina* lagoons in Australia. Western Biotechnology Ltd. operates 250 ha. of ponds (semi-intensive cultivation) at Hutt Lagoon (Western Australia); Betatene Ltd., a division of Henkel Co. (Germany), operates 460 ha. unmixed ponds (extensive cultivation) at Whyalla (South Australia). Both facilities produce biomass for β-carotene extraction. Other facilities use raceway culture ponds, such as those operated by Cyanotech Co. in Hawaii and Earthrise farms in California for the production of *Haematococcus* and *Artrosphira* biomass. In both cases large raceway ponds from 1000 to 5000 m^2 are adopted, with stirring accomplished by one large paddle wheel per pond. Raceway pond are also used for intensive cultivation of *D. salina* by Nature Beta Technologies Ltd. in Israel.

The nutrient medium for outdoor cultures is based on that used indoors, but agricultural-grade fertilizers are used instead of laboratory-grade reagents. However, fertilization of mass algal cultures in estuarine ponds and closed lagoons used for bivalve nurseries was not found to be desirable as fertilizers were expensive and it induced fluctuating algal blooms, consisting of production peaks followed by total algal crashes. In contrast, natural blooms are maintained at a reasonable cell density throughout the year and the ponds are flushed with oceanic water whenever necessary. Culture depths are typically 0.25–1 m. Cultures from indoor production may serve as inoculum for monospecific cultures. Alternatively, a phytoplankton bloom may be induced in seawater from which all zooplankton has been removed by sand filtration. Algal production in outdoor ponds is relatively inexpensive, but it cannot be maintained for prolonged period and is only suitable for a few, fast-growing species due to problems with contamination by predators, parasites, and more opportunistic algae that tend to dominate regardless of the species used as inoculum. Furthermore, outdoor production is often characterized by a poor batch to batch consistency and unpredictable culture crashes caused by changes in weather, sunlight, or water quality. As stated earlier, at present, large-scale commercial production of microalgae biomass is limited to *Dunaliella*, *Haematococcus*, *Arthrospira*, and *Chlorella*, which are cultivated in open ponds at farms located around the world (Australia, Israel, Hawaii, Mexico, China). These algae are a source for viable and inexpensive carotenoids, pigments, proteins, and vitamins that can be used for the production of nutraceuticals, pharmaceuticals, animal feed additives, and cosmetics. Mass algal cultures in outdoor ponds are applied in Taiwanese shrimp hatcheries where *Skeletonema costatum* is produced successfully in rectangular outdoor concrete ponds of 10–40 tons of water volume and a water depth of 1.5–2 m.

Photobioreactors

An alternative to open ponds for large-scale production of microalgal biomass are photobioreactors. The term "photobioreactor" is used to indicate only closed systems that do not allow direct exchange of gases or contaminants between the algal culture they contain and the atmosphere. These devices provide a protected environment for cultivated species, relatively safe from contamination by other microorganisms, in which culture parameters such as pH, oxygen and carbonic dioxide concentration, and temperature can be better controlled, and provided in known amount. Moreover, they prevent evaporation and reduce water use, lower CO_2 losses due to outgassing, permit higher cell concentration, thus reducing operating costs, and attain higher productivity. However, these systems are more expensive to build and operate than ponds, due to the need of cooling, strict control of oxygen accumulation, and biofouling, and their use must be limited to the production of very high-value compounds from algae that cannot be cultivated in open ponds. Different categories of photobioreactors exist, such as axenic photobioreactors; tubular or flat photobioreactors; horizontal, inclined, vertical, or spiral; manifold or serpentine photobioreactors; air or pump mixed; single phase, filled with culture suspension, with gas exchange taking place in a separate gas exchanger, or two-phase, with both the gas and the liquid phase contained in the photostage.

The use of these devices dates back to the late 1940s, as a consequence of the investigation on the fundamental of photosynthesis carried out with *Chlorella*. Open systems were considered inappropriate to guarantee the necessary degree of control and optimization of the continuous cultivation process. From the first vertical tubular reactors set up in the 1950s for the culture of *Chlorella* under both artificial light and sunlight, several types of photobioreactors have been designed and experimented with. Most of these are small-scale systems, for which experimentation has been conducted mainly indoors, and only few have been scaled up to commercial level. Significantly higher photosynthetic efficiencies and a higher degree of system reliability have been achieved in recent years, due in particular to the progress in understanding the growth dynamic and requirements of microalgae under mass cultivation conditions. Notwithstanding these advances, there are only few examples of photobioreactor technology that has expanded from the laboratory to the market, proving to be commercially successful. In fact, the principle obstacle remains the scaling-up phase, due to the difficulties of transferring a process developed at the laboratory scale to industrial scale in a reliable and efficient way. Two of the largest commercial systems in operation at present are the Klötze plant in Germany for the production of *Chlorella* biomass and the Algatechnologies plant in Israel for the production of *Haematococcus* biomass. Both plants utilize tubular, pump-mixed, single phase photobioreactors; in particular, the Klötze plant consists of compact and vertically arranged horizontal running glass tubes of a total length of 500,000 m and a total PBR volume of 700 m^3. In a glasshouse requiring an area of only 10,000 m^2 an annual production of 130–150 tons dry biomass was demonstrated to be economically feasible under Central European conditions.

Other industrial plants actually operating are the plant built in Maui, Hawaii (USA) by Micro-Gaia Ltd. (now BioReal, Inc. a subsidiary of Fuji Chemical Industry Co., Ltd.), which is based on a rather complex design, called BioDome^{TM}, for the cultivation of Haematococcus; the rigid, plastic tubes photobioreactor of AAPS (Addavita Ltd., UK) and the flexible, plastic tubes photobioreactor of the Mera Growth Module (Mera Pharmaceuticals, Inc., USA).

CULTURE OF SESSILE MICROALGAE

Farmers of abalone (*Haliotis* sp.) have developed special techniques to provide food for the juvenile stages, which feed in nature by scraping coralline algae and slime off the surface of rocks using their radulae. In culture operations, sessile microalgae are grown on plates of corrugated roofing plastic, which serve as substrate for settlement of abalone larvae. After metamorphosis, the spat graze on the microalgae until they become large enough to feed on macroalgae. The most common species of microalgae used on the feeder plates are pennate diatoms (e.g., *Nitzchia, Navicula*). The plates are inoculated by placing them in a current of sand filtered seawater. Depending on local conditions, the microalgae cultures on the plates take between 1 and 3 weeks to grow to a density suitable for settling of the larvae. As the spat grow, their consumption rate increases and becomes greater than the natural production of the microalgae. At this stage, the animals are too fragile to be trans-ferred to another plate and algal growth may be enhanced by increasing illumination intensity or the addition of fertilizer.

QUANTITATIVE DETERMINATIONS OF ALGAL DENSITY AND GROWTH

Although tedious, time-consuming, and requiring an excellence in taxonomic identifications, the traditional counting of phytoplankton is still unsurpassed for quantifying plankton, especially at low limits of detection. The examination and counting of preserved material also allows for direct observation and assessment of cell condition. The fundamental issue that must be addressed prior to choosing a counting technique is whether samples must be concentrated. The three routine methods used to concentrate phytoplankton samples are centrifugation and filtration for live samples and sedimentation (gravitational settling) for preserved samples. Centrifugation and sedi-mentation are the most commonly used, although filtration onto membrane filters is an effective procedure for fluorescence microscopic enumerations. If the cell density is less than 10^5 cells l^{-1}, the sample needs to be concentrated; if cell density is too high to allow direct counting, the sample will be diluted. Dilution or concentration factors need to be taken into account for calculating the final cell concentration.

Species containing gas vacuoles (cyanobacteria) are unlikely to fully sediment despite the Lugol's fixation. Two methods are available for collapsing vacuoles to assist sedimentation. Samples can either be exposed to brief ultrasonification (<1 min, but may vary for different species) or alternatively, pressure can be applied to the sample by forcing it through a syringe with a fine needle; this will collapse the vacuoles and the cells will then settle through gravity.

If the water sample contains sufficient number of algae and concentration is not required, a direct count can be undertaken. The sample is thoroughly mixed, treated with a few drops of Lugol's solution, mixed again and allowed to stand for 30–60 min. The sample is mixed again and the subsamples are used for direct counting and taxonomic identification.

When concentration is required, a simple procedure for gravitational settling is to pour a well-mixed volume of the sample of water containing the microalgae in a measuring cylinder (100 ml), add Lugol's solution to it (1% by volume), and either allow the sample to stand overnight or use a centrifuge so that the cells sink. The iodine in Lugol's solution not only preserves and stains the cells but also increases their density. When the column of water appears clear, the top 90 ml will be gently siphoned off without disturbing the sediment at the base of the cylinder. This leaves the cells concentrated in the bottom 10 ml. Subsamples of this sedimented fraction can be used for counting procedure and examined under the microscope for identification, keeping in mind that the cells of the original sample have been concentrated 10 times.

At present, there are many kinds of counting procedures available for enumerating phytoplank-ton, depending on whether counting algae in mixtures, as from field sampling, or unialgal samples, such as in growth or bioassay experiments in a laboratory. Some techniques are relatively "low-tech," and can be used also in remote locations, other are "high-tech," and require an expensive instrument.

The Sedgewick-Rafter counting chamber (Figure 6.2) is a low tech device routinely used for counting algae in mixed assemblages. This cell limits the volume and area of the sample to enable easier counting and calculation of phytoplankton numbers. It consists of a brass or poly-styrene rectangular frame mounted on a heavy glass slide from which a precise internal chamber has been cut; its dimensions are $50 \times 20 \times 1$ mm^3, with an area of 1000 mm^2, and a volume of 1.0 ml. The base is ruled in a 1 mm grid. When the liquid sample is held in the cell by its large, rectangular glass cover slips, the grid subdivides 1 ml volume into microliters.

To fill the Sedgewick-Rafter chamber, the cover slip is placed on top of the chamber, diagonally across the cell, so that the chamber is only partially covered. This helps to prevent the formation of air bubbles in the cell corners. The cover slip will rotate slowly and cover the inner portion of the cell during filling. The cover slip should not float, nor should there be any air bubbles; the former results from over-filling, the latter from under-filling the chamber. In both cases, the depth of the chamber will be different from 1 mm, and the calculations will be invalidate. A large-bore pipette should be used to transfer the sample into the chamber; after filling, the cover slip is gently pushed to cover the chamber completely. The phytoplankton sample placed into the Sedg-wick-Rafter counting chamber is allowed to stand on a flat surface for 20 min to enable the phyto-plankton to settle. It is then transferred to the stage of an upright light microscope and securely positioned, ready for counting.

Counts are done with the $4\times$ or (more usually) the $10\times$ objectives of the compound microscope (depth of field and lens length preclude the use of higher magnification objectives). A Whipple disk is inserted into one of the ocular lenses in order to provide a sample grid. It is necessary to first determine the area (A) of the Whipple field for each set of ocular and objective lenses used. This is accomplished with a stage micrometer.

There are 50 fields in the length and 20 fields in the width of the chamber (comprising a total of 1000 fields). A horizontal strip corresponds to 50 fields. All cells within randomly selected fields are counted. A convention needs to be followed for cells or units lying on a boundary line or field, such as all cells or units overlapping the right hand and top boundary are counted, but those overlapping the bottom and left hand boundary are not. The number of units per milliliter for each taxon is calculated according to following formula:

$$\text{Number of cells} * \text{mL}^{-1} = \frac{C * 1000 \text{ mm}^3}{A * D * F} \qquad (6.2)$$

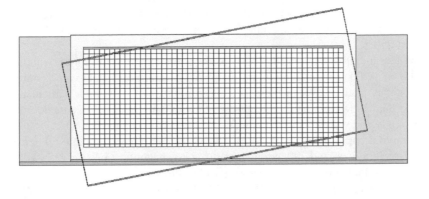

FIGURE 6.2 Schematic drawing of the Sedgewick-Rafter counting chamber.

where C is the number of cells counted, A is the area of field in mm^2, D is the depth of a field (Sedgewick-Rafter chamber depth) in mm, and F is the number of fields counted.

For colonial taxa multiply the count of units by the average number of cells per unit and use the resulting value as C in Equation (6.1). To adjust for sample concentration of dilution the result is divided or multiplied by the appropriate factor. To obtain total cell density per milliliter, sum all counts of individual taxa.

If cell density is low ($<$10 units per field) counting of long transects to cover a large proportion of the chamber floor is more appropriate. Several transects with a width of a chamber field are counted. The number of strips depends on the required precision and the phytoplankton density. The number of cells per millimeter is calculated according to the following formula:

$$\text{Number of cells} * mL^{-1} = \frac{C * 1000 \text{ mm}^3}{L * D * W * S} \tag{6.3}$$

where C is the number of cells counted, L is the length of strip in mm, D is the depth of a field (Sedgewick-Rafter chamber depth) in mm, W is the width of strip in mm, and S is the number of strips counted. To adjust for sample concentration or dilution the result is divided or multiplied by the appropriate factor.

A "high-tech" alternative to counting algae in mixed assemblages with the Sedgewick-Rafter cell is the inverted microscope method. The expensive component here is the inverted microscope, whose great advantage is that settling chamber depth does not preclude the use of high magnification objective lenses. In 1931, Utermöhl solved the problem of concentrating and enumerating algae in mixed populations when he described a one step settling and enumeration technique using the inverted microscope. The procedure involved the gravitational sedimentation of preserved phytoplankton into a counting chamber. This counting technique correctly assumed that phytoplankton would fall randomly to the bottom of the chamber and that counts would then be made on random fields or transects. The inverted microscope counting technique has gained broad popularity for phytoplankton enumeration. One of the advantages of this randomized counting technique is the capability of calculating error estimates to verify the accuracy of the enumeration. Through the years, many modified chambers have been designed and used with the inverted microscope.

Special and expensive "Utermöhl" chambers can be purchased, but cheaper ones can be constructed from large cover slips, and plastic syringe barrels. If a long focal-length lens is available, chambers may be constructed from glass slides.

A measured volume of preserved sample is added to the settling chamber and allowed to settle for at least an hour. Time periods as long as 24–48 h are preferred, especially if small algae are present in the sample (these will settle only very slowly). Upon settling, the upper portion of the chamber is removed and replaced with a glass plate. The sample is then transferred to an inverted microscope (condenser numerical aperture 0.70; objectives 25× and 40×; oculars 12.5×) with phase contrast optics.

The sample is initially enumerated at 500× using a random fields technique. A minimum of 20 random fields and 200 individual cells are enumerated. Additional fields are counted until the minimum count is attained. When there is a large number of cells of a particular taxon in a sample, fewer than 20 random fields are enumerated with a minimum of five random fields examined for this taxon. Individual cells are enumerated, whether in chains, filaments, or colonies. This allows for a more accurate estimate of biomass which is determined from the cell densities. Upon achieving 20 random fields and 200 individual cells, a low magnification scan (25×) of 20 random fields is used to estimate the rarer, larger forms within the sample.

In the case of unialgal samples (unicells, small colonies, or relatively short filaments) chambers such as the haemacytometer, the Thoma chamber (Figure 6.3), the Fuchs-Rosenthal or the Burker chambers are effective and commonly used for estimating the densities of cultures. The

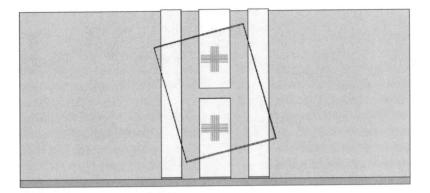

FIGURE 6.3 Schematic drawing of the Thoma counting chamber.

hematocytometer was developed for counting cells in blood samples (now this is mostly done with electronic particle counters). Most of these counting chamber have delicate, mirrored surfaces that must not be scratched. Each mirrored surface has a grid etched upon the surface. In the case of Thoma chamber each grid is composed of 16 fields of 0.2 mm side, separated by three boundary lines (Figure 6.4). These 16 fields are further subdivided into 16 smaller areas each, for a total of 256 counting fields. The

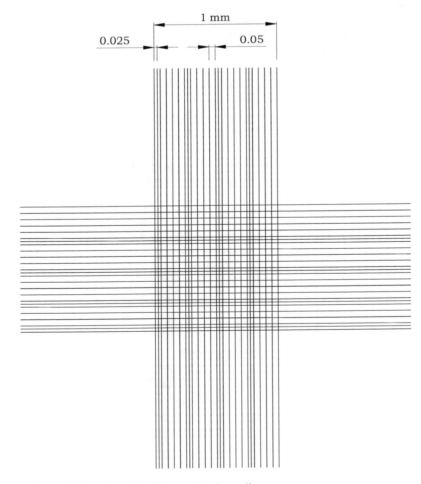

FIGURE 6.4 Schematic drawing of the Thoma chamber ruling.

chamber is 0.1 mm deep; hence each grid holds exactly 64×10^{-3} ml of sample. You have the choice of counting the algae in the entire grid; counting algae in only one of the 16 squares, then multiplying by 16; or counting algae in an even smaller area and multiplying accordingly.

When counting cell in the entire grid, that is, 16 fields, the cell concentration is calculated according to the following formula:

$$\text{Number of cells} * \text{mL}^{-1} = \frac{C * 10^3}{64} \tag{6.4}$$

where C is the number of cells counted.

Counts of about 30 cells per field are desirable for accuracy. If there are more than 30 cells per field, dilute the sample, or count algae in a lower number of fields and multiply. As for the Sedgewick-Rafter a convention needs to be followed for cells lying on a boundary line or field, such as all cells overlapping the right hand and top boundary are counted, but those overlapping the bottom and left hand boundary are not (Figure 6.5). The counting process has to be repeated at least ten times to determine an accurate mean.

"High-tech" methods for counting unialgal samples are the electronic particle counter (e.g., Coulter counter) and the digital microscopy. In spite of relatively high cost, an electronic particle counter is highly recommended for performing growth or bioassay studies that require many counts and high accuracy. In addition, the instrument will provide particle size/biovolume distributions. The principle of operation is that particles, suspended in an electrolyte solution, are sized and counted by passing them through an aperture having a particular path of current flow for a given length of time. As the particles displace an equal volume of electrolyte in the aperture, they place resistance in the path of the current, resulting in current and voltage changes. The magnitude of the change is directly proportional to the volume of the particle; the number of changes per unit time is proportional to the number of particles in the sample. When opened, the stopcock introduces vacuum into the system, draws sample through the aperture, and unbalances the mercury in the manometer. The mercury flows past the "start" contact and resets the counter to zero. When the stopcock is closed, the mercury starts to return to its balanced position and draws sample through the aperture. Variously sized aperture tubes are available for use in counting variously sized particles; the aperture size is chosen to match that of particles.

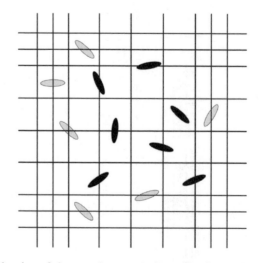

FIGURE 6.5 Schematic drawing of the counting convention: all cells overlapping the right-hand and top boundary are counted (black cells), but those overlapping the bottom and left-hand boundary are not (gray cells).

Digital microscopy necessitates of a microscope equipped with a digital TV camera connected to a personal computer running dedicated application software for cell recognition and cell counting. It represents an automatic, reliable, and very fast approach to growth determination.

GROWTH RATE AND GENERATION TIME DETERMINATIONS

Growth curves are prepared from cell density data obtained with a hemocytometer or electronic particle counter on cultures sampled at intervals, such as once per day, depending on the growth rate of the alga. Plots of number of cells against time (in days) can be made, and from these curves can be calculated specific growth rate or growth constant (μ) and division or generation time (T_g).

A typical growth curve will show a lag phase, an exponential or log phase, and a stationary or plateau phase, where increase in density has leveled off. In the stationary phase, growth is likely limited by resources such as light or nutrients. Growth rate (μ) is calculated with the following equation:

$$\mu = \frac{\ln N_2/N_1}{t_2 - t_1} \tag{6.5}$$

where N_2 and N_1 are number of cells at times t_2 and t_1.

When N_2 is twice N_1, for example, the population has undergone a doubling, the growth rate can be expressed as

$$\mu = \frac{\ln(2)}{T_g} \tag{6.6}$$

As $\ln(2)$ is equal to 0.6931, generation time can be calculated with the following equation:

$$T_g = 0.6931 * \mu^{-1} \tag{6.7}$$

SUGGESTED READING

Alverson, A. J., Manoylov, K. M., and Stevenson, R. J., Laboratory sources of error for algal community attributes during sample preparation and counting, *Journal of Applied Phycology*, 39, 1–13, 2003.

Andersen, R. A., Ed., *Algal Culturing Techniques*, Elsevier Academic Press, Burlington, 2005.

Andersen, P. and Throndsen, J., Estimating cell numbers, in Hallegraeff, G. M., Anderson, D. M., and Cembella, A. D., Eds., *Manual on Harmful Marine Algae*, IOC Manuals and Guides, No. 33, UNESCO Publishing, Paris, 2003, pp. 99–129.

Apt, K. E. and Beherens, P. W., Commercial developments in microalgal biotechnology, *Journal of Phycology*, 35, 215–226, 1999.

Becker, P., *MicroAlgae Production Manual*, http://www.mbl.edu/aquaculture

Berges, J. A., Franklin, D. J., and Harrison, P. J., Evolution of an artificial seawater medium: improvements in enriched seawater, artificial water over the last two decades, *Journal of Phycology*, 37, 1138–1145, 2001.

Chilsholm, S. W., Temporal patterns of cell division in unicellular algae, in *Physiological Bases of Phytoplankton Ecology*, Platt, T., Ed., Canadian Bulletin of Fishery and Acquatic Science, Vol. 210, pp. 150–181, 1981.

Day, J. G., Conservation strategies for algae, in: Benson, E. E., Ed., *Plant Conservation Biotechnology*, Taylor and Francis Ltd., London 1999, pp. 111–124.

Fogg, G. E., and Thake, B., *Algal Cultures and Phytoplankton Ecology*. University of Wisconsin Press, Madison, 1987.

Gualtieri, P. and Barsanti, L., Identification of cellular and subcellular features by means of digital microscopy, *International Journal of Biomedical Computing*, 20, 79–86, 1987.

Guillard, R. R. L., Culture methods. In Hallegraeff, G. M., Anderson, D. M., and A. D., Cembella, Eds., *Manual on Harmful Marine Algae*. IOC Manuals and Guides, No. 33, UNESCO Publishing, Paris, 2003, pp. 45–62.

Hamilton, P. B., Proulx, M., and Earle, C., Enumerating phytoplankton with an upright compound microscope using a modified settling chamber, *Hydrobiologia*, 444, 171–175, 2001.

Integrated Watershed Management, Ecohydrology and Phytotechnology Manual. http://www.unep.or. jp/ietc/publications/freshwater/watershed_manual/index.asp

Labeda, D. P., Ed., *Isolation of Biotechnological Organisms from Nature*, McGraw-Hill Publishing Company, New York, 1990.

Paxinos, R. and Mitchell, J. G., A rapid Utermohl method for estimating algal numbers, *Journal of Planktonic Research*, 22, 2255–2262, 2000.

Pulz, O., Photobioreactors: production system for phototropic microorganisms, *Applied Microbiology and Biotechnology*, 57, 287–293, 2001. http://link.springer.de

Richmond, A., Ed., *Handbook of Microalgae Culture*, CRC Press, Boca Raton, 1986.

Richmond, A., Ed., *Handbook of Microalgae Mass Culture and Biotechnology and Applied Phycology*, Blackwell Publishing, Malden, 2004.

Sampling and Analytic Procedure for GLNPO's Open Lake Water Quality Survey of the Great Lakes. http:// www.epa.gov/glnpo/monitoring/procedures/toc.html

Stein, R. S., Ed., *Culture Methods and Growth Measurements*, Cambridge University Press, Cambridge, 1973.

Sunda, W. G. and Huntsman, S. A., Effects of pH, light, and temperature on Fe-EDTA chelation and Fe hydrolysis in seawater, *Marine Chemistry*, 84, 35–47, 2003.

Sunda, W. G. and Huntsman, S. A., Relationship among photoperiod, carbon fixation, growth, chlorophylla, and cellular iron and zinc in a coastal diatom, *Limnology Oceanography*, 49, 1742–1753, 2004.

Tredici, M. R., Photobioreactors, in *Encyclopedia of Bioprocess Technology: Fermentation, Biocatalysis and Bioseparation*, Flinckinger, M. C., and Drew, S. W., Eds., Wiley and Sons, New York, 1999, pp. 395–419.

Tredici, M. R. and Zitelli, G. C., Efficiency of sunlight utilization: tubular versus flat photobioreactors, *Biotechnology Bioengineering*, 57, 187–197, 1998.

Warren, A., Day, J. G., and Brown, S., Cultivation of protozoa and algae, *Manual of Environmental Microbiology* in Hurtst, C. J., Crawford, R. L., Knudsen, G. R., McInerney, M. J., and Stenzenbach, L. D., Eds., ASM Press, Washington, 2002, pp. 71–83.

Wilkinson, M. H. F. and Shut, F., *Digital Images Analysis of Microbes*, Wiley and Sons, New York, 1998.

7 Algae and Men

INTRODUCTION

Microalgae and macroalgae have been utilized by man for hundreds of years as food, fodder, remedies, and fertilizers. Ancient records show that people collected macroalgae for food as long as 500 B.C. in China and one thousand of years later in Europe. Microalgae such as *Arthrospira* have a history of human consumption in Mexico and Africa. In the 14th century the Aztecs harvested *Arthrospira* from Lake Texcoco and used to make a sort of dry cake called *tecuitlatl*, and very likely the use of this cyanobacterium as food in Chad dates back to the same period, or even earlier, to the Kanem Empire (9th century A.D.).

People migrated from countries such as China, Japan, and Korea, but also from Indonesia and Malaysia, where algae have always been used as food, have brought this custom with them, so that today there are many more countries all over the world where the consumption of algae is not unusual, Europe as well.

On the east and west coasts of the U.S. and Canada, around Maine, New Brunswick, Nova Scotia and British Columbia, some companies have begun cultivating macroalgae onshore, in tanks, specifically for human consumption, and their markets are growing, both in those two countries and with exports to Japan. Ireland and Northern Ireland are showing a renewed interest in macroalgae that were once a traditional part of the diet. In addition to direct consumption, agars and carrageenans extracted from red macroalgae and alginates from brown macroalgae and microalgae have been included in a remarkable array of prepared food products, serving mostly to modify viscosity or texture. Global utilization of macroalgae is on increase, and in terms of harvested biomass per year, macroalgae are among the most important cultivated marine organisms.

Currently, there are 42 countries in the world with reports of commercial macroalgae activity. China holds first rank in macroalgae production, with *Laminaria* sp. accounting for most of its production, followed by North Korea, South Korea, Japan, Philippines, Chile, Norway, Indonesia, the U.S. and India. These top ten countries contribute about 95% of the world's commercial macroalgae volume. About 90% macroalgae production comes from culture based practices. The most cultivated macroalgae is the kelp *Laminaria japonica*, which alone accounts for over 60% of the total cultured macroalgae production while *Porphyra, Kappaphycus, Undaria, Eucheuma*, and *Gracilaria* make up most of the rest to a total of 99%. The most valuable crop is the red alga *Nori* (*Porphyra* species, mainly *Porphyra yezzoensis*), used as food in Japan, China, and Pacific. According to FAO, between 1981 and 2002, world's total harvest of macroalgae increased from 3 million tons to nearly 13 million tons (wet weight). The macroalgae that are most exploited for culture are the brown algae with 6 million tons followed by the red algae with 3 million tons and a small amount of green algae with less than 100,000 tons. East and Southeast Asian countries contribute almost 99% cultured production, with about 75% of the production (9 million tons) supplied by China.

Most output is used domestically for food, but there is a growing international trade. The *Porphyra* cultivation in Japan is the biggest macroalgae industry, with a turnover of more than 2 billion U.S. dollars per annum. Total EU imports of macroalgae in 2002 amounted to 70,000 metric tons with the Philippines, Chile, and Indonesia as the biggest suppliers. Significant quantities of *Eucheuma* are exported by the Philippines, Tanzania, and Indonesia to the U.S., Denmark, and Japan. The Philippines account for nearly 80% of the world's total *Eucheuma cottonii* production of 1,300,000 tons, roughly 35% of which is traded in dried form. They supply 14% of the world's total raw macroalgae production and hold first rank as producers of semirefined carrageenan, contributing close to 60% of the world market. Table 7.1 shows the FAO data relative to the total macroalgae harvest for the period 2000–2002 in all fishing areas of the world.

Large-scale commercial production of microalgae biomass is limited to *Dunaliella, Haematococcus, Arthrospira*, and *Chlorella*, which are cultivated in open ponds at farms located around the world.

These algae are a source for viable and inexpensive carotenoids, pigments, proteins, and vitamins that can be used for the production of nutraceuticals, pharmaceuticals, animal feed additives, and cosmetics. Examples of large-scale commercial production are the large lagoons used in Australia for *Dunaliella salina* cultivation aimed at β-carotene production, the ponds that Cyanotech Enterprise own in Australia, and Earthrise farms in California for *Haematococcus* cultivation aimed at astaxanthin production.

Cyanotech Enterprise claimed that in the 2004 their net sales were about $12 billion, with a net income only in the fourth quarter of $400,000.

SOURCES AND USES OF COMMERCIAL ALGAE

FOOD

Cyanophyta

Some *Nostoc* species are regionally being used as food and herbal ingredients. *Ge-Xian-Mi* has been regarded as *Nostoc sphaeroides*, however, its taxonomic identity remains controversial. This *Nostoc* species has been used as a delicacy for hundreds of years and is found in rice fields from December to May in Hubei, China. Colonies of Ge-Xian-Mi are dark green and pearl-shaped, develop from hormogonia, and can reach 2.5 cm in diameter. Dried *Nostoc* spp. balls are sold in Asian markets; they are stir-fried sautéed with oysters, and used in soups and as thickeners for other foods.

Another species is *Nostoc flagelliforme* Bornet et Flahault, a terrestrial cyanobacterium that naturally grows on arid and semiarid steppes in the Northern and the Northwestern parts of China, where it is considered an edible delicacy with special medical value and great economic value. The Chinese have used it as food for about 2000 years, as told in an old text dating back to the Jin Dinasty (265–316 AD). Its herbal values were recognized more than 400 years ago as recorded with other economic *Nostoc* species in "Compendium of Materia Medica" of 1578. *N. flagelliforme* is called "*Facai*" (hair vegetable) in Chinese because of its hair-like appearance. However, the pronunciation of "*Facai*" sounds like another Chinese word that means to be fortunate and get rich. Therefore, it symbolizes additionally good luck. *N. flagelliforme* has been consumed in China, especially Guangdong, and among Chinese emigrants worldwide on account of its food and herbal values as well as its spiritual image.

As this cyanobacterium has been collected and traded from old times, the resource is getting less and less as the market demand increases with economic growth. People gather it by tools, which more or less destroy the vegetation, and in addition to this loss by harvesting, pasturing

TABLE 7.1
Total Macroalgae Harvest in All Fishing Areas of the World

All Fishing Areas of the World	2000	2001	2002
Red Macroalgae	2,275,141	2,472,253	2,791,006
Brown Macroalgae	5,608,074	5,453,534	5,782,535
Green Macroalgae	96,235	93,688	76,265

of cattle is rapidly diminishing the resource. On the other hand, exploitation of land has greatly reduced the area producing *N. flagelliforme*, and the quality of the product is degrading. Hence, recently the Chinese government has prohibited further collection for the sake of environmental protection. A few phycologists in northern and western China were engaged for some years in attempts to cultivate this alga, but all in vain.

Also *Arthrospira* has a history of human consumption, which can be located essentially in Mexico and in Africa. About 1300 AD the Aztecs harvested *Arthrospira* from Lake Texcoco, near Tenochtitlan (today Mexico City) and used it to make a sort of dry cake called *tecuitlatl*, which were sold in markets all over Mexico, and were commonly eaten with maize, and other cereals, or in a sauce called *chilmolli* made with tomatoes, chilli peppers, and spices. Very likely the use of *Arthrospira* as food in Chad dates back to the same period, or even earlier, to the Kanem Empire (9th century AD), indicating that two very different and very distant populations discovered independently the food properties of *Arthrospira*. Human consumption of this cyanobacterium in Chad was reported for the first time in 1940 by the French phycologist Dangeard in the little known *Journal of the Linnean Society of Bordeaux*. He wrote about an unusual food called *dihé* and eaten by the Kanembu of Chad, and concluded that it was a purée of the filamentous cyanobacterium *Arthrospira platensis*. However, at that time, his report failed to capture the attention it deserved because of the war. In 1966 the botanist Leonard, member of the 1964–1965 Belgian Trans-Saharan Expedition, confirmed the observations of Dangeard. He reported finding a greenish, edible substance being sold as dry cakes in the market of Fort-Lamy (today N'Djamena, the capital of Chad). His investigation revealed that these greenish cakes, called *dihé*, were a common component of the diet of the Kanembu populations of Chad and Niger, and that they were almost entirely composed of *Arthrospira*, blooming naturally in the saline-soda lakes of the region. Like *tecuitlatl*, *dihé* was commonly eaten as a thick, pungent sauce made of chilli peppers, and spices, poured over millet, the staple of the region. In 1976, Delpeuch and his collaborators of ORSTOM (Office de la Recherche Scientifique et Technique Outre-mer, Paris, France) carried out a study on the nutritional and economic importance of *dihé* for the populations of the Prefectures of Kanem and Lac in Chad. The use of *Arthrospira* by Kanembu was mentioned again in 1991 in a canadian survey of food consumption and nutritional adequacy in wadi zones of Chad, which suggested that *dihé* makes a substantial contribution to vitamin A intake.

Arthrospira is still harvested and consumed by the Kanembu who live around Lake Kossorom, a soda lake at the irregular northeast fringe of Lake Chad, in the Prefecture of Lac. *Arthrospira* is harvested from Lake Kossorom throughout the year, with a minimum yield in December and January, and a maximum in the period between June and September during the rainy season. The bloom is present as a thick blue-green mat floating onto the surface of the lake only few hours a day, early in the morning. When the sun is high, the temperature of the water rises, and the bloom disperses, therefore the harvesting begins at sunrise and it is over in about 2 h.

Only Kanembu women carry out the harvesting; men are banned from entering the water, because it is a deep rooted belief that they would make the lake barren. The harvesting begins early in the morning, and the work is coordinated by an old woman (the captain) who is responsible for guarding the lake even when the harvesting is over. Just before harvesting begins, the women form a line along the shore at positions assigned to them by the captain according to the village they come from, so as to avoid overcrowding in areas where the alga bloom is more abundant and where trampling and muddying of the water would reduce the quality of *dihé*. The harvesting is carried on according to rules and procedures handed down from mother to daughter from time immemorial. With their basins, the women skim off the blue-green mat that floats at the surface of the water, especially along the shore, and pour it into twine baskets, which act as primary filters, or directly into jars (Figure 7.1). In about 2 h the harvesting is over, and the women move to sandy areas close to the lake for the filtration and drying of the alga. The women dig ·in the sand round holes, 40–50 cm in diameter and about 5 cm deep, and line them with clean sand, which is patted to obtain a smooth, firm surface. The algal suspension is then carefully poured into the

FIGURE 7.1 Harvesting of *Arthrospira* bloom from Lake Kossorom. (Courtesy of Dr. Gatugel Abdulqader.)

holes, and the surface of the biomass is smoothed with the palm of the hand. Within a few minutes, almost all of the extracellular water will have seeped out of the biomass, and the *dihé* is then cut into 8–10 cm squares, 1–1.5 cm thick, and removed from the holes as soon as it is firm enough for the squares to be handled without breaking (Figure 7.2). The drying is completed on mats in the sun. *Dihé* is traded in the local markets, and in the markets of the main towns of Chad, from where it can also be taken across the borders of Chad to Nigeria, Cameroon, and other countries. It is estimated that about 40 tons of *dihé* are harvested from this lake Kossorom every year, corresponding to a local trading value of more than US$100,000, which represents an important contribution to the economy of one of the poorest nations in the world.

Dihé is mainly used to prepare *la souce*, a kind of vegetable broth served with corn, millet, or sorghum meal, which occasionally can have fish or meat as additional ingredients. Well-dried *dihé* is crumbled in a bowl either by hand or with a mortar and pestle; cold water is then added to disperse the lumps, and the suspension is strained through a fine sieve to remove such solid impurities as sticks, grass, and leaves. The suspension is poured away from most of the sand that settles to the bottom of the bowl. The cleaned *dihé* is cooked for 1–1.5 h, which further disperses the lumps, yielding a blue-green broth that still contains small amounts of plant debris and sand. This broth is transferred into a bowl, left to settle for 5–10 min to allow sedimentation of any residual sand, strained very carefully once more, and then poured over sautéed onions. Salt, chili peppers, bouillon cubes, and *gombo* (*Hibiscus esculentus*) are added, and *la souce* is then simmered and occasionally stirred until cooked.

A minor utilization of *dihé* is as remedy applied onto wounds to speed up the healing process, or as a poultice to soothe the pain and reduce the swelling of mumps.

FIGURE 7.2 Drying of *Arthrospira* and preparation of *dihé* on the shore of Lake Kossorom. (Courtesy of Dr. Gatugel Abdulqader.)

Arthrospira biomass has a high content of protein, about 55–60% of the dry matter, with respect to other foods such as milk, eggs, etc.; the proteins are low in lysine and sulfured aminoacids such as methionine and cystein, but their amount is much higher than in other vegetables, including legumina. Phycobiliproteins represent a major portion of proteins, and among them phycocyanin can reach 7–13% of the dry matter; carbohydrates reach 10–20% of the dry weight, and consist mainly of reserve products, while lipids account for 9–14% of the dry weight. The mineral fraction represents 6–9% of the dry biomass, rich in K, P, Na, Ca, Mn, and Fe. Group A, B, and C vitamins are also present, with an average β-carotene content of 1.5 mg g^{-1} of *Arthrospira*, corresponding to 0.25 μg of vitamin A. Considering a recommended dietary allowance (RDA) of vitamin A of about 800 μg, and taking into account a natural 20–30% decrease in β-carotene level due to *dihé* storing conditions, we can say that a daily consumption of 5 g of *dihé* would provide about 100% of RDA.

Rhodophyta

Porphyra (Bangiophyceae) is popularly known as *Nori* in Japan, *Kim* in Korea, and *Zicai* in China, (see Chapter 1, Figure 1.24). It is among the most nutritious macroalgae, with a protein content of 25–50%, and about 75% of which is digestible. This alga is an excellent source of iodine, other trace minerals, and dietary fibers. Sugars are low (0.1%), and the vitamin content very high, with significant amounts of vitamins A, complex B, and C, but the shelf life of vitamin C can be short in the dried product. During processing to produce the sheets of *nori*, most salt is washed away, so the sodium content is low. The characteristic taste of *nori* is caused by the large amounts of three amino acids: alanine, glutamic acid, and glycine. It also contains taurine, which controls blood cholesterol levels. The alga is a preferred source of the red pigment *r*-phycoerythrin, which is utilized as a fluorescent "tag" in the medical diagnostic industry.

Porphyra has been cultivated in Japan and the Republic of Korea since the 17th century, because even at that time natural stocks were insufficient to meet demand. Today *Porphyra* is one of the largest aquaculture industries in Japan, Korea, and China. Because of its economic importance and other health benefits, *Porphyra* cultivation is now being expanded to other countries.

Porphyra species are primarily intertidal, occurring mainly in temperate zones around the world, but also in subtropical and sub-Artic regions, as confirmed by its history of being eaten by the indigenous peoples of northwest America (Alaska) and Canada, Hawaii, New Zealand, and parts of the British Isles. *Porphyra abbottae* Krishnamurthy is a nutritionally and culturally important species of red alga used by First Peoples of coastal British Columbia and neighboring areas, down to northern California. This species, along with *Porphyra torta* and possibly others, is a highly nutritious food, still gathered in quantity today by the Coast Tsimshian, Haida, Heiltsuk, Kwakwaka'wakw, and other coastal peoples from wild populations in large quantities, dried and processed, and served in a variety of ways: toasted as a snack, cooked with clams, salmon eggs, or fish in soup, or sprinkled on other foods as a condiment. Common linguistic origin of the majority of names for this species among some 16 language groups in five language families indicates widespread exchange of knowledge about this seaweed from southern Vancouver Island north to Alaska. The harvesting and preparation of this seaweed is exacting and time-intensive. It necessitates a wide range of knowledge and skills, including an understanding of weather patterns, tides, and currents; an appreciation of the growth and usable life stages of the seaweed; and a knowledge of the optimum drying locations and techniques and of the procedures for secondary moistening, chopping, and drying to achieve the best flavors and greatest nutritional value. *P. abbottae* is generally harvested in May. Though formerly a women's activity, as for *dihé* harvesting by Kanembu women, both genders now participate in seaweed gathering. The postharvest preparation and handling of the seaweed is fairly labor-intensive and detailed. Once processed, seaweed is considered "an expensive and prestigious food" and is valued as a gift or trade item that is often exchanged for equally valuable products from other groups, especially on the central and northern coasts of British Columbia and Alaska. *P. abbottae* is valued also for its medicinal properties as gastrointestinal aid, taken as decoction or applied as a poultice for any kind of sickness in the stomach, and as orthopedic aid applied on broken collarbones.

Nearly 133 species of *Porphyra* have been reported from all over the world, which includes 28 species from Japan, 30 from North Atlantic coasts of Europe and America, and 27 species from the Pacific coast of Canada and United States. Seven species have been reported from the Indian coast, but they are not exploited commercially.

Porphyra grows as a very thin, flat, blade, which can be yellow, olive, pink, or purple. It can be either round, round to ovate, obovate, linear or linear lanceolate, from 5 to 35 cm in length. The thalli are either one or two cells thick, and each cell has one or two stellate chloroplasts with a pyrenoid. *Porphyra* has a heteromorphic life cycle with an alternation between an aploid gametophyte consisting of a macroscopic foliose thallus, which is eaten, and a filamentous diploid sporophyte called conchocelis phase. This diploid conchocelis phase in the life cycle was earlier thought to be *Conchocelis rosea*, a shell-boring independent organism. Understanding that these two phases were connected was a major research advance made in 1949 in Britain, when Drew demonstrated in culture that *Porphyra umbilicalis* (L.) Kütz had a diploid conchocelis phase. Until this landmark work, cultivation of *Porphyra* was developed intuitively, by observing the seasonal appearance of spores, but nobody knew where the spores came from, so there was little control over the whole cultivation process. Drew's findings completely revolutionized and transformed the *Porphyra* industry in Japan and subsequently throughout Asia, allowing indoor mass cultivation of the filamentous form in sterilized oyster shells and the seeding of conchospores directly onto nets for outplanting in the sea. All Japanese species of *Porphyra* investigated so far produce the conchocelis phase, which can be maintained for long periods of time in free culture. It grows vegetatively under a wide range of temperature, irradiances, and photoperiods, and it is probably a perennial persistent stage in the life history of many species in nature as well.

Since Drew's time, cultivation has flourished, and now accounts for virtually all the production in China, Japan, and the Republic of Korea. Japan produces about 600,000 wet tons of edible macroalgae annually, around 75% of which is for *nori*. In 1999, the combined annual production from these three countries was just over 1,000,000 wet tons. *Nori* is a high value product, worth approximately US$16,000/dry tons. Japanese cultivation of *Porphyra* yields about 40,000 wet tons/year and this is processed into ca. 10 billion *nori* sheets (each 20 × 20 cm, 3.5–4.0 g), representing an annual income of 1500 million U.S. dollars. In the Republic of Korea, cultivation produces 270,000 wet tons, while China produces 210,000 wet tons.

Processing of wet *Porphyra* into dried sheets of *nori* has become highly mechanized, by an adaptation of the paper-making process. Wet *Porphyra* is rinsed, chopped into small pieces, and stirred in a slurry. It is then poured onto mats or frames, most of the water drains away, and the mats run through a dryer. The sheets are peeled from the mats and packed in bundles of ten for sale. This product is called *hoshi-nori*, which distinguishes it from *yaki-nori*, which is toasted. *Nori* is used mainly as a luxury food. It is often wrapped around the rice ball of sushi, a typical Japanese food consisting of a small handful of boiled rice with a slice of raw fish on the top. It can be incorporated into soy sauce and boiled down to give an appetizing luxury sauce. It is also used as a raw material for jam and wine. In China it is mostly used in soups and for seasoning fried foods. In the Republic of Korea it has uses similar to Japan.

Dried *nori* is in constant oversupply in Japan and producers and dealers are trying to encourage its use in the U.S. and other countries. Production and markets in China are expanding, although the quality of the product is not always as good as that from the Republic of Korea and Japan.

The fronds of the red alga *Palmaria* (*Rodimenia*) *palmata* (Florideophyceae) are known as "*dulse*" (see Chapter 1, Figure 1.15); they are eaten raw as a vegetable substitute or dried and eaten as a condiment in North America and Europe (Brittany, Ireland, and Iceland). Natives of Alaska consume the fronds fresh or singed on a hot stove, or add the air-dried fronds to soups and fish head stews. *Palmaria* is found in the eulittoral zone and sometimes the upper sublittoral. It is collected by hand by harvesters plucking it from the rocks at low tide. It is perennial and when either plucked or cut, new growth appears from the edge of the previous season's leaf. It is harvested mainly in Ireland and the shores of the Bay of Fundy in eastern Canada, and is especially abundant around Grand Manan Island, situated in the Bay of Fundy, in a line with the Canadian-the U.S. border between New Brunswick and Maine. The harvest season here is from mid-May to mid-October. After picking, fronds are laid out to sun dry for 6–8 h; if the weather is not suitable, it can be stored in seawater for a few days, but it soon deteriorates. Whole *dulse* is packed for sale in plastic bags, 50 g per bag. Inferior *dulse*, usually because of poor drying, is broken into flakes or ground into powder for use as a seasoning. In Ireland, it is sold in packages and looks like dark-red bundles of flat leaves. It is eaten raw in Ireland, like chewing tobacco, or is cooked with potatoes, in soups and fish dishes.

Dulse is a good source of minerals, being very high in iron and containing all the trace elements needed in human nutrition, and has also a high vitamin content. In Canada, one company has cultivated it in land-based systems (tanks) and promotes it as a sea vegetable with the trade name "Sea Parsley." It is a variant of normal *dulse* plants, but with small frilly outgrowths from the normally flat plant. It was found by staff at the National Research Council of Canada's laboratories in Halifax, Nova Scotia, among samples from a commercial *dulse* harvester.

Chondrus crispus (Florideophyceae), the *Irish moss* or *carrageenan moss*, has a long history of use in foods in Ireland and some parts of Europe (Figure 7.3). It is not eaten as such, but used for its

FIGURE 7.3 Frond of *Chondrus crispus.*

thickening powers when boiled in water, a result of its carrageenan content. One example is its use in making blancmange, a traditional vanilla-flavored pudding. In eastern Canada, a company is cultivating a strain of *C. crispus* and marketing it in Japan as *hana nori*, a yellow macroalga that resembles the more traditional Japanese *nori* that is in limited supply from natural resources because of overharvesting and pollution. First introduced to the Japanese market in 1996, the dried product, to be reconstituted by the user, was reported to be selling well at the end of 1999, with forecasts of a market valued at tens of millions of U.S. dollars. It is used in macroalgae salads, sushimi garnishes, and as a soup ingredient.

Fresh *Gracilaria* (Florideophyceae) has been collected and sold as a salad vegetable in Hawaii (USA) for several decades. It is known as *Ogo*, *ogonori*, or *sea moss*. The mixture of ethnic groups in Hawaii (Hawaiians, Filipinos, Koreans, Japanese, and Chinese) creates an unusual demand and supply has at times been limited by the stocks available from natural sources. Now it is being successfully cultivated in Hawaii using an aerated tank system, producing up to 6 tons fresh weight per week. *Limu manauea* and *limu ogo* are both sold as fresh vegetables, the latter usually mixed with raw fish. In Indonesia, Malaysia, Philippines, and Vietnam, species of *Gracilaria* are collected by coastal people for food. In southern Thailand, an education programme was undertaken to show people how it could be used to make jellies by boiling and making use of the extracted agar. In the West Indies, *Gracilaria* is sold in markets as "*sea moss*"; it is reputed to have aphrodisiac properties and is also used as a base for a non-alcoholic drink. *Gracilaria* sp. contains (wet weight basis): $6.9 \pm 0.1\%$ total proteins, $24.7 \pm 0.7\%$ crude fiber, $3.3 \pm 0.2\%$ total lipids, and $22.7 \pm 0.6\%$ ash. It contains 28.5 ± 0.1 mg of vitamin C per 100 g of wet biomass, 5.2 ± 0.4 % mg of β-carotene per 100 g of dry weight, which corresponds to a vitamin A activity of 865 μg. According to standard classification adopted by AOAC (Association of Official Analytical Chemists), this can be considered a very high value of vitamin A activity for a food item, which makes *Gracilaria* a potential source of β-carotene for human consumption.

In Chile, the demand for edible macroalgae has increased and *Callophyllis variegata* (*carola*) (Florideophyceae) is one of the most popular (Figure 7.4). Its consumption has risen from zero in 1995 to 84

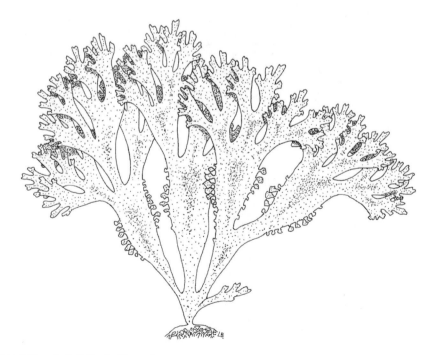

FIGURE 7.4 Frond of *Callophyllis variegata*.

wet tons in 1999. This red macroalgae has a promising future due to its high commercial value, which in 2001 was at about US$30 kg^{-1}d.w. However, knowledge of its biology is limited and research project have been funded for the management of the natural resources and opportunities for cultivation.

Heterokontophyta

Alaria esculenta, is a large brown kelp, which grows in the upper limit of the sublittoral zone (Figure 7.5). It is known as *winged kelp*. It has a wide distribution in cold waters and does not survive above 16°C. It is found in areas such as Ireland, Scotland (U.K.), Iceland, Brittany (France), Norway, Nova Scotia (Canada), Sakhalin (Russia), and northern Hokkaido (Japan). In Ireland it grows up to 4 m in length and favors wave-exposed rocky reefs all around the Irish coast. Eaten in Ireland, Scotland (U.K.), and Iceland either fresh or cooked, it is said to have the best protein among the kelps and is also rich in trace metals and vitamins, especially niacin. It is usually collected from the wild and eaten by local people, and while it has been successfully cultivated, this has not been extended to a commercial scale.

China is the largest producer of edible macroalgae, harvesting about 5 million wet tons annually. The greater part of this is for *haidai*, produced from hundreds of hectares of the brown macroalga *Laminaria japonica* (Figure 7.6). It is a large macroalga, usually 2–5 m long, but it

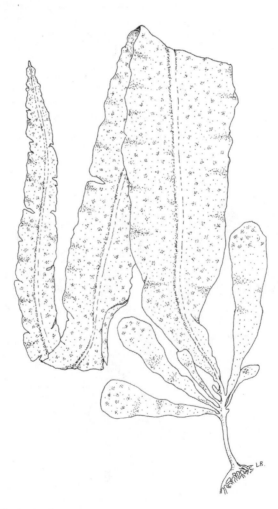

FIGURE 7.5 Frond of *Alaria esculenta.*

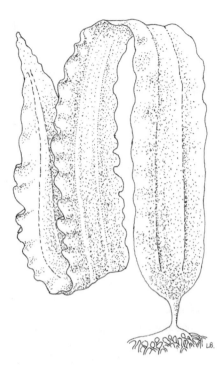

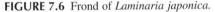

FIGURE 7.6 Frond of *Laminaria japonica*.

can grow up to 10 m in favorable conditions. It requires water temperatures below 20°C. *Laminaria* was originally native to Japan and the Republic of Korea, where it is cultivated since 1730. It was introduced accidentally to China in 1927 probably by the importation of logs from Hokkaido in Japan. Prior to that, China had imported its needs from the naturally growing resources in Japan and the Republic of Korea. In the 1950s, China developed a method for cultivating *Laminaria*; sporelings ("seedlings") are grown in cooled water in greenhouses and later planted out in long ropes suspended in the ocean. This activity became a widespread source of income for large numbers of coastal families. By 1981, 1,200,000 wet tons of macroalgae were being produced annually. In the late 1980s, production fell as some farmers switched to the more lucrative but risky farming of shrimp. By the mid-1990s, production had started to rise again and the reported annual harvest in 1999 was 4,500,000 wet tons. China is now self-sufficient in *Laminaria* and has a strong export market. In the Republic of Korea, the demand for *Laminaria* is much lower and most is now provided from cultivation. *Kombu* is the Japanese name for the dried macroalgae that is derived from a mixture of *Laminaria* species. These include *L. longissima*, *L. japonica*, *L. angustata*, *L. coriacea* and *L. ochotensis*. These are all harvested from natural sources, mainly on the northern island of Hokkaido, with about 10% coming from the northern shores of Honshu. The first three of the above are the main components of the harvest. The plants grow on rocks and reefs in the sublittoral zone, from 2 to 15 m deep. They prefer calm water at temperatures between 3 and 20°C. The naturally growing plants are biennial and are ready for harvesting after 20 months. Harvesting is from June to October, from boats. Hooks of various types are attached to long poles and used to twist and break the macroalgae from the rocky bottom. As demand grew in the 1960s, attempts were made to develop artificial cultivation methods, but the 2 yr cycle meant the costs were too high. In the 1970s, forced cultivation was introduced, reducing the cultivation period to 1 yr, similar to the system developed in China in the early 1950s. Today, about one third of Japan's requirements come from cultivation, with the remaining two thirds still coming from natural resources.

For cultivation, *Laminaria* must go through its life cycle, and this involves an alternation of generations. Seedstock is produced from meiospores released from the sori of wild or cultivated sporophytes, cleaned by wiping or brief immersion in bleach, and incubated in a cool, dark place for up to 24 h. Released spore attach to a substratum within 24 h and develop into gametophytes. Release of gametes, fertilization, and growth into sporophytes about 6 mm long require a couple of months. In the original two-years method, seedstock was produced in late autumn, when the sporophyte produce their sori, and was available for out-planting from December to February, and the crop was ready to harvest in 20 months. In the forced-cultivation method, seedstock is produced in the summer, because sporophytes that spent 3 months in the autumn in the field prior to their second growth season behaved as second-year plants. This method saves 1 yr.

Seedstock is reared on horizontal or vertical strings, placed in sheltered waters for about 10 days; after this period the strings are cut into small pieces that are inserted into the warp of the culture rope. The ropes with the young sporophytes are attached to floating rafts, which belong to two basic types. The first type is the vertical-line raft, consisting of a large diameter rope, 60 m long, kept floating by buoys fixed every 2–3 m. Each end of the rope is anchored to a wooden peg driven into the sea bottom. The ropes with the young sporophytes attached hang down from this rope at 50 cm intervals. The second type is the horizontal-line raft, consisting of three ropes laid out parallel, 5 m apart. The ropes with the young sporophytes are tied across two ropes so that they are more or less horizontal, and each of them has equal access to light.

In China, the largest region for *Laminaria* cultivation is in the Yellow Sea, which has been found to be low in nitrogen fertilizer. Yields are increased when the floating raft areas, which are usually set out in rectangles, are sprayed with a nitrate solution using a powerful pump mounted in a boat. The plants take up the nitrate quickly and very little is lost in the sea. In Japan, the cultivation is mainly in the waters between Honshu and Hokkaido islands and fertilizing is not necessary. Harvesting takes place in the summer, from mid-June to early July.

The kelp is usually laid out in the Sun to dry and then packed into bales. In Japan, the whole macroalgae is washed thoroughly with seawater, cut into 1 m lengths, folded and dried; the product is *suboshi kombu* and is delivered to the local fisheries cooperative.

Laminaria species contain about 10% protein, 2% fat, and useful amounts of minerals and vitamins, though generally lower than those found in *nori*. For example, it has one tenth the amounts of vitamins but three times the amount of iron compared with *nori*. Brown macroalgae also contain iodine, which is lacking in *nori* and other red macroalgae. In China, *haidai* is regarded as a health vegetable because of its mineral and vitamin content, especially in the north, where green vegetables are scarce in winter. It is usually cooked in soups with other ingredients. In Japan, it is used in everyday food, such as a seasoned and cooked *kombu* that is served with herring or sliced salmon.

Another exploited kelp is *Undaria* sp., which together with *Laminaria* sp. is one of the two most economically important edible algae. This alga has been a food item of high value and importance in Japan since 700 A.D. Cultivation began in Japan at the beginning of this century, when the demand exceeded the wild stock harvest, and was followed later in China. The Republic of Korea began the cultivation around 1970 and today it is the largest producer of *wakame/quandai-cai* from *Undaria pinnatifida*. This cultivation accounts for about 50% of the about 800,000 wet tons of edible macroalgae produced annually in this country. Some of this is exported to Japan, where production is only about 80,000 wet tons yr^{-1}. *Undaria* is less popular than *Laminaria* in China; by the mid-1990s China was harvesting about 100,000 wet tons yr^{-1} of *Undaria* from cultivation, compared with 3 million wet tons yr^{-1} of *Laminaria* at that time.

U. pinnatifida is the main species cultivated; it grows on rocky shores and bays in the sublittoral zone, down to about 7 m, in the temperate zones of Japan, the Republic of Korea, and China (Figure 7.7). It grows best between 5 and 15°C, and stops growing if the water temperature rises above 25°C. It has been spread, probably via ship ballast water, to France, New Zealand, and Australia.

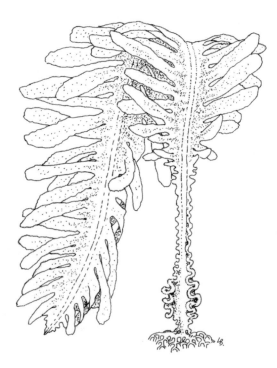

FIGURE 7.7 Frond of *Undaria pinnatifida.*

Undaria is an annual plant with a life cycle similar to *Laminaria*. It has an alternation of generations with the large macroalga as the sporophyte and a microscopic gametophyte as the alternate generation.

The ropes are immersed in seawater tanks containing fertile sporophytes in April–May, when the water temperature is 17–20°C, and let stand until about 100 spores per centimetre of twine have become attached. The ropes are then lashed to frames submerged in seawater tanks until September–November. When the young plants are 1–2 cm long, and the temperature falls below 15°C, the ropes are removed from the frames and wound around a rope that is suspended by floats and anchored to the bottom at each end. However, a variation from the *Laminaria* cultivation is that the rope long-lines are suspended 2–3 m below the surface. In sheltered bays, the ropes are placed 10 m apart; in open waters, where there is more movement, the single ropes are assembled into a grid pattern using connecting ropes to hold the long-lines about 2 m apart. Harvesting is in two stages. First the plants are thinned out by cutting them off at a point close to the rope. This is done by pulling the rope over the edge of a boat, cutting and dragging the plant into the boat. The remaining plants on the rope have plenty of space and continue to grow. Harvesting finishes in April.

In Japan, the seeded strings are often cut into small lengths and inserted in the twist of a rope that is then hung vertically from a floating rope, much the same as is done with *Laminaria*. Harvesting in southern Japan is from March to May, but around Hokkaido it is from May to July.

After its accidental introduction, cultivation of *Undaria* has been undertaken also in France. Here the above methods were found to be inappropriate because the high nutrient concentrations in the water allowed a large variety of other plant and animal life (epiphytes) to grow on the frames holding the strings. In the seeding method adopted by the French, the gametophytes are formed and maintained in a sterile laboratory medium. One month before out-planting the gametophytes are brought to maturation. After fertilized eggs (zygotes) are formed, the solution with the suspended zygotes is sprayed onto a nylon line that is wound around a frame. The zygotes germinate and young sporophytes begin to grow on the frames, which are free of epiphytes. The sporophytes are out-planted on floating ropes in the usual way.

Undaria is processed into a variety of food products. After harvesting, the plants are washed with seawater, then freshwater, the central midrib is removed, and the pieces are dried in the sun or a hot air dryer; this is *suboshi wakame*. However, as this product often fades during storage because various enzymes are still active, another process can be used in which the fresh fronds are mixed with ash from wood or straw, so that the alkalinity of the ash inactivates the enzymes. This mixture is spread on the ground for 2–3 days, placed in plastic bags and kept in the dark. The plants are washed with seawater, then freshwater to remove the salt and ash, the midrib is removed and the pieces are dried. This is *haiboshi wakame*, which keeps its deep green color for a long time, and retains the elasticity of the fresh fronds.

The major *wakame* product is salted *wakame*. Fresh fronds are heated into water at 80°C for 1 min and quickly cooled. The mixture is then dehydrated after 24 h incubation with salt (30% w/w) and then stored at −10°C. When ready for packaging, it is taken from storage, the midribs are removed, and the pieces placed in plastic bags for sale. It is a fresh green color and can be preserved for long periods when stored at low temperatures.

Cut *wakame* is made from salted *wakame*, which is washed with freshwater to remove salt, cut into small pieces, dried in a flow-through dryer and passed through sieves to sort the different sized pieces. It has a long storage life and is a fresh green color when rehydrated. It is one of the most popular dried *wakame* products used for various instant foods such as noodles and soups, and its consumption is very popular.

The crude protein content of *wakame* and *kombu* is 16.3 and 6.2 g (g/100 g), respectively, and both algae contain all essential amino acids, which account for 47.1% of the total amino acid content in *wakame* and for 50.7% in *kombu*. Table 7.2–Table 7.4 summarize the vitamins, minerals, and fiber contents of the two edible algae. This data shows that *wakame* and *kombu* have high contents of β-carotene, that is, 1.30 and 2.99 mg $(100 \text{ g d.w.})^{-1}$ or 217 and 481 μg retinol $(100 \text{ g d.w.})^{-1}$, respectively. The basic component in sea vegetables is iodine, an essential trace element and an integral part of two hormones released by the thyroid gland. According to the results in Table 7.3, *wakame* and *kombu* contain 26 and 170 mg $(100 \text{ g d.w.})^{-1}$ of iodine, respectively. The recommended daily dose for adults is 150 μg, meaning that the consumption of 557 mg of *wakame* and 88 mg of *kombu* would satisfy the daily requirement for iodine. The toxic dose of iodine for adults is thought to be over 2000 μg day^{-1}. The intake of recommended amount of *wakame* and *kombu* per day would supply 0.94 and 6.29 mg of iodine, respectively, meaning that 1.18 g of *kombu* a day would not exceed the recommended safe dose of iodine. However, it is claimed that iodine supplements can be toxic only if taken in excess, while eating sea vegetables should cause no concern.

TABLE 7.2

Vitamin Contents of Marine Algae *Wakame* (*U. pinnatifida*) and *Kombu* (*L. japonica*) (in mg [100 g d.w.]$^{-1}$)

Vitamins	Kombu	Wakame
β-carotene	2.99 ± 0.09	1.30 ± 0.12
Retinol equivalent	0.481 ± 0.015	0.217 ± 0.006
Vitamin B$_1$	0.24 ± 0.02	0.30 ± 0.04
Vitamin B$_2$	0.85 ± 0.08	1.35 ± 0.09
Vitamin B$_6$	0.09 ± 0.01	0.18 ± 0.02
Niacin	1.58 ± 0.14	2.56 ± 0.11

TABLE 7.3
Mineral Composition of *Wakame* (*U. pinnatifida*) and *Kombu* (*L. japonica*) (in mg [100 g d.w.]$^{-1}$)

Minerals	Kombu	Wakame
Ca	880 ± 20	950 ± 30
Mg	550 ± 15	405 ± 10
P	300 ± 10	450 ± 12
I	170 ± 5.5	26 ± 2.4
Na	2532 ± 120	6494 ± 254
K	5951 ± 305	5691 ± 215
Ni	0.325 ± 0.020	0.265 ± 0.015
Cr	0.227 ± 0.073	0.072 ± 0.026
Se	<0.05	<0.05
Fe	1.19 ± 0.03	1.54 ± 0.07
Zn	0.886 ± 0.330	0.944 ± 0.038
Mg	0.294 ± 0.017	0.332 ± 0.039
Cu	0.247 ± 0.076	0.185 ± 0.016
Pb	0.087 ± 0.021	0.079 ± 0.015
Cd	0.017 ± 0.007	0.028 ± 0.006
Hg	0.054 ± 0.005	0.022 ± 0.003
As	0.087 ± 0.006	0.055 ± 0.008

Hizikia fusiforme is another brown algae popular as food in Japan and the Republic of Korea known as *Hiziki*. Up to 20,000 wet tons were harvested from natural beds in the Republic of Korea in 1984, when cultivation began. Since then, cultivation has steadily increased, on the southwest coast, such that in 1995 about 37,000 wet tons were farmed and only 6000 wet tons were harvested from the wild. A large proportion of the production is exported to Japan, where there is little activity in *Hizikia* cultivation. This medium-dark brown macroalga grow to about 20–30 cm in length; the many branched along central stipes give an appearance slightly reminiscent of conifer leaves, with a finer frond structure than *wakame* and *kombu*. It is collected from the wild in Japan and cultivated in the Republic of Korea, grows at the bottom of the eulittoral and top of the sublittoral zones, and is on the southern shore of Hokkaido, all around Honshu, on the Korean peninsula and most coasts of the China Sea. About 90% of the Republic of Korea production is processed and exported to Japan.

The protein, fat, carbohydrate, and vitamin contents are similar to those found in *kombu*, although most of the vitamins are destroyed in the processing of the raw macroalgae. The iron, copper, and manganese contents are relatively high, certainly higher than in *kombu*. Like most

TABLE 7.4
Dietary Fiber Content of *Wakame* (*U. pinnatifida*) and *Kombu* (*L. japonica*) (% d.w.)

	Soluble	Insoluble	Total
Kombu	32.6	4.7	37.3
Wakame	30	5.3	35.3

brown macroalgae, its fat content is low (1.5%) but 20–25% of the fatty acid is eicosapentaenoic acid (EPA).

Cultivation process is very similar to that of *Undaria* and *Laminaria*. Young fronds collected from natural beds are inserted in a rope at 10 cm intervals. Seeding ropes are attached to the main cultivation rope, which is kept at a depth of 2–3 m using flotation buoys along the rope and anchoring it to the seabed at each end. Cultivation is from November to May, and harvesting is in May–June.

The harvested fronds are washed with seawater, dried in the sun, and boiled with the addition of other brown macroalgae such as *Eisenia bicyclis* or *Ecklonia cava* (Figure 7.8), which helps removing phlorotannin. This pigment gives *Hizikia* its astringent bitter taste. The resulting product is cut into short pieces, sun-dried, and sold packaged as *hoshi hiziki*. Typically it is cooked in stir fries, with fried bean curd and vegetables such as carrot or it may be simmered with other vegetables.

Japan produces also *Cladosiphon okamuranus* cultivated around Okinawa Island. This brown macroalga is also harvested from natural populations around the southern islands of Japan and consumed as *Mozuku*. It is characterized by a thallus with a stringy not turgid fronds and it can exceed 50 cm in length. *Cladosiphon* grows in the sublittoral, mainly at depths of 1–3 m. As *Undaria* and *Laminaria*, its life history involves an alternation of generations, with the fronds being the

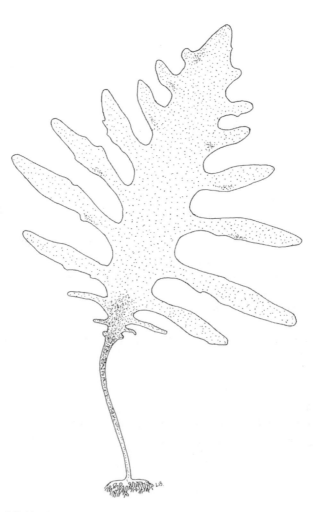

FIGURE 7.8 Frond of *Ecklonia cava*.

sporophyte. Spores collected from the sporophytes are stored during the summer and used for seeding nets in fall. When young sporophytes have grown to 2–5 cm, the nets are moved to the main cultivation sites. The fronds are harvested after about 90 days, when they have grown to 50 cm. Harvesting is done by divers using a suction pump that draws the macroalgae up and into a floating basket besides the attending boat. The harvested macroalgae are washed, salted with 20–25% salt, and let to dehydrate for about 15 days. Drained fronds are sold in wet, salted form in packages.

Chlorophyta

Monostroma (Figure 7.9) and *Enteromorpha* (Figure 7.10) are the two green macroalgae genera cultivated in Japan, and known as *aonori* or *green laver*.

Monostroma latissimum occurs naturally in the bays and gulfs of southern areas of Japan, usually in the upper eulittoral zone. The fronds are bright green in color, flat and leafy, consisting of a single cell layer. They are slender at the holdfast and growing wider toward the apex, often with a slight funnel shape that has splits down the side. *Monostroma* reproduces seasonally, usually during tropical dry season or temperate spring. It is found in shallow sea water usually less than 1 m in depth; generally grows on rocks, coral, mollusk shells, or other hard substrates, but also grows as an epiphyte on sea plants including crops such as *Kappaphycus* and *Eucheuma*. It averages 20% protein and has a useful vitamin and mineral content. It has a life cycle involving an alternation of generations, one generation being the familiar leafy plant, the other microscopic and approximately spherical. It is this latter generation that releases spores that germinate into the leafy frond. For cultivation, these spores are collected on rope nets by submerging the nets in areas where natural *Monostroma* populations grow. The seeded nets are then placed in the bay or estuary, fixed to poles so that they are under water at high tide and exposed for about 4 h at low tide, or using floating rafts in deeper water. The nets are harvested every 3–4 weeks and the

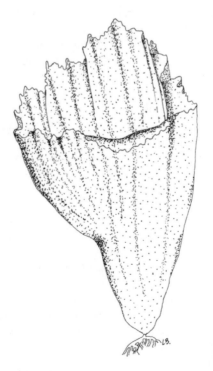

FIGURE 7.9 Frond of *Monostroma latissimum.*

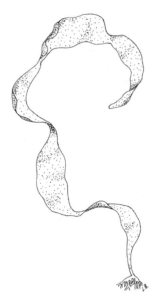

FIGURE 7.10 Frond of *Enteromorpha* sp.

growing season allows about three to four harvests. The harvested macroalga is washed well in seawater and freshwater. It can then either be processed into sheets and dried, for sale in shops, or dried, either outside or in dryers, and then boiled with sugar, soy sauce, and other ingredients to make "*nori-jam*."

Enteromorpha prolifera and *Enteromorpha intestinalis* are found in bays and river mouths around Japan, and are also found in many other parts of the world, including Europe, North America, and Hawaii. Fronds usually flat, narrow, and bright green in color can be seen waving gently with water movement. They can be attached to firm substrate in clear, shallow waters, and also occur as epiphyte on cultured red seaweeds such as *Kappaphycus, Eucheuma, Gracilaria, Gelidiella*, and others.

It can thrive in both salt and brackish waters and is usually found at the top of the sublittoral zone. It contains about 20% protein, little fat, low sodium, and high iron and calcium. Its vitamin B-group content is generally higher than most vegetables, and while its vitamin A is high, it is only half of that found in spinach. Its life history involves an alternation of generations with the same appearance of long, tubular filaments. As for *Monostroma*, rope nets are seeded with spores by submerging them in areas where *Enteromorpha* is growing naturally.

In the Republic of Korea, seed collection is from June to August and the strings or ropes are taken to culture sites in September; in Japan, seeding is done in September, and by early November young plants are visible. The nets are placed in calm bays or estuaries using either fixed poles in shallow waters or floating rafts in deeper waters. Harvesting can be done two to three times during the growing period, either by hand picking from the nets or by machine. Harvested fronds are washed in freshwater and dried in large trays.

Ulva sp. (see Chapter 1, Figure 1.22) is known as sea lettuce, as fronds may be convoluted and have an appearance rather like lettuce. It can be collected from the wild and added to *Monostroma* and *Enteromorpha* as part of aonori. It has a higher protein content than the other two, but much lower vitamin content, except for niacin, which is double that of *Enteromorpha*. Bright green in color, it has a double or multiple cell layer. Slender at the holdfast and growing wider toward the apex, it reproduces seasonally, usually during tropical dry season or temperate spring. It is naturally found in shallow sea water usually less than 1 m in depth, where it grows on rocks, coral, mollusk shells, or other hard substrate, but also as an epiphyte on other sea macroalgae. It was used as flavoring with other seaweed by Kashaya Pomo natives of northern California.

Caulerpa lentillifera (Figure 7.11) and *Caulerpa racemosa* are the two edible green algae used in fresh salads and known as sea grapes or green caviar. As the common name suggests, their appearance looks like bunches of green grapes. These algae often produce "runners" under the substrate, which can produce several vertical branches that extend above the substrate. They naturally grow on sandy or muddy bottom in shallow protected waters.

Caulerpa lentillifera has been very successfully cultivated in enclosures similar to prawn ponds in the central Philippines, where about 400 ha of ponds are under cultivation, producing 12–15 tons of fresh macroalgae per hectare per year. Water temperature can range between 25 and 30°C. Pond depth should be about 0.5 m and areas of about 0.5 ha are usual. Also some strains of *C. racemosa* give good yields under pond cultivation conditions.

Planting is done by hand; about 100 g lots are pushed into the soft bottom at 0.5–1 m intervals. Harvesting can commence about 2 months after the first planting; fronds are pulled out of the muddy bottom, but about 25% of the plants are left as seed for the next harvest. Depending on growth rates, harvesting can then be done every 2 weeks. The harvested plants are washed thoroughly in seawater to remove all sand and mud, then sorted and placed in 100–200 g packages; these will stay fresh for 7 days if chilled and kept moist.

Table 7.5 summarizes edible algae and the corresponding food item.

Extracts

Agar, alginate (derivative of alginic acid), and carrageenan are three hydrocolloids that are extracted from various red and brown macroalgae. A hydrocolloid is a non-crystalline substance with very large molecules, which dissolves in water to give a thickened (viscous) solution. Agar, alginate, and carrageenan are water-soluble carbohydrates used to thicken aqueous solutions, to form gels (jellies) of varying degrees of firmness, to form water-soluble films, and to stabilize certain products, such as ice-cream (they inhibit the formation of large ice crystals, allowing the ice-cream to retain a smooth texture). The use of macroalgae as a source of these hydrocolloids

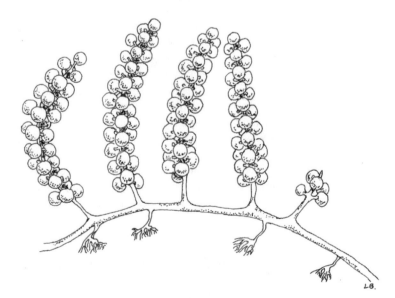

FIGURE 7.11 Frond of *Caulerpa lentillifera.*

TABLE 7.5

Summary of Edible Algae and the Corresponding Food Item

Scientific Name	Common Name	Class
Nostoc flagelliforme	Facai	Cyanophyceae
Arthrospira sp.	Dihé/Tecuitlatl	Cyanophyceae
Chondrus crispus	Pioca/Irish moss	Floridophyceae
Porphyra spp.	Nori/Laber/Zicai	Bangiophyceae
Palmaria (Rodimenia) palmata	Dulse	Floridophyceae
Callophyllis variegata	Carola	Floridophyceae
Asparagopsis taxiformis	Limu kohu	Floridophyceae
Gigartina spp.	Botelhas	Floridophyceae
Gracilaria coronopifolia	Limu manauea	Floridophyceae
Gracilaria parvisipora	Ogo	Floridophyceae
Gracilaria verucosa	Ogo-nori/Sea moss	Floridophyceae
Sargassum echinocarpum	Limu kala	Pheophyceae
Dictyopteris plagiogramma	Limu lipoa	Pheophyceae
Undaria pinnatifida	Wakame	Pheophyceae
Laminaria spp.	Kombu	Pheophyceae
Nereocystis spp.	Black kelp	Pheophyceae
Hizikia fusiforme	Hiziki/Hijiki	Pheophyceae
Alaria esculenta	Oni-wakame	Pheophyceae
Cladosiphon okamuranus	Mozuku	Pheophyceae
Codium edule	Limu wawale'iole	Bryopsidophyceae
Enteromorpha prolifera	Limu 'ele'ele/green laver	Ulvophyceae
Ulva fasciata	Limu palahalaha	Ulvophyceae
Caulerpa lentillifera	Limu Eka	Charophyceae
Monostroma nitidum	Aonori	Ulvophyceae

dates back to 1658, when the gelling properties of agar, extracted with hot water from a red macro-algae, were first discovered in Japan. Extracts of Irish moss (*Chondrus crispus*), another red macro-algae, contain carrageenan and were popular as thickening agents in the 19th century. It was not until the 1930s that extracts of brown macroalgae, containing alginate, were produced commercially and sold as thickening and gelling agents. Industrial uses of macroalgae extracts expanded rapidly after the Second World War, but were sometimes limited by the availability of raw materials. Once again, research into life cycles has led to the development of cultivation industries that now supply a high proportion of the raw materials for some hydrocolloids. Today, approximately 1 million tons of wet macroalgae are harvested annually and extracted to produce the above three hydrocolloids. Total hydrocolloid production is in the region of 55,000 tons/yr, with a value of 585 million U.S. dollars.

There are a number of artificial products reputed to be suitable replacements for macroalgae gums but none have the exact gelling and viscosity properties of macroalgae gums and it is very unlikely that macroalgae will be replaced as the source of these polysaccharides in the near future.

Agar

Agar, a general name for polysaccharides extracted from some red algae, is built up of alternating D- and L-galactopyranose units. The name agar is derived from a Malaysian word "*agar-agar*," which literally means "macroalgae." As the gelling agent "*kanten*," it is known from Japan since the 17th century; extracts from red macroalgae were carried up the mountains to freeze overnight so that water and other impurities could be extracted from the material. Agar finds its widest use as a solid microbiological culture substrate. Modern agar is a purified form consisting largely of the neutral fraction known as agarose; the non-ionic nature of the latter makes it more suitable for a range of laboratory applications. Agar in a crude or purified form also finds wide usage in the food industry where it is used in various kinds of ices, canned foods, and bakery products.

The higher-quality agar (bacteriological-grade agar) is extracted from species of the red algal genera *Pterocladia* and *Gelidium* (Figure 7.12) that are harvested by hand from natural populations in Spain, France, Portugal, Morocco, the Azores, California, Mexico, New Zealand, South Africa, India, Indonesia, the Republic of Korea, Chile, and Japan. Agars of lesser quality are extracted from *Gracilaria* and *Hypnea* species. Food-grade agar is seasonal in *Pterocladia* species, being low in the colder months and high in the warmer. *Gelidium* is a small, slow-growing alga and, while efforts to cultivate it in tanks and ponds have been biologically successful, they have generally proved uneconomic. *Gracilaria* species were once considered unsuitable for agar production because the quality of the agar was poor. In the 1950s, it was found that pre-treatment of the macro-algae with alkali before extraction lowered the yield but gave a good-quality agar. This allowed expansion of the agar industry, which had been previously limited by the available supply of *Gelidium*, and led to the harvesting of a variety of wild species of *Gracilaria* in countries such as Argentina, Chile, Indonesia, and Namibia. Chilean *Gracilaria* was especially useful, but evidence of over-harvesting of the wild crop soon emerged. Cultivation methods were then developed, both

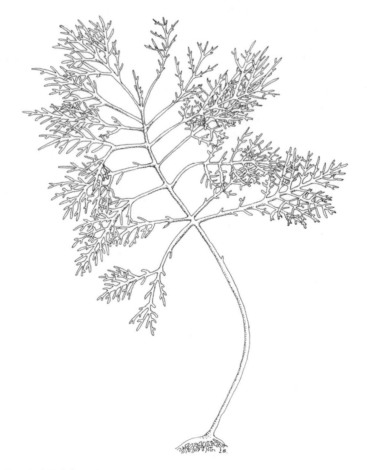

FIGURE 7.12 Frond of *Gelidium* sp.

in ponds and in the open waters of protected bays. These methods have since spread beyond Chile to other countries, such as China, Indonesia, the Republic of Korea, Namibia, the Philippines and Vietnam, usually using species of *Gracilaria* native to each particular country. *Gracilaria* species can be grown in both cold and warm waters. Today, the supply of *Gracilaria* still derives mainly from the wild, with the extent of cultivation depending on price fluctuations. Agar production is valued at U.S. $132 million annually.

Alginate

Alginates are cell-wall constituents of brown algae (Phaeophyceae). They are chain-forming heteropolysaccharides made up of blocks of mannuronic acid and guluronic acid. Composition of the blocks depends on the species being used for extraction and the part of the thallus from which extraction is made. Extraction procedures probably also affect alginate quality. Alginates of one kind or another seem to be present in most species of brown algae but they occur in exploitable quantities (30–45% d. w.) only in the larger kelps and wracks (Laminariales and Fucales). The more useful brown macroalgae grow in cold waters, thriving best in waters up to about 20°C. Brown macroalgae are also found in warmer waters, but these are less suitable for alginate production and are rarely used as food. A wide variety of species are used, harvested in both the northern and southern hemispheres.

The ammonia and alkali metal salts of alginic acid readily dissolve in cold water at low concentrations to give viscous solutions. Alginates, especially sodium alginate, are widely used in the textile industry because they form an excellent dressing and polishing material. Calcium alginate, which is insoluble in water, has been used in the manufacture of a medical dressing very suitable for burns and extensive wounds where a normal dressing would be extremely difficult to remove; the calcium alginate is extruded to make a fiber which is then woven into a gauze-like product; alginates with a high proportion of guluronic acid blocks are most suitable for this purpose. When applied to either a wound or burn, a network is formed around which a healthy scab may form; the bandage may be removed with a sodium chloride solution, which renders the alginate soluble in water. Alginates are also used as a thickening paste for colors in printing textiles, as a hardener and thickener for joining threads in weaving; the alginates may subsequently be dissolved away, giving special effects to the material. Other uses include glazing and sizing paper, special printers' inks, paints, cosmetics, insecticides, and pharmaceutical preparations. In the U.S. alginates are frequently used as stabilizers in ice cream, giving a smooth texture and body, and also as a suspending agent in milk shakes. Alginates take up atomically heavy metals in a series of affinities; for example, lead and other heavy metals will be taken up in preference to sodium, potassium, and other "lighter" metals; accordingly, alginates are useful in lead and strontium-90 poisoning.

Approximately 32,000–39,000 tons of alginic acid per annum are extracted worldwide. The main producers are Scotland, Norway, China, and the U.S., with smaller amounts being produced in Japan, Chile, Argentina, South Africa, Australia, Canada, Chile, the U.K., Mexico, and France. In the U.S., the giant kelp, *Macrocystis pyrifera* is used (Figure 7.13); it is harvested from large offshore beds off the coasts of California and Mexico. Around 50,000 tons wet weight are gathered each year using ships equipped with cutting machinery. *Macrocystis* has the distinction of being the largest macroalgae in the world; the largest attached plant recorded was 65 m long and the plants are capable of growing at up to 50 cm day^{-1}. *Ascophyllum nodosum* and *Laminaria hyperborea* are used in Norway and Scotland. *Ascophyllum* is sustainably harvested in Ireland to produce macroalgae meal that is exported to Scotland for alginate extraction. A decrease in the demand for alginates in recent years has reduced this harvest to about 15,000 tons in 2001, yielding about 3000 tons of meal for alginates. About 6000–8000 tons of alginates are produced from this and other sources (kelp from Chile and Australia) in Scotland, but the amounts of weed used are not known. Norway processes *L. hyperborea*, manufacturing about 5000 tons of high-quality alginates.

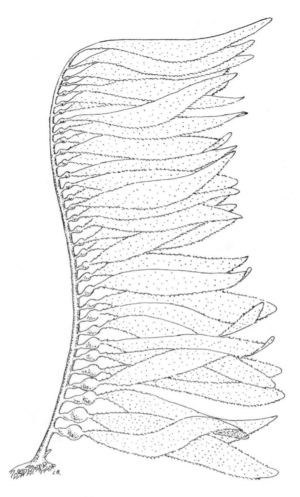

FIGURE 7.13 Frond of *Macrocystis pyrifera*.

Japan produces about 1500–2500 tons of alginates from kelps. China, a relatively recent addition to the alginate manufacturers of the world, produces about 8000–10000 tons of alginates from cultivated *L. japonica*. Alginate production is valued at U.S. $213 million annually.

Carrageenan

Carrageenan is a general name for polysaccharides extracted from certain kinds of algae which are built up, in contrast to agar, from D-galactopyranose units only. The word carrageenan is derived from the colloquial Irish name for this macroalgae, *carrageen* (from the Irish place name, probably Carrigeen Head in Co. Donegal, Carraigín; "little rock"); the use of this macroalgae to extract a gel is known in Ireland since 1810. *Chondrus crispus* used to be the sole source of carrageenan, but species of *Eucheuma, Ahnfeltia*, and *Gigartina* are now commonly used. About 28,000 tons of carrageenan are manufactured worldwide, and although *Chondrus* is no longer the unique source, it is still the principal one. Modern carrageenan is a branded product designed by mixing various types of carrageenan, to give a gel with particular qualities. Most of the *Chondrus* that is used in the carrageenan industry comes from the Maritime Provinces of Canada (Nova Scotia, etc.), where about 50,000 wet tons of *Chondrus crispus* are harvested each year from natural populations. The bulk of the harvest is collected using long-handled rakes and dredges from small boats. The macroalgae is then dried, either by spreading and air-drying or by using rotary dryers, and exported to the U.S. and Denmark for processing.

Carrageenan production (valued at U.S. $240 million annually) was originally dependent on wild macroalgae, especially *Chondrus crispus* (Irish moss), a small macroalgae growing in cold waters, with a limited resource base in France, Ireland, Portugal, Spain, and the east coast provinces of Canada. As the carrageenan industry expanded, the demand for raw material began to strain the supply from natural resources. However, since the early 1970s the industry has expanded rapidly following the availability of other carrageenan-containing macroalgae that have been successfully cultivated in warmwater countries with low labor costs. Today, most raw material comes from two species originally cultivated in the Philippines, *Kappaphycus alvarezii* and *Eucheuma denticulatum*, but which are now also cultivated in other warm-water countries such as Indonesia and the United Republic of Tanzania. Limited quantities of wild *Chondrus* are still used; attempts to cultivate *Chondrus* in tanks have been successful biologically, but it has proved uneconomic as a raw material for carrageenan. Wild species of *Gigartina* and *Iridaea* from Chile are also being harvested and efforts are being made to find cultivation methods for these.

Table 7.6 summarizes commercially exploited algae and the corresponding extract.

ANIMAL FEED

Microalgae are utilized in aquaculture as live feeds for all growth stages of bivalve molluscs (e.g., oysters, scallops, clams, and mussels), for the larval/early juvenile stages of abalone, crustaceans,

TABLE 7.6
Summary of Commercially Exploited Algae and the Corresponding Extracts

Scientific Name	Class	Extracts
Gracilaria chilensis	Floridophyceae	Agar
Ahnfeltia plicata	Floridophyceae	Agar/Carrageenan
Gelidium lingulatum	Floridophyceae	Agar
Pterocladia spp.	Floridophyceae	Agar
Hypnea spp.	Floridophyceae	Agar
Chondrus crispus	Floridophyceae	Carrageenan
Gigartina skottsbergii	Floridophyceae	Carrageenan
Gigartina canaliculata	Floridophyceae	Carrageenan
Mazzaella laminaroides	Floridophyceae	Carrageenan
Sarcothalia crispata	Floridophyceae	Carrageenan
Kappaphycus alvarezii	Floridophyceae	Carrageenan
Eucheuma denticulatum	Floridophyceae	Carrageenan
Iridaea spp.	Floridophyceae	Carrageenan
Laminaria hyperborea	Phaeophyceae	Alginate
Laminaria digitata	Phaeophyceae	Alginate
Laminaria japonica	Phaeophyceae	Alginate
Laminaria saccarina	Phaeophyceae	Alginate
Macrocystis pyrifera	Phaeophyceae	Alginate
Ascophyllum nodosum	Phaeophyceae	Alginate
Durvillea potatorum	Phaeophyceae	Alginate
Ecklonia spp.	Phaeophyceae	Alginate
Lessonia nigrescens	Phaeophyceae	Alginate
Lessonia trabiculata	Phaeophyceae	Alginate

and some fish species, and for zooplankton used in aquaculture food chains. Over the last four decades, several hundred microalgae species have been tested as food, but probably less than 20 have gained widespread use in aquaculture. Microalgae must possess a number of key attributes to be useful aquaculture species. They must be of an appropriate size for ingestion, for example, from 1 to 15 μm for filter feeders; 10 to 100 μm for grazers and readily digested. They must have rapid growth rates, be amenable to mass culture, and also be stable in culture to any fluctuations in temperature, light, and nutrients as may occur in hatchery systems. Finally, they must have a good nutrient composition, including an absence of toxins that might be transferred up the food chain.

Successful strains for bivalve culture included *Isochrysis galbana*, *Isochrysis* sp. (T.ISO), *Pavlova lutheri*, *Tetraselmis suecica*, *Pseudoisochrysis paradoxa*, *Chaetoceros calcitrans*, and *Skeletonema costatum*.

Isochrysis sp. (T.ISO), *P. lutheri*, and *C. calcitrans* are the most common species used to feed the larval, early juvenile, and broodstock (during hatchery conditioning) stages of bivalve molluscs; these are usually fed together as a mixed diet. Many of the strains successfully used for bivalves are also used as direct feed for crustaceans (especially shrimp) during the early larval stages, especially diatoms such as *Skeletonema* spp. and *Chaetoceros* spp. Benthic diatoms such as *Navicula* spp. and *Nitzschia* are commonly mass-cultured and then settled onto plates as a diet for grazing juvenile abalone. *Isochrysis* sp. (T.ISO), *P. lutheri*, *T. suecica*, or *Nannochloropsis* spp. are commonly fed to *Artemia* or rotifers, which are then fed on to later larval stages of crustacean and fish larvae.

Microalgal species can vary significantly in their nutritional value, and this may also change under different culture conditions. Nevertheless, a carefully selected mixture of microalgae can offer an excellent nutritional package for larval animals, either directly or indirectly (through enrichment of zooplankton). Microalgae that have been found to have good nutritional properties — either as monospecies or within a mixed diet — include *C. calcitrans*, *C. muelleri*, *P. lutheri*, *Isochrysis* sp. (T.ISO), *T. suecica*, *S. costatum*, and *Thalassiosira pseudonana*. Several factors can contribute to the nutritional value of a microalga, including its size and shape, digestibility (related to cell wall structure and composition), biochemical composition (e.g., nutrients, enzymes, and toxins if present), and the requirements of the animal feeding on the alga. As the early reports demonstrated biochemical differences in gross composition between microalgae and fatty acids, many studies have attempted to correlate the nutritional value of microalgae with their biochemical profile. However, results from feeding experiments that have tested microalgae differing in a specific nutrient are often difficult to interpret because of the confounding effects of other microalgal nutrients. Nevertheless, from examining all the literature data, including experiments where algal diets have been supplemented with compounded diets or emulsions, some general conclusions can be reached.

Microalgae grown to late-logarithmic growth phase typically contain 30–40% proteins, 10–20% lipids and 5–15% carbohydrates. When cultured through to stationary phase, the proximate composition of microalgae can change significantly; for example, when nitrate is limiting, carbohydrate levels can double at the expense of protein. There does not appear to be a strong correlation between the proximate composition of microalgae and nutritional value, though algal diets with high levels of carbohydrate are reported to produce the best growth for juvenile oysters (*Ostrea edulis*) and larval scallops (*Patinopecten yessoensis*), provided polyunsaturated fatty acids (PUFAs) are also present in adequate proportions. In contrast, high dietary protein provided best growth for juvenile mussels (*Mytilus trossulus*) and Pacific oysters (*Crassostrea gigas*). PUFAs derived from microalgae, that is, docosahexanoic acid (DHA), eicosapentanoic acid (EPA) and arachidonic acid (AA) are known to be essential for various larvae.

The fatty acid content showed systematic differences according to taxonomic group, although there were examples of significant differences between microalgae from the same class. Most microalgal species have moderate to high percentages of EPA (7–34%). Prymnesiophytes (e.g., *Pavlova* spp. and *Isochrysis* sp. [T.ISO]) and cryptomonads are relatively rich in DHA (0.2–11%), whereas eustigmatophytes (*Nannochloropsis* spp.) and diatoms have the highest percentages of AA (0–4%). Chlorophytes (*Dunaliella* spp. and *Chlorella* spp.) are deficient in both C20 and C22 PUFAs, although some species have small amounts of EPA (up to 3.2%). Because of this PUFA deficiency, chlorophytes generally have low nutritional value and are not suitable as a single species diet. Prasinophyte species contain significant proportions of C_{20} (*Tetraselmis* spp.) or C_{22} (*Micromonas* spp.) — but rarely both. In late-logarithmic phase, prymnesiophytes, on average, contain the highest percentages of saturated fats (33% of total fatty acids), followed by diatoms and eustigmatophytes (27%), prasinophytes and chlorophytes (23%), and cryptomonads (18%). The content of saturated fats in microalgae can also be improved by culturing under high light conditions.

The content of vitamins can vary between microalgae. Ascorbic acid shows the greatest variation, that is, 16-fold ($1-16$ mg g^{-1} d. w.). Concentrations of other vitamins typically show a two- to four-fold difference between species that is, β-carotene $0.5-1.1$ mg g^{-1}, niacin $0.11-0.47$ mg g^{-1}, α-tocopherol $0.07-0.29$ mg g^{-1}, thiamin 29 to 109 μg g^{-1}, riboflavin $25-50$ μg g^{-1}, pantothenic acid $14-38$ μg g^{-1}, folates $17-24$ μg g^{-1}, pyridoxine $3.6-17$ μg g^{-1}, cobalamin $1.8-7.4$ μg g^{-1}, biotin $1.1-1.9$ μg g^{-1}, retinol ≤ 2.2 μg g^{-1} and vitamin D <0.45 μg g^{-1}. To put the vitamin content of the microalgae into context, data should be compared with the nutritional requirements of the consuming animal. Unfortunately, nutritional requirements of larval or juvenile animals that feed directly on microalgae are, at best, poorly understood. However, the requirements of the adult are far better known and, in the absence of information to the contrary, will have to serve as a guide for the larval animal. These data suggest that a carefully selected, mixed-algal diet should provide adequate concentrations of the vitamins for aquaculture food chains.

The amino acid composition of the protein of microalgae is very similar between species and relatively unaffected by the growth phase and light conditions. Further, the composition of essential amino acids in microalgae is very similar to that of protein from oyster larvae (*Crassostrea gigas*). This indicates that it is unlikely the protein quality is a factor contributing to the differences in nutritional value of microalgal species. Sterols, minerals, and pigments also may contribute to nutritional differences of microalgae.

A common procedure during the culture of both larval fish and prawns is to add microalgae (i.e., "green water") to intensive culture systems together with the zooplankton prey. Addition of the microalgae to larval tanks can improve the production of larvae, though the exact mechanism of action is unclear. Theories advanced include (a) light attenuation (i.e., shading effects), which have a beneficial effect on larvae, (b) maintenance of the nutritional quality of the zooplankton (c) an excretion of vitamins or other growth-promoting substances by algae, and (d) a probiotic effect of the algae. Most likely, the mechanism may be a combination of several of these possibilities. Maintenance of NH_3 and O_2 balance has also been proposed, though this has not been supported by experimental evidence. The most popular algae species used for green water applications are *N. oculata* and *T. suecica*. More research is needed on the application of other microalgae, especially those species rich in DHA, to green water systems. Green water may also be applied to extensive outdoor production systems by fertilizing ponds to stimulate microalgal growth, and correspondingly, zooplankton production, as food for larvae introduced into the ponds.

For a long time, animals such as sheep, cattle, and horses that lived in coastal areas have eaten macroalgae, especially in those European countries where large brown macroalgae were washed ashore. Today the availability of macroalgae for animals has been increased with the production of macroalgae meal: dried macroalgae that has been milled to a fine powder. In the early 1960s, Norway was among the early producers of macroalgae meal, using *Ascophyllum nodosum*, a macroalga that grows in the eulittoral zone so that it can be cut and collected when exposed at low tide. France has used *Laminaria digitata*, Iceland both *Ascophyllum* and *Laminaria* species, and the U.K., *Ascophyllum*.

Because *Ascophyllum* is so accessible, it is the main raw material for macroalgae meal and most experimental work to measure the effectiveness of macroalgae meal has been done on this macroalgae. The macroalgae used for meal must be freshly cut, as drift macroalgae is low in minerals and usually becomes infected with mould. The wet macroalgae is passed through hammer mills with progressively smaller screens to reduce it to fine particles. These are passed through a drum dryer starting at 700–800°C and exiting at no more than 70°C. It should have a moisture level of about 15%. It is milled and stored in sealed bags because it picks up moisture if exposed to air. It can be stored for about a year.

Analysis shows that it contains useful amounts of minerals (potassium, phosphorus, magnesium, calcium, sodium, chlorine, and sulfur), trace elements, and vitamins. Trace elements are essential elements needed by humans and other mammals in smaller quantities than iron (approximately 50 mg/kg body weight), and include zinc, cobalt, chromium, molybdenum, nickel, tin, vanadium, fluorine, and iodine. Because most of the carbohydrates and proteins are not digestible, the nutritional value of macroalgae has traditionally been assumed to be in its contribution of minerals, trace elements, and vitamins to the diet of animals. In Norway, it has been assessed as having only 30% of the feeding value of grains.

Ascophyllum is a very dark macroalga due to a high content of phenolic compounds. It is likely that the protein is bound to the phenols, giving insoluble compounds that are not attacked by bacteria in the stomach or enzymes in the intestine. *Alaria esculenta* is another large brown macroalgae, but much lighter in color and in some experimental trials it has been found to be more effective than *Ascophyllum* meal. It is this lack of protein digestibility that is a distinct drawback to *Ascophyllum* meal providing useful energy content. In preparing compound feedstuffs, farmers may be less concerned about the price per kilogram of an additive; the decisive factor is more likely to be the digestibility and nutritive value of the additive.

In feeding trials with poultry, adding *Ascophyllum* meal had no benefit except to increase the iodine content of the eggs. With pigs, addition of 3% *Ascophyllum* meal had no effect on the meat yield. However, there have been some positive results reported with cattle and sheep. An experiment for 7 yr with dairy cows (seven pairs of identical twins) showed an average increase in milk production of 6.8% that lead to 13% more income. A trial involving two groups each of 900 ewes showed that those fed macroalgae meal over a 2 yr period maintained their weight much better during winter feeding and also gave greater wool production.

The results of trials reported above and in the suggested reading below leave the impression that macroalgae meal is probably only really beneficial to sheep and cattle. Certainly the size of the industry has diminished since the late 1960s and early 1970s, when Norway alone was producing about 15,000 tons of macroalgae meal annually. Nevertheless, a Web search for "macroalgae meal" shows that there are companies in at least Australia, Canada, Ireland, Norway, the U.K. and the U.S. advocating the use of macroalgae meal as a feed additive for sheep, cattle, horses, poultry, goats, dogs, cats, emus, and alpacas. The horse racing industry seems to be especially targeted. An interesting report from a U.S. university states that the immune system of some animals is boosted by feeding a particular Canadian macroalgae meal. Obviously the industry is still active, pursuing niche markets and fostering research that might lead back to further expansion.

In fish farming, wet feed usually consists of meat waste and fish waste mixed with dry additives containing extra nutrients, all formed together in a doughy mass. When thrown into the fish ponds or cages it must hold together and not disintegrate or dissolve in the water. A binder is needed; sometimes a technical grade of alginate is used. It has also been used to bind formulated feeds for shrimp and abalone. However, cheaper still is the use of finely ground macroalgae meal made from brown macroalgae; the alginate in the macroalgae acts as the binder. The binder may be a significant proportion of the price of the feed so macroalgae meal is a much better choice. However, as the trend is to move to dry feed rather than wet, this market is not expected to expand.

There is also a market for fresh macroalgae as a feed for abalone. In Australia, the brown macroalgae *Macrocystis pyrifera* and the red macroalgae *Gracilaria edulis* have been used. In

South Africa, *Porphyra* is in demand for abalone feed and recommendations have been made for the management of the wild population of the macroalgae. Pacific *dulse* (*Palmaria mollis*) has been found to be a valuable food for the red abalone, *Haliotis rufescens*, and development of land-based cultivation has been undertaken with a view to producing commercial quantities of the macroalgae. The green macroalgae, *Ulva lactuca*, has been fed to *Haliotis tuberculata* and *Haliotis discus*. Feeding trials showed that abalone growth is greatly improved by high protein content, and this is attained by culturing the macroalgae with high levels of ammonia present.

FERTILIZERS

There is a long history of coastal people using macroalgae, especially the large brown macroalgae, to fertilize nearby land. Wet macroalgae is heavy, so it was not usually carried very far inland, although on the west coast of Ireland enthusiasm was such that it was transported several kilometres from the shore. Generally drift macroalgae or beach-washed macroalgae is collected, although in Scotland farmers sometimes cut *Ascophyllum* exposed at low tide. In Cornwall (U.K.), the practice was to mix the macroalgae with sand, let it rot, and then dig it in. For over a few hundred kilometres of the coast line around Brittany (France), the beach-cast, brown macroalgae is regularly collected by farmers and used on fields up to a kilometre inland. Similar practices can be reported for many countries around the world. For example, in a more tropical climate like the Philippines, large quantities of *Sargassum* have been collected, used wet locally, but also sun-dried and transported to other areas. In Puerto Madryn (Argentina), large quantities of green macroalgae are cast ashore every summer and interfere with recreational uses of beaches. Part of this algal mass has been composted and then used in trials for growing tomato plants in various types of soil. In all cases, the addition of the compost increased water holding capacity and plant growth, so composting simultaneously solved environmental pollution problems and produced a useful organic fertilizer.

Macroalgae meal is dried, milled macroalgae, and again it is usually based on the brown macroalgae because they are the most readily available in large quantities. Species of *Ascophyllum, Ecklonia*, and *Fucus* are the common ones. They are sold as soil additives and function as both fertilizer and soil conditioner. They have a suitable content of nitrogen and potassium, but are much lower in phosphorus than traditional animal manures and the typical N : P : K ratios in chemical fertilizers. The large amounts of insoluble carbohydrates in brown macroalgae act as soil conditioners (improve aeration and soil structure, especially in clay soils) and have good moisture retention properties. Their effectiveness as fertilizers is also sometimes attributed to the trace elements they contain, but the actual contribution they make is very small compared with normal plant requirements. A company in Ireland that produces milled macroalgae for the alginate industry is developing applications for macroalgae meal in Mediterranean fruit and vegetable cultivation. "Afrikelp" is another example of a commercially available dried macroalgae, sold as a fertilizer and soil conditioner; it is based on the brown macroalgae *Ecklonia maxima* that is washed up on the beaches of the west coast of Africa and Namibia. Like all brown macroalgae, *Ascophyllum* contains alginate, a carbohydrate composed of long chains. When calcium is added to alginate, it forms strong gels. By composting the dried, powdered *Ascophyllum* under controlled conditions for 11–12 days, the alginate chains are broken into smaller chains and these chains still form gels with calcium but they are weaker. The composted product is a dark brown, granular material containing 20–25% water and it can be easily stored and used in this form. Steep slopes are difficult to cultivate with conventional equipment and are likely to suffer soil loss by runoff. Spraying such slopes with composted *Ascophyllum*, clay, fertilizer, seed, mulch, and water has given good results, even on bare rock. Plants quickly grow and topsoil forms after a few years. The spray is thixotropic, that is, it is fluid when a force is applied to spread it but it sets to a weak gel when standing for a time and sticks to the sloping surface. It holds any soil in place and retains enough moisture to allow the seeds to germinate. Composted *Ascophyllum* has been used after the construction of roads in a number of countries, and has found other uses as well.

Maerl is the common name of a fertilizer derived from calcareous red algae; *maerl* beds are characterized by accumulations of living and dead unattached non-geniculate calcareous rhodophytes (mostly Corallinaceae but also Peyssonneliaceae, such as *Phymatolithon calcareum* and *Lithothamnion corallioides*). Also known as rhodolith beds, these habitats occur in tropical, temperate, and polar environments. In Europe they are known from throughout the Mediterranean and along most of the Western Atlantic coast from Portugal to Norway, although they are rare in the English Channel, Irish Sea, North Sea, and Baltic Sea. *Maerl* beds are often found in subdued light conditions and their depth limit depends primarily on the degree of light penetration. In the northeast Atlantic, *maerl* beds occur from low in the intertidal to ca. 30 m depth; in the West Mediterranean they are found down to 90–100 m, while in the East they occur down to depths of ca. 180 m. *Maerl* beds in subtidal waters have been utilized over a long period in Britain, with early references dating back to 1690. In France also, *maerl* has been used as a soil fertilizer for several centuries. Extraction of *maerl*, either from beds where live thalli are present or where the *maerl* is dead or semifossilized, has been carried out in Europe for hundreds of years. Initially, the quantities extracted were small, being dug by hand from intertidal banks, but in the 1970s about 600,000 tons of *maerl* was extracted per annum in France alone. Amounts have declined to about 500,000 tons per annum since then, though *maerl* extraction still forms a major part of the French seaweed industry, both in terms of tonnage and value of harvest. Live *maerl* extraction is obviously very problematic with regard to growth rates for replacement. Dead *maerl* extraction is liable to lead to muddy plumes and excessive sediment load in water that later settles out and smothers surrounding communities.

The *maerl* is marketed mainly for use as an agricultural fertilizer, for soil improvement in horticulture, mainly to replace lime as an agricultural soil conditioner. There are conflicting reports on the benefits of *maerl* use as opposed to the use of dolomite or calcium carbonate limestone. Other uses include: as an animal food additive, for biological denitrification, and in neutralization of acidic water in the production of drinking water, aquarium gravel as well as in the pharmaceutical, cosmetics, nuclear, and medical industries. These uses are all related to the chemical composition of *maerl*, which is primarily composed of calcium and magnesium carbonates. It is occasionally used for miscellaneous purposes such as hardcore for filling roads, and surfacing garden paths.

Maerl beds are analogous to the sea-grass beds or kelp forests in that they are structurally and functionally complex perennial habitats formed by marine algae that support a very rich biodiversity. The high biodiversity associated with *maerl* grounds is generally attributed to their complex architecture. Long-lived *maerl* thalli and their dead remains build up on underlying sediments to produce deposits with a three-dimensional structure that is intermediate in character between hard and soft grounds. *Maerl* thalli grow very slowly such that *maerl* deposits may take hundreds of years to develop, especially in high latitudes. One of the most obvious threats is commercial extraction, as this has led to the wholesale removal of *maerl* habitats (e.g., from five sites around the coasts of Brittany) while areas adjacent to extraction sites show significant reductions in diversity and abundance. Even if the proportion of living *maerl* in commercially collected material is low, extraction has major effects on the wide range of species present in both live and dead *maerl* deposits. Brittany is the main area for *maerl* extraction with about 500,000 tons extracted annually; smaller amounts are extracted in southwest England and southwest Ireland. *Maerl* beds represent a non-renewable resource as extraction and disruption far out-strips their slow rate of accumulation. In France, *maerl* extraction is now considered to be "mining," which implies more constraints for the extractors and more controls on the impact of extraction. Scientists, managers, and policy makers have been slow to react to an escalating degradation of these habitats such that there is now an urgent need to protect these systems from severe human impacts.

Macroalgae extracts and suspensions have achieved a broader use and market than macroalgae and macroalgae meal. They are sold in concentrated form, are easy to transport, dilute, and apply, and act more rapidly. They are all made from brown macroalgae, although the species varies between countries. Some are made by alkaline extraction of the macroalgae and anything that does not dissolve is removed by filtration; others are suspensions of very fine particles of macroalgae.

Macroalgae extracts have given positive results in many applications. There are probably other applications where they have not made significant improvements, but these receive less, if any, publicity. However, there is no doubt that macroalgae extracts are now widely accepted in the horticultural industry. When applied to fruit, vegetable, and flower crops, some improvements have included higher yields, increased uptake of soil nutrients, increased resistance to some pests such as red spider mite and aphids, improved seed germination, and more resistance to frost. There have been many controlled studies to show the value of using macroalgae extracts, with mixed results. For example, they may improve the yield of one cultivar of potato but not another grown under the same conditions. No one is really sure about why they are effective, despite many studies having being made. The trace element content is insufficient to account for the improved yields, etc. It has been shown that most of the extracts contain several types of plant growth regulators such as cytokinins, auxins, and betaines, but even here there is no clear evidence that these alone are responsible for the improvements. Finally there is the question, are macroalgae extracts an economically attractive alternative to NPK fertilizers? Perhaps not when used on their own, but when used with NPK fertilizers they improve the effectiveness of the fertilizers, so less can be used, with a lowering of costs. Then there are always those who prefer an "organic" or "natural" fertilizer, especially in horticulture, so macroalgae extracts probably have a bright future.

COSMETICS

Extract of algae is often found on the list of ingredients on cosmetic packages, particularly in face, hand, and body creams or lotions, but the use of algae themselves in cosmetics, rather than extracts of them, is rather limited.

Milled macroalgae, packed in sachets, is sold as an additive to bath water, sometimes with essential oils added. Bath salts with macroalgae meal are also sold. Thalassotherapy has come into fashion in recent years, especially in France. In thalassotherapy, macroalgae pastes, made by cold-grinding or freeze-crushing, are applied to the person's body and then warmed under infrared radiation. This treatment, in conjunction with seawater hydrotherapy, is said to provide relief for rheumatism and osteoporosis. Mineral-rich seawater is used in a range of therapies, including hydrotherapy, massage, and a variety of marine mud and algae treatments. One of the treatments is to cover a person's body with a paste of fine particles of macroalgae, sometimes wrap them in cling wrap, and warm the body with infrared lamps. It is said to be useful in various ways, including relief of rheumatic pain or the removal of cellulite. Paste mixtures are also used in massage creams, with promises to rapidly restore elasticity and suppleness to the skin. The macroalgae pastes are made by freeze grinding or crushing. The macroalgae is washed, cleaned, and then frozen in slabs. The slabs are either pressed against a grinding wheel or crushed, sometimes with additional freezing with liquid nitrogen that makes the frozen material more brittle and easier to grind or crush. The result is a fine green paste of macroalgae.

There appears to be no shortage of products with ingredients and claims linked to macroalgae: creams, face masks, shampoos, body gels, bath salts, and even a do-it-yourself body wrap kit. The efficacy of these products must be judged by the user. A company recently pointed out that the lifetime of cosmetic products has reduced over the years and now rarely exceeds 3 or 4 years. Perhaps the macroalgae products that are really effective will live longer than this.

Cosmetic products, such as creams and lotions, sometimes show on their labels that the contents include "marine extract," "extract of alga," "macroalgae extract" or similar. This usually means that one of the hydrocolloids extracted from macroalgae has been added. Alginate or carrageenan could improve the skin moisture retention properties of the product.

Therapeutic Supplements

Microalgae are a unique source of high-value compounds and systematic screening for therapeutic substances, particularly from cyanobacteria, has received great attention. Among cyanobacteria *Spirulina* sp. has undergone numerous and rigorous toxicological studies that have highlighted its potential therapeutic applications in the area of immunomodulation, anticancer, antiviral, and cholesterol reduction effects.

A number of extracts were found to be remarkably active in protecting human lymphoblastoid T-cells from the cytopathic effects of HIV infection. Active agents consisting of sulfolipids with different fatty acid esters were isolated from *Lyngbya lagerheimii* and *Phormidium tenue*. Additional cultured cyanobacterial extracts with inhibitory properties were also found in *Phormidium cebennse, Oscillatoria raciborskii, Scytonema burmanicum, Calotrix elenkinii*, and *Anabena variabilis*. A protein called cyanovirin, isolated from an aqueous cellular extract of *Nostoc elipsosporum* prevents the *in vitro* replication and citopathicity of primate retroviruses.

Cryptophycin 1, an active compound isolated from *Nostoc* strain GSV224, exerts antiproliferative and antimitotic activities by binding to the ends of the microtubules, thus blocking the cell cycle at the metaphase of mitosis. Cryptophycin 1 is the most potent suppressor of microtubule dynamics yet described. Research has been focused on its potent antitumor activity and a synthetic analogue, cryptophycin-52, is at present in Phase II clinical trials by Eli Lilly & Co, Inc. (Indianapolis, IN). Other studies using water soluble extracts of cyanobacteria have found a novel sulfated polysaccharide, calcium spirulan to be an antiviral agent. This compound appears to be selectively inhibiting the penetration of enveloped viruses into host cells, thereby preventing the replication. The effect was described for many different viruses like herpex simplex, measles, and even HIV-1.

Among eukaryotic microalgae, a glycoprotein prepared from *Chlorella vulgaris* culture supernatant exhibited protective activity against tumor metastasis and chemotherapy-induced immunosuppression in mice.

Extracts from several macroalgae may prove to be a source of effective antiviral agents; although the tests have been either *in vitro* (in test-tubes or similar) or on animals, with few advancing to trials involving people. A notable exception is Carraguard, a mixture of carrageenans similar to those extracted from Irish moss. Carraguard has been shown to be effective against human immunodeficiency virus (HIV) *in vitro* and against herpes simplex virus in animals. Testing has advanced to the stage where the international research organization, the Population Council, is supervising large-scale HIV trials of Carraguard, involving 6000 women over 4 yr. Extracts from the brown macroalgae, *Undaria pinnatifida*, have also shown antiviral activity; an Australian company is involved in several clinical trials, in Australia and the U.S., of such an extract against HIV and cancer. The Population Council's trials against HIV involve the vaginal application of a gel containing carrageenan. Because antiviral substances in macroalgae are composed of very large molecules, it was thought they would not be absorbed by eating macroalgae. However, it has been found in one survey that the rate of HIV infection in macroalgae-eating communities can be markedly lower than it is elsewhere. This has led to some small-scale trials in which people infected with HIV ate powdered *Undaria*, with a resulting decrease of 25% in the viral load.

Fucoxanthin, a carotenoid commonly distributed in brown algae, such as *U. pinnatifida, Scytosiphon lomentaria, Petalonia binghamiae*, and *Laminaria religiosa*, is a potent drug candidate and can be utilized as an excellent supplement like astaxanthin, because it acts as an antioxidant and inhibits GOTO cells of neuroblastoma and colon cancer cells. Recently, the apoptosis activity against HL-60 (human leukema) and Caco-2 (cancer colon) cells has been reported for fucoxanthin.

Among the microalgal high-value compounds there are carotenoids such as β-carotene, astaxanthin, PUFA such as DHA and EPA, and polysaccharides such as β-glucan. Carotenoids by their quenching action on reactive oxygen species carry intrinsic anti-inflammatory properties, PUFA exhibit antioxidant activity and polysaccharides act as immunostimulators.

A major bottleneck in the exploitation of microalgal biomass for the production of high-value compounds is low productivity of the culture, both in terms of biomass and product formation. A fundamental reason for this is slow cell growth rates owing to inefficient use of strong light. Furthermore, most microalgal products are secondary metabolites that are produced when growth is limited. A solution to this bottleneck could be to milk the secondary metabolites from the microalgae. This involves continuous removal of secondary metabolites from cells, thereby enabling the biomass to be reused for the continuous production of high-value compounds.

Recently, a new method was developed for milking β-carotene from *Dunaliella salina* in a two-phase bioreactor. In this technique, cells are first grown under normal growth conditions and then stressed by excess light to produce larger amounts of β-carotene. At this stage, the second, biocompatible organic phase is added and the β-carotene is extracted selectively via continuous recirculation of a biocompatible organic solvent (lipophilic compound) through the aqueous phase containing the cells. Because the cells continue to produce β-carotene, the extracted product is continuously replaced by newly produced molecules. Therefore, the cells are continuously reused and. do not need to be grown again. In contrast to existing commercial processes, this method does not require the harvesting, concentrating, and destroying of cells for extraction of the desired product. Furthermore, purification of the product is simple owing to the selectivity of the extraction process. The general application of this process would facilitate the commercialization of microalgal biotechnology and development of microalgal products.

The properties of the cell membrane play an important role in the contact between biocompatible lipophilic solvents and hydrophobic parts of the cell membrane might be prevented by presence of a cell wall or hydrophilic parts of the outer membrane. Physiological properties of the cells, such as their capacity for continuous endo- and exocytosis, might also play a role in the milking process. Other considerations are the location and way in which the product accumulates inside the cells and the function of that product inside the cells. A product like chlorophyll would be difficult to extract owing to its location in thylakoid membranes and because it is bound strongly to other cell components. The extraction of a product with a protective effect on the cells (e.g., β-carotene) will enhance its synthesis. The milking process can be applied also to other algae and other products besides *D. salina*, for example in *Haematococcus pluvialis* for the recovery astaxanthin, and marine microalgae for PUFA.

In addition to its use in aquaculture (e.g., to give salmon a pink color), astaxanthin has also been described as having nutraceutical importance related to free-radical scavenging, immunomodulation, and cancer prevention. *H. pluvialis* can produce and accumulate astaxanthin to concentrations of 1–8% of the dry weight. This concentration within the cell would make milking of *H. pluvialis* more successful compared with *D. salina*. However, cultivation of *H. pluvialis* is more complex than *D. salina*, and productivity is lower. Furthermore, extraction, purification, and concentration are a heavy burden on the production cost. As the final product cost is also sensitive to algal productivity and duration of the growth period, this process is not economically feasible at present.

PUFAs are gaining increasing importance as valuable pharmaceutical products and ingredients of food owing to their beneficial effect on human health. DHA ($22:6\ \omega3$) and EPA, ($20:5\ \omega3$), in particular, are important in the development and functioning of brain, retina, and reproductive tissues both in adults and infants. They can also be used in the treatment of various diseases and disorders, including cardiovascular problems, a variety of cancers, and inflammatory disease. At present, PUFAs are produced commercially from fish oil, but this is an insufficient source of these products and microalgae provide an optimal lipid source of PUFAs.

The heterotrophic marine dinoflagellate *Crypthecodinium cohnii* has a lipid content greater than 20% dry weight and is known for its ability to accumulate fatty acids that have a high fraction (30–50%) of DHA. Lipids are important components of algal cell membranes but also accumulate in globules in other parts of the cells. Microalgal growth and fatty acid formation is affected by medium composition and environmental conditions (e.g., carbon sources). Lipid production occurs under growth-limiting conditions; during linear growth, the cells are stressed owing to nutrient limitation and therefore produce more lipids. Also the concentration of DHA, hence the lipid quality, is negatively affected by increases in lipid concentration. The highest quality lipid it obtained when glucose is used as the carbon source, and when the cell concentration and lipid content of the cells are the lowest.

Milking can be used also for DHA production by *C. cohnii*. In this process, cells are first grown under the correct conditions for growth, after which they are stressed to produce higher concentrations of DHA. A biocompatible organic solvent is added during the DHA production stage to extract the product. This process enables the production of high-quality lipid, thereby reducing extraction and purification costs. Furthermore, higher amounts of DHA are produced by substitution of extracted lipids by newly synthesized lipid, increasing the productivity of the system.

Microalgae represent also one of the most promising EPA producers, as the purification of this PUFA from fish oil, which remains the main commercial source of EPA, involves many drawbacks. Many Eustigmatophyceae, such as *Nannochloropsis* sp. and *Monodus subterraneus*, and Bacillariophyceae species contain a considerable amount of EPA. An EPA production potential has been found in the genus *Nitzschia* (especially *N. alba* and *N. laevis*). It was reported that the oil content of *N. alba* was as high as 50% of cell dry weight and the EPA comprises 4–5% of the oil. *N. laevis* could utilize glucose or glutamate as single substrate for heterotrophic growth, and the cellular EPA content of the alga in heterotrophic conditions was also higher than that in photoautotrophic conditions suggesting that this diatom is a good heterotrophic EPA producer.

The bioprocess engineering aspect of heterotrophic EPA production by *N. laevis* has been extensively studied. Major achievements include:

- Optimization of various medium components (including silicate, glucose, nitrogen sources, salts and trace elements) and environmental factors (including pH, temperature) for the alga culture
- Investigation of detailed physiological behavior of the alga (cell growth, nutrient consumption, fatty acid compositions, etc.) by a continuous culture (the dilution rate and glucose concentration in the feed medium were optimized in terms of EPA productivity and glucose utilization efficiency)
- Development of high cell density and high productivity techniques, which led to an EPA yield of 1112 mg l^{-1} and an EPA productivity of 174 mg l^{-1} day^{-1}, both of which are the highest ever reported in microalgal cultures.

Microalgae produce many different types of polysaccharides, which may be a costituent of the cell wall as in unicellular red algae as *Porphyridium* sp. and *Rhodella* sp., or be present inside the cell, as in the Euglenophyceae. The polysaccharides of Rhodophyta are highly sulfated and consist mainly of xylose, glucose, and galactose. These compounds selectively inhibit reverse transcriptase (RT) enzyme of human immunodeficiency virus (HIV) and its replication *in vitro*.

Rodents fed with a diet supplemented with biomass and polysaccharides derived from *Porphyridium* results in a decrease in blood cholesterol concentration (by 22% and 29%, respectively) and triglyceride levels, increased feces weight (by 130% and 196%, respectively) and bile acid excretion (5.1- and 3.2-fold or more). Moreover, algal biomass or polysaccharide increased the length of both the small intestine (by 17% and 30%, respectively) and the colon (by 8.5% and 32%, respectively).

Paramylon is the term used for the reserve polysaccharide of *Euglena* and euglenoids in general. Its granules appearing in various locations inside the cell; in many species they are scattered throughout the cytoplasm, but others can be massed together or few, but large, and located in a fairly constant position. Their shape and size differ markedly, and together with their distribution inside the cell, represent a taxonomic feature. Paramylon consists of β-1,3-glucan, a linear poly-saccharide found in the cell walls of many bacteria, plants, and yeasts. It belongs to a group of natu-rally occurring polysaccharides such as lentinan, fungal glucans, sizofiran, and pachyman, which are considered bioactive compounds. Lentinan and sizofiran are known to inhibit the growth of various tumors, whereas fungal glucans are used clinically for their stimulatory effect on the immune system, in particular on the macrophages. The main interest in β-glucans stems from their ability to act as nonspecific immune system stimulants, by binding to a specific site on mono-cytes/macrophages and granulocytes. They have been successfully used in aquaculture to strengthen the nonspecific defence of many important species of fishes and shrimps by injection, immersion, or in the feed. Sulfated derivatives of *Euglena* paramylon, in particular, have shown anti-HIV (human immunodeficiency virus) activity. Moreover, it has been suggested that this poly-saccharide has a cholesterol-lowering effect when incorporated in the diet of either humans or animals, and moderates the postprandial blood glucose and insulin response in humans. Since *Euglena gracilis* can accumulate large quantities of paramylon when grown in the presence of an utilizable carbon source, it could represent an alternative source of this compound to *Saccharomyces cerevisiae* (baker's yeast), which is currently exploited industrially for its extraction. More-over, *Euglena* has been investigated as a potential protein source, and a promising dietary

TABLE 7.7
Summary of Commercially Exploited Algae and the Corresponding Products or Applications

Scientific Name	Class	Products/Applications
Lyngbya lagerheimii	Cyanophyceae	Sulpholipids/spirulan
Nostoc spp.	Cyanophyceae	Cryptophycin 1
Arthrospira spp.	Cyanophyceae	Health food
Palmaria mollis	Floridophyceae	Abalone feed
Phymatolithon calcareum	Floridophyceae	Fertilizers
Lithothamnion coralloides	Floridophyceae	Fertilizers
Nannochloropsis spp.	Eustigmatophyceae	EPA/fish fry feed
Monodus subterraneus	Eustigmatophyceae	EPA
Skeletonema spp.	Bacillariophyceae	Fish fry feed
Chaetoceros spp.	Bacillariophyceae	Fish fry feed
Nitzschia alba	Bacillariophyceae	EPA
Nitzschia laevis	Bacillariophyceae	EPA
Petalonia binghamiae	Phaeophyceae	fucoxanthin
Scytosiphon lomentaria	Phaeophyceae	fucoxanthin
Ascophyllum nodosum	Phaeophyceae	Fertilizers
Sargassum spp.	Phaeophyceae	Fertilizers
Laminaria digitata	Phaeophyceae	Animal feed
Macrocystis pyrifera	Phaeophyceae	Abalone feed
Isochrysis spp.	Haptophyceae	DHA/fish fry feed
Tetraselmis spp.	Haptophyceae	Fish fry feed
Pavlova spp.	Haptophyceae	Fish fry feed
Crypthecodinium cohni	Dinophyceae	DHA
Euglena gracilis	Euglenophyceae	β-1,3-glucan
Haematococcus pluvialis	Chlorophyceae	astaxanthin
Dunaliella salina	Chlorophyceae	β-carotene
Chlorella spp.	Chlorophyceae	Health food/fish fry feed

supplement due to its content of highly nutritious proteins and PUFAs, and to the simultaneous production of antioxidant compounds, such as β-carotene, vitamin C, and vitamin E.

Table 7.7 summarizes commercially exploited algae and the corresponding nutraceutical.

TOXIN

Toxins are all compounds that are either synthesized by the algae or formed by the composition of metabolic products and hence represent an intrinsic characteristic of the organism. Of the millions of species of microalgae those that produce specific toxins scarcely exceed a hundred. These occur in both salt and freshwaters, and while most are planktonics some are benthic or floating at water surface. Toxins can attract particular attention when they cause the death of livestock that has drunk water containing them or fish and shellfish in the sea, or humans that consume these. Toxins have been divided into different classes based on the syndromes associated with exposure to them, such as paralytic shellfish poisoning (PSP), diarrheic shellfish poisoning (DSP), neurotoxic shellfish poisoning (NSP), ciguatera fish poisoning (CFP), and amnesic shellfish poisoning (ASP).

Algae which seem to be directly producer of toxic substances mostly belong to three taxonomic groups: Cyanophyta, Haptophyta, and Dinophyta. In addition to these there are some groups which include one or two toxic members. Species of *Chattonella* and *Heterosygma*, belonging to the Raphidophyceae, form toxic red tides in Japanese waters and a few diatoms of the genus *Peusdonitzschia* produce domoic acid, a low molecular amino acid causing amnesic shellfish poisoning.

Among the 50 freshwater existing cyanobacteria genera, 12 are capable of producing toxins. While blue-green algae have significant taste and odor constituents, representing a moldy smell, their toxic metabolites have no taste, odor, or color. The risk of exposure to algal toxins may come from drinking water, recreational water, dietary supplements, or residue on produce irrigated with contaminated water and consumption of animal tissue. Avoiding cyanobacteria toxins is not as easy as avoiding a harmful algal bloom as toxins may be present in fish, shellfish and water even after the bloom has dissipated. Cyanobacterial toxins are responsible for a variety of health effects such as skin irritations, respiratory ailments, neurological effects, and carcinogenic effects.

The three major classes of these compounds are:

- Cyclic peptides (nodularins, microcystins). *Nodularia*, a well-known cyanobacterium, produce nodularins that are primarily a concern in marine and brackish waters thus creating a risk to recreational swimmers. The 65 variants of microcystins, however, are isolated from freshwaters worldwide and are produced by *Microcystis* (the most commonly identified cyanobacteria in human and animal poisonings), *Anabaena*, and other algae. They are very stable in the environment and resistant to heat, hydrolysis, and oxidation. Both toxins have an affinity for the liver. Other symptoms of exposure to microcystins may range from weakness, loss of appetite, vomiting, and diarrhea to cancer.
- Alkaloids (anatoxin, saxitoxin). Anatoxins may affect the nervous system, skin, liver, or gastrointestinal tract. These neurotoxins can cause symptoms of diarrhea, shortness of breath, convulsions and death, in high doses, due to respiratory failure. Saxitoxins are the cause of paralytic shellfish poisonings in humans consuming contaminated shellfish. There are no reports of similar poisonings via the drinking water route.
- Lipopolysaccharides (endotoxins). A similar cell wall toxin as found in *Salmonella* bacteria, but appears to be less toxic.

The physicochemical nature of the water source can have an effect on not only the growth but also the toxicity of the algal bloom. For example, some algae increase in toxicity when blooms are iron-deficient. In general, temperature, sunlight, and nutrient loads have a substantial impact on the proliferation of the bloom.

Haptophyta contain a few toxic species. *Prymnesium parvum* produces a potent toxin that causes extensive fish mortality in brackish water. Another flagellate *Chrysochromulina polylepis* produces and excretes glycolipids with hemolytic and ichtytoxic properties causing osmoregolatory failure similar to that brought about by *P. parvum*. Another widely distributed marine haptophyte *Phaeocystis* sp. is familiar to fishermen in the form of extensive blooms of mucilaginous colonies, which are avoided by herring. It produces large quantities of acrylic acid, which has strong bacteriocidal properties.

Already 70 years ago, *P. pouchetii* was suspected to cause avoidance of herring. Later it was demonstrated that copepods avoided gracing on healthy *P. pouchetii* colonies, food intake and growth were reduced in sea cage cultivated salmon during the spring bloom of *P. pouchetii* and water from *P. pouchetii* cultures acted toxic towards cod larvae.

Species belonging to the genus *Phaeocystis* are important in all oceans, and *P. pouchetii* is an important component of the spring bloom of phytoplankton in northern waters. Its life cycle is only partly resolved but is known to be polymorphic consisting of at least two solitary and one colonial stage.

P. pouchetii has been reported to produce a polyunsaturated aldehyde (PUA) as diatoms, namely the 2-*trans*-4-*trans*-decadienal. This compound is known to interfere with the proliferation of different cell types, both prokaryotic and eukaryotic. As mechanical stress is known to induce the release of PUAs in other phytoplankton species, the release of 2-*trans*-4-*trans*-decadienal has been suggested to be a mean of deterring grazers, for example, zooplankton or fish larvae. *P. pouchetii* is a common component of northern and temperate spring blooms; it is grazed by zooplankton at normal rates and can also be a diet preferable to diatoms. Copepods may avoid gracing on healthy colonies of *P. pouchetii*, thus, the production and excretion of PUAs seems to be depending on the state of the cells or on environmental factors, or both. It has been reported that this PUA can be released into the sea in the absence of grazers, indicating that it may serve as an allelochemical, that is, a compound which gives *P. pouchetii* a competitive advantage over phytoplankton species blooming at the same time by inhibiting their growth.

The Dinophyta includes about a dozen genera, with at least 30 species, producing water and lipid soluble, low molecular weight, neuroactive secondary metabolites that are among the most potent nonproteinaceous poisons known. In general, these toxins have been shown to block the influx of sodium through excitable nerve membranes, thus preventing the formation of action potentials. The toxins are accumulated and sometimes metabolized by the shellfish which feed upon these dinoflagellates, causing different types of shellfish poisoning, such as PSP, DSP, and NSP.

Ingestion of contaminated shellfish results in a wide variety of symptoms, depending on the toxins(s) present, their concentrations in the shellfish and the amount of contaminated shellfish consumed. In the case of PSP, caused by toxin of *Alexandrium* spp., the effects are predominantly neurological and include tingling, burning, numbness, drowsiness, incoherent speech, and respiratory paralysis. Less well characterized are the symptoms associated with DSP and NSP.

DSP is caused by okadaic acid, a diarrhoric shellfish toxin and tumor promoter found in many dinoflagellates of the genera *Dinophysis* and in *Prorocentrum lima*. DSP is primarily observed as a generally mild gastrointestinal disorder, that is, nausea, vomiting, diarrhoea, and abdominal pain accompanied by chills, headache, and fever.

Both gastrointestinal and neurological symptoms characterize NSP, including tingling and numbness of lips, tongue, and throat, muscular aches, dizziness, reversal of the sensations of hot and cold, diarrhea, and vomiting. The NSP toxins are produced by *Gymnodinium breve*, also denominated as *Ptychodiscus brevis*.

Another syndrome caused by dinoflagellate toxins is the ciguatera poisoning, connected with eating contaminated tropical reef fish. Ciguatoxins that cause ciguatera poisoning are actually produced by *Gambierdiscus toxicus*, a photosynthetic dinoflagellate that normally grows as an epiphyte and has a relatively slow growth rate of approximately one division every 3 days. In its coral reef habitat, *G. toxicus* is biflagellate and swims if disturbed, but is usually motionless and attached to certain macroalgae. The dinoflagellate may also be associated with macroalgal detritus on the sea floor. Some scientists believe that the diverse symptoms of CFP are a result of a combination of several toxins or their metabolites, produced by one or more dinoflagellates. However, *G. toxicus*, which is found on a variety of macroalgae eaten by herbivorous fish, is now widely considered the single-celled source of ciguatoxins and the potential cause of CFP.

G. toxicus produces two classes of polyether toxins, the ciguatoxins (CTX) and maitotoxins (MTX). The CTX are lipophilic and are accumulated in fish through food web transfer. More than 20 CTX congeners have been isolated; however, only a few have been fully characterized structurally. The maitotoxins are transfused ladder-like polyether toxins, but are somewhat more polar, due to the presence of multiple sulfate groups. MTX was originally identified as a water soluble toxin in the viscera of surgeonfishes, and later found to be the principal toxin produced by *G. toxicus*. MTXs have not been demonstrated to bioaccumulate is fish tissues, possibly due to their more polar structure. Thus, if MTX is involved in ciguatera poisoning, it may be implicated only in ciguatera poisonings derived from herbivorous fishes.

The toxic potency of MTX exceeds that of CTX (respectively 10 ng/kg and 50 ng/kg in mice). Its mode of action has not been fully elucidated. Its biological activity is strictly calcium dependent and causes both membrane depolarization and calcium influx in many different cell types.

These toxins become progressively concentrated as they move up the food chain from small fish to large fish that eat them, and reach particularly high concentrations in large predatory tropical reef fish. Barracuda are commonly associated with ciguatoxin poisoning, but eating grouper, sea bass, snapper, mullet, and a number of other fish that live in oceans between latitude 35°N and 35°S has caused the disease. These fish are typically caught by sport fishermen on reefs in Hawaii, Guam and other South Pacific islands, the Virgin Islands, and Puerto Rico. Ciguatoxin usually causes symptoms within a few minutes to 30 h after eating contaminated fish, and occasionally it may take up to 6 h. Common nonspecific symptoms include nausea, vomiting, diarrhea, cramps, excessive sweating, headache, and muscle aches. The sensation of burning or "pins-and-needles," weakness, itching, and dizziness can occur. Patients may experience reversal of temperature sensation in their mouth (hot surfaces feeling cold and cold, hot), unusual taste sensations, nightmares, or hallucinations. Ciguatera poisoning is rarely fatal. Symptoms usually clear in 1–4 weeks.

In its typical form, CFP is characterized initially by the onset of intense vomiting, diarrhoea, and abdominal pain within hours of ingestion of toxic fish. Within 12–14 h of onset, a prominent neurological disturbance develops, characterized by intense paraesthesia (tingling, crawling, or burning sensation of the skin) and dysaesthesia (painful sensation) in the arms, legs, and perioral region, myalgia, muscle cramping, and weakness. During this stage of the illness, pruritus and sweating are commonly experienced.

Pseudonitzschia spp. are among the several other marine algae that can produce domoic acid, the cause of ASP. Domoic acid was first isolated in Japan from the macroalgae species *Chondria armata* in 1958, and was consequently called after the Japanese word for macroalgae, which is *domoi*. Its identification in 1987 as a neurotoxin was first treated with scepticism, because this water-soluble amino acid was known as a folk medicine in Japan to treat intestinal pinworm infestations when used in very small doses. Production of domoic acid by algae seems to be a genetic property for a secondary metabolite with no known function in defence or primary metabolism.

Domoic acid can enter the marine food chain via uptake by molluscan shellfish such as mussels that filter their food out of the water. This water can contain both diatoms and the toxin, which is released to the water column (although there is no evidence yet that the toxin can be taken up directly). The toxin accumulates in the digestive gland and certain tissues of shellfish, and it appears to have no effect on the animals. Domoic acid may be metabolized by bacteria (e.g., of the genera *Alteromonas* and *Pseudomonas*) present in tissue of blue mussels (*Mytilus edulis*). Scallops are reported not to contain these elimination bacteria. Anchovies can also contain domoic acid in their guts, by feeding on toxic *Pseudonitzschia* spp.; this toxin affects their behavior and survival. Effects are also seen in seals. In humans, consumption of contaminated seafood mostly affected the elderly or infirm. Heat does not destroy domoic acid, although shellfish toxicity can decrease during cooking or freezing via domoic acid transfer from the meats to the surrounding liquid. The sea otter is the only animal known to be able to avoid intoxication, probably recognizing toxic prey by their odor. The mechanism of domoic acid toxicity is explained by its structural similarity with the excitatory neurotransmitter glutamic acid and its analogs, but with a much stronger receptor affinity. Domoic acid is three times more potent than its analog kainic acid and up to 100 times more potent than glutamic acid. After exposure, domoic acid binds predominately to N-methyl-D-aspartate (NMDA) receptors in the central nervous system, causing depolarization of the neurones. Subsequently, the intercellular calcium concentration increases, resulting in sustained activation of calcium-sensitive enzymes, eventually leading to depletion of energy, neuronal swelling, and cell death. The affected neurones are mainly located in the hippocampus, explaining the most striking effect of domoic acid poisoning, which is short-term memory loss, observed in 25% of the affected persons in the 1987 contaminated mussel event. Other symptoms are confusion, nausea, vomiting, gastroenteritis, cramps and diarrhea, all within 24 h. Neurological complaints, including ataxia, headaches, disorientation, difficulty in breathing and coma, develop 48 h after consumption. Permanent brain damage can also be caused by domoic acid intoxication. Effects of chronic low level ingestion are unknown. Domoic acid from mussels is more neurotoxic than the chemically pure compound. This increase is due to domoic acid potentiation, caused by high concentrations of glutamic and aspartic acids present in mussel tissue. This neurotoxic synergism occurs through a reduction in the voltage-dependent Mg_2 block at the receptor associated channel, following activation of non-NMDA receptors, in addition to the NMDA receptor activation by domoic acid itself.

Intensive research over the last years has revealed a new class of phytotoxins produced by diatoms with more subtle and less specific effects, a discovery that has drawn a lot of attention as the diatoms have traditionally been regarded a key component of the food chain. Three

TABLE 7.8
Summary of Toxic Algae and the Corresponding Metabolites

Scientific Name	Class	Toxin
Nodularia spp.	Cyanophyceae	Nodularin
Microcystis spp.	Cyanophyceae	Microcystin
Chondrus armata	Floridophyceae	ASP domoic acid
Prymnesium parvum	Haptophyceae	Fish toxin
Phaeocystis pouchetii	Haptophyceae	PUA
Chrysochromulina polylepis	Haptophyceae	Ichtytoxic glycolipids
Alexandrium spp.	Dinophyceae	PSP
Dinophysis spp.	Dinophyceae	DSP
Prorocentrum lima	Dinophyceae	DSP
Gymnodinium breve	Dinophyceae	NSP
Ptychodiscus brevis	Dinophyceae	NSP
Gambierdiscus toxicus	Dinophyceae	Ciguatoxin/maitotoxin
Pseudonizschia spp.	Bacillariophyceae	ASP domoic acid
Thalassiosira rotula	Bacillariophyceae	PUA
Skeletonema costatum	Bacillariophyceae	PUA

closely-related PUAs have been isolated from *Thalassiosira rotula, Skeletonema costatum*, and *Pseudonitzschia delicatissima*, namely 2-*trans*-4-*cis*-7-*cis*-decatrienal, 2-*trans*-4-*trans*-7-*cis*-deca-trienal and 2-*trans*-4-*trans*-decadienal. In the same study these aldehydes were found to inhibit cleavage of sea urchin embryos, reduce growth of Caco-2 cells and hatching of copepod eggs. The structural element shared by these compounds, the unsaturated aldehyde group, is able to form adducts with nucleophiles and is thus capable of inducing reactions that are toxic to the cell. The harmful effects of PUAs have been demonstrated at the organism level as inducers of apoptosis in sea urchin embryos, at the cell level as cytotoxicity in human cell lines and at the protein level by deactivation of enzymes.

Table 7.8 summarizes toxic algae and the corresponding metabolites.

SUGGESTED READING

Abdulqader, G., Barsanti, L., and Tredici, M. R., Harvest of *Artrosphira platensis* from Lake Kossorom (Chad) and its household usage among the Nanembu, *Journal of Applied Phycology*, 12, 493–498, 2000.

Andersen, R. A., Ed., *Algal Culturing Techniques*, Elsevier Academic Press, Burlington, 2005.

Amin, I. and Hong, T. S. S. Antioxidant activity of selected commercial seaweeds. *Malaysian Journal of Nutrition*, 8, 167–177, 2002.

Barbera, C., Bordehore, C., Borg, J. A., Glemarec, M., Grall, J., Hall-Spencer, J. M., De La Huz, Ch., Lanfranco, E., Lastra, M., Moore, P. G., Mora, J., Pita, M. E., Ramos-Espla, A. A., Rizzo, M., Sanchez-Mata, A., Seva, A., Schembri, P. J., and Valle, C., Conservation and management of northeast Atlantic and Mediterranean *maerl* beds, *Aquatic Conservation: Marine and Freshwater Ecosystems*, 13, s65–s76, 2003.

Barsanti, L., Bastianini, A., Passarelli, V., Tredici, M. R., and Gualtieri, P., Fatty acid content in wild type and WZSL mutant of *Euglena gracilis, Journal of Applied Phycology*, 12, 515–520, 2000.

Barsanti, L., Vismara, R., Passarelli, V., and Gualtieri, P., Paramylon (β-1,3-glucan) content in wild type and WZSL mutant of *Euglena gracilis*. Effects of growth conditions, *Journal of Applied Phycology*, 13, 59–65, 2001.

Becker, W., Microalgae in human and animal nutrition, in *Handbook of Microalgae Mass Culture and Biotechnology and Applied Phycology*. Richmond, A., Ed., Blackwell Publishing, Malden, 2004.

Borowitzka, M. A., Microalgae for aquaculture: opportunities and constraints, *Journal of Applied Phycology*, 9, 393–401, 1997.

Borowitzka, M. A., Commercial production of microalgae: ponds, tanks, tubes and fermentors, *Journal of Biotechnology*, 70, 313–3121, 1999.

Brown, M. R., Mular, M., Miller, I., Farmer, C., and Trenerry, C., The vitamin content of microalgae used in aquaculture, *Journal of Applied Phycology*, 11, 247–255, 1999.

Buschmann, A. H., Correaa, J. A., Westermeier, R., Hernandez-Gonzales, M. C., Norambuena, R., Red algal farming in Chile: a review, *Aquaculture*, 194, 203–220, 2001.

But, P. P., Cheng, L., Chan, P. K., Lau, D. T., and But, J. W., *Nostoc flagelliforme* and faked items retailed in Hong Kong, *Journal of Applied Phycology*, 14, 143–145, 2002.

Chini-Zittelli, G., Rodolfi, L., and Tredici, M. R., Mass cultivation of *Nannochloropsis* sp. in annular reactors, *Journal of Applied Phycology*, 15, 107–114, 2003.

Cohen, Z., Ed., *Chemical from Microalgae*, Taylor and Francis, London, 1999.

Daranas, A. H., Norte, M., and Fernandez, J. J., Toxic marine microalgae, *Toxicon*, 39, 1101–1132, 2001.

Delisle, H., Allasoumgue, F. B., Nandjingar, K., and Lasorsa, C., Household food consumption and nutritional adequacy in wadi zones of Chad. *Ecology, Food and Nutrition*, 25, 229–248, 1991.

Delpeuch, F. A. J. and Cavalier, C., Consommation alimentaire et apport nutritionnel des algues bleues chez quelques populations du Kanem (Chad), *Annales Nutrition Alimentation*, 29, 497–516, 1976.

Dungeard, P., Su rune algue bleue alimentaire pour l'homme, *Actes Society Linnean Bordeaux*, 91, 39–41, 1940.

Faulkner, D. J., Marine natural products, *Natural Products Reports*, 18, 1–49, 2001.

Farra, W. V., Tecuitlatl: a glimpse of Atzec food technology, *Nature*, 211, 341–342, 1966.

Funaki, M., Nishizawa, M., Sawaya, T., Inoue, S., and Yamagishi, T. Mineral composition in the holdfast of three brown algae of the genus *Laminaria, Fisheries Science*, 67, 295–300, 2001.

Gao, K., Chinese studies on the edible blue-green alga, *Nostoc flagelliforme:* a review, *Journal of Applied Phycology*, 10, 37–49, 1998.

Gao, K. and Ai, H., Relationship of growth and photosynthesis with colony size in an edible cyanobacterium, Ge-Xian-Mi (*Nostoc*), *Journal of Phycology*, 40, 523–526, 2004.

Hallegraeff, G. M., Anderson, D. M., and Cembella, A. D., Eds. *Manual on Harmful Marine Algae*. IOC Manuals and Guides, No. 33, UNESCO Publishing, Paris, 2003.

Harel, M., Koven, W., Lein, I., Bar, Y., Behrens, P., Stubblefield, J., Zohar, Y., and Place, A. R., Advanced DHA, EPA and ArA enrichment materials for marine aquaculture using single cell heterotrophs, *Aquaculture*, 213, 347–362, 2002.

Hetland, G., Ohno, N., Aaberge, I. S., and Lovik, M., Protective effect of beta-glucan against systemic *Streptococcus pneumoniae* infection in mice, *FEMS*, 27, 111–116, 2000.

Irianto, A. and Austin, B., Probiotics in aquaculture, *Journal of Fish Diseases*, 25, 633–642, 2002.

Jyonouchi, H., Sun, S., Iijima, K., and Gross, M. D., Antitumor activity of Astaxanthin and its mode of action, *Nutrition and Cancer*, 36, 59–65, 2000.

Khan, S. and Satam, S. B., Seaweed marineculture: scope and potential in India, *Aquculture Asia*, 8, 26–29, 2003.

Kusmic, C., Barsacchi, R., Barsanti, L., Gualtieri, P., and Passarelli, V., *Euglena gracilis* as source of the antioxidant vitamin E. Effects of culture conditions in the wild type strain in the natural mutant WZSL, *Journal of Applied Phycology*, 10, 555–559, 1999.

Lee, Y., Microalgal mass culture systems and methods: their limitation and potential, *Journal of Applied Phycology*, 13, 307–315, 2001.

Leonard, J., The 1964–1965 Belgian Trans-Saharan expedition, *Nature*, 209, 126–128, 1966.

Leonard, J. and Compere, P., *Spirulina platensis* algue bleue de grande valeur alimentatire par sa richesse en proteins, *Bulletin Jarden Botanique Naturelle Belgique*, 37, 3–23, 1967.

Lourenço, S. O., Barbarino, E., De-Paula, J. C., Da, L. O., Pereira, S., and Marquez, U. M. L., Amino acid composition, protein content and calculation of nitrogen-to-protein conversion factors for 19 tropical seaweeds. *Phycological Research*, 50, 233–241, 2002.

Lu, C., Rao, K., Hall, D., and Vonshak, A., Production of eicosapentaenoic acid (EPA) in *Monodus subterraneus* grown in a helical tubular photobioreactor as affected by cell density and light intensity, *Journal of Applied Phycology*, 13, 517–522, 2001.

Luning, K. and Pang, S., Mass cultivation of seaweeds: current aspects and approaches, *Journal of Applied Phycology*, 15, 115–119, 2003.

Malloch, S., Marine plant management and opportunities in British Columbia. Fisheries, Sustainable Economic Development Branch, Victoria (British Columbia), 2000. http://www.agf.gov.bc.ca/fisheries/pdf/Marine_Plant_Management_and_Opportunities_Report.pdf

McHugh, D. J., *A Guide to the Seaweed Industry*, Fao Fisheries Technical Paper, N. 441, Fao, Rome, 2003.

Murakami, T., Ogawa, H., Hayashi, M., and Yoshizumi, H., Effect of *Euglena* cells on blood pressure, cerebral peripheral vascular changes and life-span in stroke-prone spontaneously hypertensive rats. *Journal Japanese Society Nutrition Food Science*, 41, 115–125, 1998.

Nagai, T. and Yukimoto, T., Preparation and functional properties of beverages made from sea algae, *Food Chemistry*, 81, 327–332, 2003.

Norziah, M. H. and Ching, C. Y., Nutritional composition of edible seaweeds *Gracilaria changgi*, *Food Chemistry*, 68, 69–76, 2000.

Parker, N. S., Negri, A. P., Frampton, D. M. F., Rodolfi, L., Tredici, M. R., and Blackburn, S. I., Growth of the toxic dinoflagellate *Alexandrium minutum* (Dinophyceae) using high biomass culture systems, *Journal of Applied Phycology*, 14, 313–324, 2002.

Richmond, A., Microalgal biotechnology at the turn of the millennium: a personal view, *Journal of Applied Phycology*, 12, 441–451, 2000.

Ridolfi, L., Chini-Zitelli, G., Barsanti, L., Rosati, G., and Tredici, M. R., Growth medium recycling in *Nannochloropsis* sp. mass cultivation, *Biomolecular Engineering*, 20, 243–248, 2003.

Steneck, R. S., Graham, M. H., Bourque, B. J., Corbett, D., Erlandson, J. M., Estes, J. A., and Tegner, M. J., Kelp forest ecosystems: biodiversity, stability, resilience and future, *Environmental Conservation*, 29, 436–459, 2002.

Stepp, J. R. and Moerman, D. E., The importance of weed in ethnopharmacology, *Journal of Ethnopharmacology*, 75, 19–23, 2001.

Takeyama, H., Kanamaru, A., Yoshino, Y., Kakuta, H., Kawamura, Y., and Matsunaga, T., Production of antioxidant vitamins, β-carotene, vitamin C, vitamin E, by two-step culture of *Euglena gracilis* Z. *Biotechnology and Bioengineering*, 53, 185–190, 1997.

Talyshinsky, M. M., Souprun, Y. Y., and Huleihel, M. M., Anti-viral activity of red microalgal polysaccharides against retroviruses, *Cancer Cell International*, 2, 8–15, 2002.

Tani, Y. and Tsumura, H., Screening for tocopherol-producing microorganisms and α-tocopherol production by *Euglena gracilis* Z, *Agriculture Biological Chemistry*, 53; 305–312, 1989.

Tseng, C. K., Algal biotechnology industries and research activities in China, *Journal of Applied Phycology*, 13, 375–380, 2001.

Turner, N. J., The ethnobotany of edible seaweeds (*Porphyra abbotae*) and its use by First Nations on the Pacific Coast of Canada, *Canadian Journal of Botany*, 81, 283–293, 2003.

Vicente, M. F., Basilio, A., Cabello, A., and Pelaez, F., Microbial natural products as a source of antifungals, *Clinical Microbiology and Infectious Disease*, 9, 15–32, 2003.

Vismara, R., Vestri, S., Barsanti, L., and Gualtieri, P., Diet-induced variations in fatty acid content and composition of two on-grown stages of *Artemia* sp., *Journal Applied Phycology*, 15, 477–483, 2003a.

Vismara, R., Vestri, S., Kusmic, C., Barsanti, L., and Gualtieri, P., Natural vitamin E enrichment of *Artemia salina* fed freshwater and marine microalgae, *Journal of Applied Phycology*, 15, 75–80, 2003b.

Vismara, R., Vestri, S., Frassanito, A. M., Barsanti, L., and Gualtieri, P., Stress resistance induced by paramylon treatment in *Artemia* sp., *Journal of Applied Phycology*, 16, 61–67, 2004.

Wang, B., Zarka, A., Trebst, A., and Boussiba, S., Astaxanthin accumulation in *Haematococcus pluvialis* as an active photoprotective process under high irradiance, *Journal of Phycology*, 39, 1116–1124, 2003.

Wilson, S., Blake, C., Berges, J. A., and Maggs, C. A., Environmental tolerances of free-living coralline algae (maerl): implication for European marine conservation, *Biological Conservation*, 120, 283–293, 2004.

Wikfors, G. H. and Ohno, M., Impact of algal research in aquaculture, *Journal of Phycology*, 37, 968–974, 2001.

Zemke-White, W. L. and Ohno, M. World seaweed utilisation: an end-of-century summary, *Journal of Applied Phycology*, 11, 369–376, 1999.

http://www.aqua-in-tech.com
http://www.english-nature.org.uk/uk-marine/reports/pdfs/maerl.pdf
http://www.herb.umd.umich.edu/
http://www.marlin.ac.uk/species/Lithothamnioncorallioides.htm
http://www.mumm.ac.be/SUMARE
http://www.scotland.gov.uk/cru/kd01/orange/sdsp-06.asp

Index